Frontiers in Stem Cell and Regenerative Medicine Research

(Volume 4)

Edited By

Atta-ur-Rahman, ***FRS***

Honorary Life Fellow, Kings College
University of Cambridge, Cambridge, UK

&

Shazia Anjum

Department of Chemistry, Cholistan Institute of Desert
Studies, The Islamia University of Bahawalpur, Pakistan

Bentham Science Publishers Ltd.
Executive Suite Y - 2
PO Box 7917, Saif Zone
Sharjah, U.A.E.
Email: subscriptions@benthamscience.org

CONTENTS

PREFACE

There continues to be tremendous progress in the field of stem cell and regenerative medicine research. The developments promise to change the face of medicine. This 4th volume of '*Frontiers in Stem Cell and Regenerative Medicine Research*' should be of considerable interest to the readers as it presents state-of-the art reviews written by renowned experts in this fast moving field.

Ilham Saleh Abduljadayel presented a comprehensive review in chapter 1 on epimorphic regeneration and retrodifferentiation. Both processes have the capacity to recreate and reconstruct tissue with precise positional integration of cells in such a way that will enable healing without scars. In chapter 2 Fraser *et al.* extensively reviewed the unique genetic programmes that lead to mesendoderm formation and the pathways leading to mesoderm and endoderm specification. They also present examples where mature cell types from both germ layers interact to support their mutual development. These programmes are being employed to direct the differentiation of pluripotent cells *in vitro* into mesendoderm derived cells and tissues. Fraser *et al.* also reviewed the role of stem, progenitor and supportive cells within the hematopoietic tissues as essential elements of regenerative medicine in chapter 3.

Cell-based therapy is an emerging field in veterinary medicine that has been used for developing new therapies for degenerative diseases. In chapter 4 Izadyar *et al.* described different cell based therapies, their risks and benefits and their possible therapeutic use for veterinary medical applications. Mesenchymal stem cells (MSCs) are the most favored cellular candidates for regenerative therapeutics. Bhat *et al.* discussed how MSCs contribute to therapeutic efficiency include facilitating secretion of bioactive factors, induction of cellular recruitment and retention of progenitor faculties in the last chapter.

Knowledge of stem cell and regenerative medicine research continues to move ahead on many fronts. We hope that the readers will enjoy reading about the latest and stimulating development in this hot area in the 4th volume of this series.

We are pleased to place on record our heartfelt thanks to all the authors for their contributions. We are also grateful to the editorial staff of Bentham Science Publishers, particularly Dr. Faryal Sami, Mr. Shehzad Naqvi and Mr. Mahmood Alam for their constant support and great help.

Prof. Atta-ur-Rahman, FRS
Honorary Life Fellow
Kings College, University of Cambridge
Cambridge
UK

&

Dr. Shazia Anjum
Department of Chemistry
Cholistan Institute of Desert Studies
The Islamia University of Bahawalpur
Pakistan

List of Contributors

Badwi B. Boumelhem	Discipline of Physiology, School of Medical Science, University of Sydney, Camperdown, Australia
Chad Maki	VetCell Therapeutics, 2917 Daimler Street, USA
Chanukya K. Colonne	Discipline of Physiology, School of Medical Science, University of Sydney, Camperdown, Australia
Chelsea Pilgrim	Discipline of Physiology, School of Medical Science, University of Sydney, Camperdown, Australia
Christoph Hammans	Discipline of Physiology, School of Medical Science, University of Sydney, Camperdown, Australia
Damien T. M. Le	Discipline of Physiology, School of Medical Science, University of Sydney, Camperdown, Australia
Fariborz Izadyar	VetCell Therapeutics, 2917 Daimler Street, USA
Ganapathi M. Bhat	Jaslok Hospital and Research centre, 15, Dr. G. Deshmukh Marg, Mumbai 400026, India
Henry Williams	Discipline of Physiology, School of Medical Science, University of Sydney, Camperdown, Australia
Ilham Saleh Abuljadayel	TriStem UK Limited, 571 Finchley Road, Unit 320, London, England
Jia Hao Yeo	Discipline of Physiology, School of Medical Science, University of Sydney, Camperdown, Australia
Kurt W. L. Brigden	Discipline of Physiology, School of Medical Science, University of Sydney, Camperdown, Australia
Marietta P. Tanudisastro	Discipline of Physiology, School of Medical Science, University of Sydney, Camperdown, Australia
Pooja Shahapurkar	Jaslok Hospital and Research centre, 15, Dr. G. Deshmukh Marg, Mumbai 400026, India
Rasika G. Bhat	Department of Medical Genetics, College of Nursing, Bombay Hospital, Mumbai 400026, India
Stephen J. Assinder	Discipline of Physiology, School of Medical Science, University of Sydney, Camperdown, Australia
Stuart T. Fraser	Discipline of Physiology, School of Medical Science, University of Sydney, Camperdown, Australia Discipline of Anatomy & Histology, School of Medical Science, University of Sydney, Camperdown, Australia
Thomas Ramos	VetCell Therapeutics, 2917 Daimler Street, USA
Veronica I. Antas	Discipline of Physiology, School of Medical Science, University of Sydney, Camperdown, Australia

Frontiers in Stem Cell and Regenerative Medicine Research

Frontiers in Stem Cell and Regenerative Medicine Research

Volume # 4

Editors: Prof. Atta-ur-Rahman and Dr. Shazia Anjum

eISBN (Online): 978-1-68108-435-0

ISBN (Print): 978-1-68108-436-7

eISSN (Online): 2352-7633

ISSN (Print): 2467-9593

First published in 2017.

CHAPTER 1

Retrodifferentiation: From Concept to Bedside Stem Cell Therapy

Ilham Saleh Abuljadayel*

TriStem UK Ltd., London, England

Abstract: Epimorphic regeneration is a process by which damaged tissues or severed body parts are restored to the original. This type of sophisticated regeneration is observed in urodeles and fetal mammals. For example, through this process, an amputated limb of a salamander can be restored, by re-growing an exact replica, irrespective of its age. During limb epimorphic regeneration: committed mesenchymal cells at the stump site dedifferentiate, forming a cluster of heterogeneous population of stem cells, known as the blastema. Upon blastema integration, positioning and expansion, constituent cells embark on redifferentiation and remorphogenesis to restore the lost appendage. Similar to epimorphic regeneration is retrodifferentiation in human leukocytes. In response to ligation of monomorphic regions of MHC class II antigens with monoclonal antibody CR3/43, human leukocytes retrodifferentiate into a variety of heterogeneous stem cell types belonging to the mesoderm, ectoderm or endoderm lineage, depending on culture media and conditions. During this process, leukocytes lose lineage-associated markers home and undergo homocytic aggregation, upregulate expression of stem cell antigens, and subsequently redifferentiate to give rise original tissue or, transdifferentiate into a different tissue altogether. The hematopoietic retrodifferentiated stem cells have been shown to engraft an animal host in two proofs of principle clinical studies, demonstrating long-term engraftment and safety in acquired aplastic anaemia, while transient amelioration of beta thalassemia major was also observed. Binding of MHC class II antigens on leukocytes with the monoclonal antibody CR3/43 appears to emulate stress and injury in human tissue *in vitro*, similar to limb amputation in salamander. The ease by which various stem cell types can be generated from human peripheral blood has allowed the design of various kits to guarantee the specificity, sterility and efficacy of stem cells production for various

* **Corresponding author Ilham Saleh Abuljadayel:** TriStem UK Limited, 571 Finchley Road, Unit 320, London NW3 7BN, England; Tel: + 44 20 7722 5449; Fax: 44 20 7586 5290; E-mail: iabuljadayel@tristemcorp.com

Atta-ur-Rahman & Shazia Anjum (Eds.)

clinical and research applications. The robustness and efficacy of the retrodifferentiation process in generating unprecedented quantities of stem cells belonging to the three germ layers will enable organ and tissue reconstruction *ex vivo*, using bio-printing and various scaffold materials. Epimorphic regeneration and retrodifferentiation both have the capacity to recreate and reconstruct tissue with precise positional integration of cells in such a way that will enable us to heal without scars and to understand how to maintain tissue integrity and architecture in the face of a hostile environment.

Keywords: Axolotl, Bioprinting, Blastema, Bone, Dedifferentiation, Ectoderm Endoderm, Epimorphic Regeneration, Hematopoietic, Hepatocytes, Histogenesis, Human leukocytes, Mesenchymal, Morphogenesis, Neurons, Pluripotent, Positional integration, Regeneration, Retrodifferentiation, Salamander, Scaffolding, Stem Cells, Tissue Repair, Transdifferentiation.

INTRODUCTION

The ability to regenerate an entire complex tissue, organ or appendage upon damage is almost nonexistent in higher vertebrates, for instance in an adult human. This is because neither all tissues in the human body are endowed with stem cells that have the ability to proliferate and differentiate to replenish damaged or spent cells nor the adult human body possess sophisticated regenerative processes that enable replacement of body parts lost to severe injury. Most injuries are dealt with simply utilizing repair mechanism which entails closure of the injured site by deposition of fibrous tissue instead of cells. This leads to altering organ geometry and architecture including deterioration in function. In stark contrast, amphibians and particularly a selected group of urodele salamander can re-grow body parts, such as an amputated limb, even when old, in a process known as epimorphic regeneration. In this process, complex mechanisms, including dedifferentiation, innervation, positional integration and re-morphogenesis occur to restore the severed appendage. During the dedifferentiation phase, fully mature specialised cells from various positions around the circumference of the amputate, belonging to the mesenchymal lineage home to the stump site and dedifferentiate. This leads to the formation of a cluster of heterogeneous population of stem cells, known as the blastema. Homing and integration of mesenchymal cells occur according to positional values. This facilitates alignment and arrangement of cells of different lineages, in a configuration that upon re-morphogenesis permits restoration of the original geometry and architecture of the limb. Elucidation and understanding of epimorphic regeneration, including harnessing similar mechanisms in human, has tremendous applications in regenerative medicine than mere production of stem cells which at best proliferate and differentiate *ex-vivo*, but may fall short of positional integration or morph into tumors when transplanted into humans. The process which has been termed retrodifferentiation [1] is similar to the process of cellular dedifferentiation which facilitates blastema formation in salamander [2].

Retrodifferentiation of human leukocytes into a variety of pluripotent stem cell classes occurs in response to ligation of the monomorphic region of the major histocompatibility complex beta chain, using a monoclonal antibody (clone CR3/43) [3]. Each stem cell type generated is determined by the type of culture media and conditions utilized during retrodifferentiation. Similar to mesenchymal cell dedifferentiation in salamanders, leukocytes lose lineage associated markers, undergo homing and homocytic aggregation, and form heterogeneous stem cell colonies which subsequently redifferentiate into cellular components of the original tissue. Unlike salamander regeneration, retrodifferentiation is capable of transdifferentiation and histogenesis giving rise to entirely different tissues. In this process, mature mononuclear leukocytes can be converted into a variety of stem cell types belonging to the three germ layers: mesoderm, endoderm or ectoderm. Three hour human male retrodifferentiated haematopoietic stem cells (RHSC) have been shown to engraft in minimally irradiated NOD/SCID female mice [4]. Furthermore, 3 hr autologous RHSC were capable of long term engraftment in severe acquired aplastic anemia patients without any form of pre-conditioning therapy [5]. While in beta thalassemia major [6], a genetic blood disorder, the autologous RHSC were only able to ameliorate the course of the disease for six months. The ease by which any stem cell type can be prepared from human peripheral blood *via* retrodifferentiation, enabled the development of kits for the treatment of haematological and degenerative disorders, as a bed side stem cell therapy. In this manner, and in combination with leukopheresis and washing devices, the automation of stem cell production, will guarantee efficiency, sterility and specificity of stem cell infusate. Most importantly, retrodifferentiation recapitulates the process of histogenesis of various human tissues *ex-vivo*, thus paving the way to reconstructing organs by utilizing bio-printing devices in combination with various scaffold materials. Furthermore, the elucidation of the genetic events partaking in retrodifferentiation, transdifferentiation and histogenesis is paramount to regenerative medicine [7]. For example, the elaboration of cellular and extracellular pathways and signaling occurring during positional integration and histogenesis of a heterogeneous conglomerate of stem cells will enable us heal without scars, and mend tissue and organs with utmost precision and fidelity.

HEALING MECHANISMS IN RESPONSE TO TISSUE INJURY IN ADULT HUMAN

Post cellular and tissue injuries, the human body mends or heals itself by utilizing two mechanisms: regeneration and repair. During regeneration, restitution of damaged cellular components of a given tissue, tend to be restored to normal configuration and function. On the other hand, the repair mechanism involves containment and closure of damaged tissue by deposition of collagen, which leads to scar formation with loss of normal architecture and deficit in function [8, 9].

Tissues capable of regeneration are those of the hematopoietic system [10] and epithelia of the skin and gastrointestinal tract. This is because they are endowed with stem cells capable of self-renewal and differentiation during injury or infection to replenish aged or damaged cells as long as the stem cell compartment is intact [11]. Tissues containing stem cells are known as labile tissues; for example, throughout the life of the haematopoietic system, there is a continuous renewal of this tissue, where injured, or aged specialized cells are replaced by stem cells, through proliferation and differentiation. In the process known as asymmetric cell division, the stem cells divide to generate daughter cells. One daughter cell embarks on differentiation to replace the damaged or spent specialized cell, while the other daughter cell, remains quiescent to prevent dwindling of the stem cell compartments in that tissue. The haematopoietic stem cells (HSCs) reside in the bone marrow of an adult human and consist of a heterogeneous population of stem cells with hierarchical development potential. For example, pluripotent/multipotent HSCs can replace granulocytes, megakaryocytes and erythrocytes, including platelets, through division and differentiation [12]. Other HSCs are either bipotent or monopotent, such as those giving rise to granulocytes / monocytes and erythrocytes only, respectively. As the HSCs divide and differentiate, apoptosis mechanisms operate to keep the population size of different lineage in check and in a specific proportion to each other [13, 14]. In other words, the haematopoietic system, though a proliferative tissue, renews at different rates during steady state and post injury, as long as the damage is not impacting various mechanisms associated with its proliferation development and apoptosis [15, 16].

The generation of various blood lineages, termed hematopoiesis, is regulated by the bone marrow microenvironment or niche. This niche regulates the quiescence, proliferation and differentiation of HSCs. The niche containing HSCs is widely distributed in the bone marrow and consists of perivascular and endosteal. The former consists of vascular endothelial cells, mesenchymal stem cells (MSCs), Cxcl12-abundant reticular (CAR) cells, perivascular stromal cells and Schwann cells, while the latter consists of osteoblasts. Niche factors include Stem cell factor (SCF; c-Kit ligand), thrombopoietin (Thpo or TPO) and Cxcl12. Conditional deletion of SCF in endothelial cells and perivascular stromal cells results in a decrease in HSC number, suggesting an important role of these cells in HSC maintenance. In addition the physical environment, such as oxygen tension, temperature and contractile force also exerts a regulatory effect. During homeostasis, the majority of HSCs are relatively more quiescent than during infection stress, both proliferation and differentiation accelerate to neutralize and sequester pathogens [17].

Dysregulation and injury to the haematopoietic stem cells will either lead to over expansion or depletion of committed haematopoietic cell lineages, respectively. Cell loss has been observed in acquired aplastic anemia [18]; a hematological condition believed to be autoimmune in nature [19]. In this condition, bone marrow becomes gradually hollow or hypo-cellular, devoid of haematopoietic cells which are replaced by fat cells resulting in (i) low leukocyte count, (ii) deficiency in platelets, and (iii) red blood cell production leading to hemorrhage; and thus a propensity to contracting infection and, if left untreated, to morbidity. Alternatively, malignancy of the hematopoietic system results in clonal expansion of cells stuck at certain stages of development. Continuous proliferation of tumor cells leads to accumulation of genetic mutation(s) [20, 21], which transform the cells into aggressive clones capable of metastasis and obstruction of organs. In either scenario, cellular deficiency or over production, the underlying pathogenesis cannot be rectified spontaneously by the bone marrow and will require procedural intervention to replace defective cells [22].

COMPENSATORY REGENERATION

The ability of the liver to regenerate is recorded in the myths of ancient Greeks [23] despite there is no convincing evidence that they had any specific knowledge about liver regeneration. As recorded by the poet Hesiod (8th century BC), Titan Prometheus was subjected to the most severe punishment for his twice deception of Zeus for protecting the welfare of human kind. Firstly, he offered mortals the best meat from a slaughtered cow and gave the fat and bones to the gods. The second time, when an infuriated Zeus punished humans by taking fire, Prometheus stole it back for humankind. Prometheus was bound to the mountain Caucasus and subjected to perpetual torture by an eagle who fed from his liver each day, but the liver regenerated overnight. Peter Paul Rubens' (1618) grand painting depicting a giant eagle piercing through Prometheus' right hypochondrium and holding a portion of his liver in its beak is probably the most familiar illustration of the punishment of Prometheus used by lecturers on hepatic regeneration. Nearly 3000 years later, the work of Higgins and Anderson (1931) provided the first experimental evidence of liver regeneration following partial hepatectomy. Subsequently, many studies have shed light on how the liver regenerate at the cellular and molecular level.

While injuries to a stable or quiescent tissue such as the liver are met with compensatory regeneration, liver parenchyma exists in a non-dividing state and will enter the cell cycle in response to injury. For example, when a liver lobe is dissected for donation, the remaining liver cells will proliferate at such a rate that the liver reaches its original size [24] (before resection) albeit with similar cellular composition but without preservation of the original architecture or geometry of the liver master plan. While this kind of liver re-growth is a result of cell division, as in regeneration, but since it does not achieve return to the original form; it is called compensatory regeneration, although normal function is restored. Tissues falling into this category include parenchymal cells of the kidney, pancreas, lymphocytes, fibroblasts, endothelial cells and smooth muscle cells. Nonetheless, after persistent injury to the liver such as viral hepatitis, fibrosis is the end result of wound healing. This involves excessive accumulation of extracellular matrix proteins (ECM) including collagen, orchestrated by activated hepatic stellate cells, portal fibroblasts and bone marrow myofibroblasts. In advanced stages of fibrosis, the liver contains approximately six times more extracellular matrix (ECM) than normal, including collagens (I, III, and IV), fibronectin, undulin, elastin, laminin, hyaluronan, and proteoglycans [25]. The accumulation of ECM results from both increased synthesis and decreased degradation; and advanced liver fibrosis results in cirrhosis, liver failure, and portal hypertension and often requires a liver transplant [26].

In contrast, a nondividing tissue such as the myocardium is unable to regenerate; instead it is repaired by replacement of damaged cells by deposition of fibrous material, in order to reconnect the injured site. The adult heart has limited endogenous regenerative capacity [27] (stem cells), and in this instance, repair is achieved by patching the injured site with fibrous tissue [28]. This leads to a profound change in heart geometry and tensile strength. Prior to repair by connective tissue, an influx of inflammatory cells removes debris, leads to an exacerbation of inflammation and may augment matrix degradation resulting in cardiac rupture and weak scar formation, with reduced tensile strength and chamber dilation. Continued expression of pro-inflammatory mediators may impact non infarcted regions of the heart due to activation of pro-apoptotic pathways, enhancing fibrosis and deterioration in diastolic function [29].

On the other hand, adult wounds apart from minor cuts; injuries to fetal mammalian skin during early gestation result in non-scar wound healing. In humans, the window of non-scar wound healing terminates at approximately 24 weeks of gestation and is diminished at an earlier gestation age if wound size is large, as observed in fetal limbs [30]. In response to tissue injury, the fetal skin is capable of regenerating a seamless layer of collagen matrix that is identical to that of the original tissue [31]. In addition, dermal structures, such as sebaceous glands and hair follicles, form normally after fetal injury. The differences between fetal and adult skin wound healing appear to reflect the processes inherent to fetal tissue, such as the unique fetal fibroblasts, a more rapid and ordered deposition and turnover of tissue components, and particularly, a markedly reduced inflammatory infiltrate and cytokine profile.

Therefore, non-scar fetal wounds are relatively deficient in the inflammatory cytokine, Transforming Growth Factor β (TGF-β), including reduced expression of TGF-β1 and higher levels of hyaluronan in the extracellular matrix. In contrast, the fibrosis characteristic of adult wound repair may be associated with excess TGF-β. Experimental studies suggest that specific anti-TGF-β therapeutic strategies can ameliorate scar formation in adult wound repair and fibrotic diseases [32].

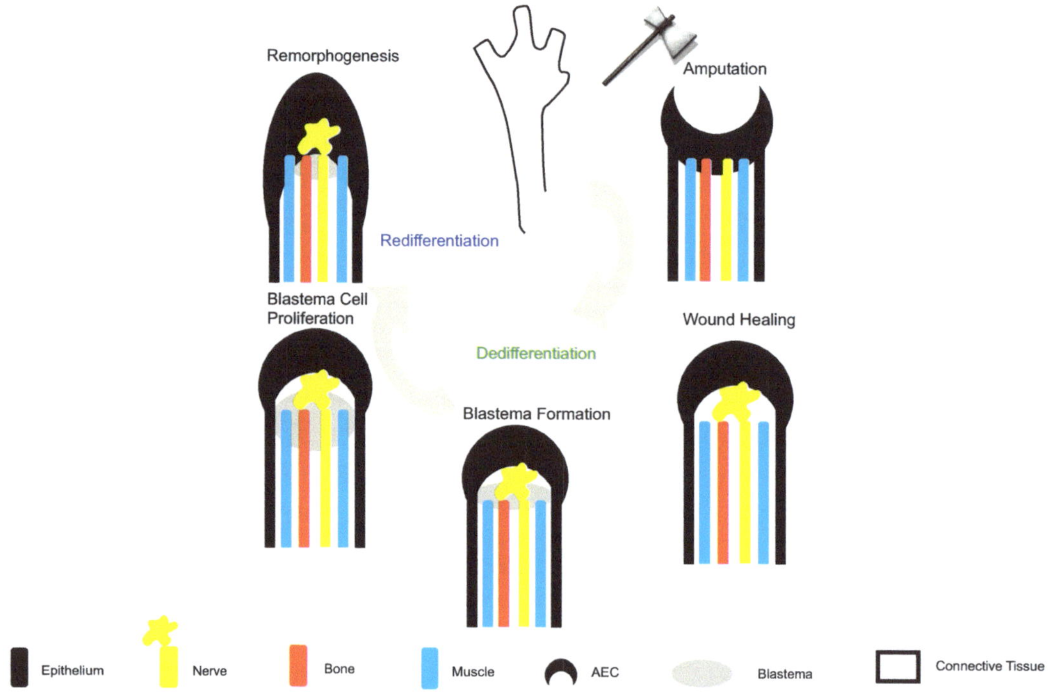

Fig. (1). Stages of epimorphic regeneration in Salamander in response to amputation: wound healing, innervation, establishment of AEC signalling centre, induction of Blastema formation (dedifferentiation) and proliferation, leading to remorphogenesis (redifferentiation) including rapid reappearance of digits when compared to normal development.

EPIMORPHIC REGENERATION

Humans are inept of more complex forms of regeneration, such as the replacement of an entire tissue, an amputated limb or dysfunctional organs; which are exact histological and morphological replica of the original. Urodele amphibians, such as axolotls salamanders and the eastern newt, *Notophthalmus viridescens*; have the unique ability to regenerate limbs, spinal cord, eye structures, and many vital organs through a process called regeneration. Two centuries ago, Spallanzani [33] was the first to describe regeneration in worms, snails, tadpoles and salamanders. The elucidation of his pioneering work is more relevant to regenerative medicine than the mere production of stem cells. The exact molecular mechanism behind epimorphic regeneration is not fully explored, though the cellular sequence of events involved in such a process are much clearer.

For example, in salamanders, after limb amputation and bleeding, the stump becomes covered by a fibrin clot within minutes of the injury. Subsequently, the blood clot becomes covered by migrating epidermal cells which originate by detachment from the basement membrane at the edge of the wound surface (Fig. **1**). A layer of epithelium covers the wound within roughly 9 hours post injury [34]. This layer of cells, known as wound epithelium, (WE) [35] in subsequent regenerations becomes thickened to form the apical epithelial cap (AEC) [36]. Dedifferentiation ensues post epithelial wound healing, during which fully differentiated mesenchymal cells, including muscle cartilage and connective tissue proximal to the injury site; become undifferentiated by losing specialized characteristics [37 - 39]. Loss of differentiation is morphologically marked by rounding of nuclei, development of a prominent nucleoli, an increase in nuclear to cytoplasmic ratio and an increase in number of ribosomes [40]. The dedifferentiated cells [41 - 49], including resident stem cells, accumulate at the apex of the regenerate following proliferation to form mesenchymal cell aggregate known as the blastema. Blastema formation proceeds the establishment of the thickened AEC; a derivative of WE to facilitate the interaction of both cellular structures and to confer division signals to the blastema before embarking on redifferentiation [50 - 52]. In addition, before transition into AEC, the WE becomes innervated from the regenerating ends of an underlying nerve.

Innervation induces the wound epithelium to become a specialized signaling centre that produces a number of growth or neuroptrophic factors and morphogens which act as chemoattractants and mitogens to attract and expand the blastema cell population. It is important to stress here that in the absence or obliteration of innervation, regeneration is stunted [53 - 57].

Furthermore, as regeneration progresses, the blastema undergoes morphogenesis and redifferentiation by transforming from a cluster of cells or small bud into elongated flattened palette-like structure. Redifferentiation occurs along a proximodistal axis from the amputation site, during which cartilage condensation and myogenesis occur first, followed by the differentiation of connective tissue, blood vessels and the nervous network [58]. Continued morphogenesis and growth of the regenerate give rise to a perfect functional replica, indistinguishable from the original limb; including the restitution of pigments which become fully formed shortly after. It is important to stress here that the initial patterning of the developing limb involves the sequential appearance of digits 1 and 2, followed by digits 3 and 4 [59]. During regeneration, all the digits seem to appear at once [60]. This rapid tissue regeneration is also observed during retrodifferentiation which will be discussed in more detail later on in the chapter.

POSITIONAL INTEGRATION AND MOLECULAR MECHANISMS OF REGENERATION

Pattern formation of a limb or an organ, during development from an embryonic stem cell or during regeneration from blastema, occurs in an organized manner along a developmental axis during which cells cement with other cells in a specific spatial configuration [61]. This is because, cells possess positional identity of variable magnitude, across which they align themselves relative to each other to shape and form organs, appendages and various body parts in an organized three dimensional plan. This cellular property is epigenetically determined in human [62] and salamander [63] cells, and persists throughout adult life and is known as positional information. For example, grafting experiment performed by exchanging (host/recipient) blastemas from the same position between donor and recipient amputations on the limbs gives rise to an appendage at the same site [64]. However, exchanging blastemas from different positions on a limb will result in ectopic regeneration of an extra limb, not in the same location [65]. No graft grows by confronting a blastema from a completely different site on the body; such as head to limb exchange [66, 67], due to positional discontinuity and non-resolution through intercalary response.

Molecular insight into the mechanism of regeneration resembles the recapitulation of genes involved in limb development, examples are members of the Hox family [68], retinoic acid receptors (RARs) [69, 70], sonic hedgehog (shh) [71] and Tbox genes [72, 73]. Fibroblast growth factors and receptors are essential during wound healing and are secreted by the AEC and nerves, which upon denervation cause regeneration to fail, and this can be restored by exogenous application of FGF-2, FGF8 and FGF10 [74 - 80]. On the other hand, during the dedifferentiation phase, genes involved in extracellular degradation and remodeling appear essential, such as members of the matrix metalloproteinase (MMP) [81, 82] family; which may be key players in preventing scar formation and promoting destabilisation of the differentiated state. MMPs are essential for mammalian cell migration and proliferation, which may be important during migration and subsequent proliferation of mesenchymal cells, to form the blastema. Other extracellular proteases, such as thrombin and elastase [83 - 85], have also been implicated in dedifferentiation and extracellular remodeling. The homeobox gene, msx-1 [86, 87], has also been shown to have a key role in dedifferentiation. Surprisingly, ectopic expression of msx1, coupled with serum stimulation, can induce C2C12 [88] mouse myotubes to fragment and dedifferentiate *in vitro* into mononucleated cells which subsequently transdifferentiate into cells expressing chondrogenic, adipogenic, myogeneic and osteogeneic antigens.

ORIGIN OF BLASTEMA

Genetic lineage tracing using a green fluorescent compound shows dedifferentiation of myofibers by fragmentation and conversion into proliferating Pax7 mononuclear cells [89], which integrates in the regenerating limb in one species of salamander *Notophthalmus viridescens* (newt). Whereas, in another specie, *Ambystoma mexicanum* (axolotl), myofibres neither generate proliferating cells, nor they contribute to newly regenerated muscle, instead resident PAX7+ cells (skeletal stem cells) provide the regeneration activity. Surprisingly, evidence of newt-like dedifferentiation was observed in specialized secretory cells of the epithelium of mouse trachea [90], which was demonstrated by ablation of CK5-expressing stem cells with doxyclycine, through inhalation and labeling of secretory cells expressing SCG1A with YFP. Lineage tracing of secretory cells, demonstrated that the luminal secretory cells had dedifferentiated into basal stem cells expressing CK5. Single secretory cells clonally dedifferentiated into multipotent stem cells, when cultured *ex vivo* without basal stem cells. In contrast, direct contact with a single basal stem cell was sufficient to prevent secretory cell dedifferentiation, albeit the propensity of committed cells to dedifferentiate was inversely correlated to their state of maturity.

REPROGRAMMING THE DIFFERENTIATED STATE

Reprogramming differentiated cells into pluripotent stem cells can be achieved through nuclear transfer technology [91] or by forced expression of transcription factors inside specialized cells. During nuclear reprogramming, the nucleus of a differentiated cell is introduced into an enucleated oocyte, at a specific stage in the cell cycle, to re-jump start development to a totipotent stem cell state, capable of giving rise to an entire organism. Sir Gurdon was the first to demonstrate the plasticity of the differentiated state of gut epithelial cell nuclei which on transfer into enucleated eggs gave rise to adult frogs. His pioneering nuclear transfer experiment laid the basic background of cloning techniques, which started to gain interest after the birth of Dolly the sheep [92]. On the other hand, direct reprogramming of differentiated cells into an induced pluripotent stem cell (IPS) [93] state requires forced ectopic expression of genetic reprogramming factors. This is achieved by integration of genetic reprogramming material into differentiated cells: belonging to a single lineage; such as fibroblasts, using viruses, plasmids and modified RNA [94]. The difference between IPS and dedifferentiated blastema cells is that, reprogramming factors KLF4 (Kruppel-like factor) and C-myc (myelocytomatosis oncogene) are re-expressed in the latter, but without expression of embryonic stem cell antigens, as observed in the former [95]. In contrast to IPS produced from a single fibroblast, ablastema formed from a heterogeneous population of committed mesenchymal cells consists of a mixed population of stem cells, committed to various mesenchymal cell lineages. More similar to the process of blastema formation is retrodifferentiation. In this process, human leukocytes consisting of heterogeneous populations of mononuclear cells; when engaged with a monoclonal antibody against the homologous region of major histocompatibility complex (MHC) class II antigen beta chain revert to heterogeneous colonies of pluripotent stem cells, with hierarchical developmental potential. These retrodifferentiated stem cells can embark on two developmental routes, either redifferentiating into cellular components of the tissue of origin, or transdifferentiating into a new tissue altogether (Fig. **2**).

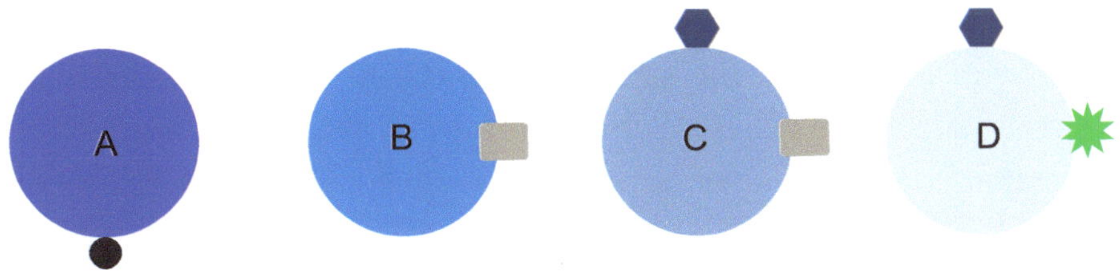

Fig. (2). Characteristics acquired and lost during retrodifferentiation or differentiation, respectively: Where (**D**), (**C**), (**B**) and (**A**) each represent a set of characteristics distinctive in cell development. Represented here as star and hexagon (more mature markers), rectangle (intermediate stage markers) and circle (stem cells markers) for simplicity.

RETRODIFFERENTIATION OF HUMAN LEUKOCYTES

Retrodifferentiation is a process by which mature adult cells are converted into stem cells by cell surface receptor contact. This process is highly dynamic, during which differentiation or cell ontogeny is rewound backwards in a stepwise fashion in order for specialized cells to attain a stem cell state. Events involved in the gradual reversion from the differentiated state to stemness equate to rewinding a video clip backwards. For example, if development of a given stem cell involves acquiring sequentially characteristics A, B, C and D during forward differentiation, reversion to a stem cell stage involves losing characteristic D first, then C, then B before reacquiring characteristic A, which is a stem cell marker in this case (Fig. **3**). This dynamic change in recapitulating development occurs rapidly, requiring sequential analysis of the reversion process, in order to account for every step in the conversion. Time lapse analysis of a retrodifferentiating heterogeneous population of cells belonging to different lineages and of known ontogonies are paramount to the elucidation of how differentiation is reversed to generate a heterogeneous population of stem cells. The most characterized tissue in the human body is the hematopoietic system, therefore deciphering retrodifferentiation in such complex tissue is much easier than others. Dedifferentiation (blastema formation) and retrodifferentiation are similar in many aspects. For example, specialized cell lose lineage-associated markers, migrate, and undergo homocytic aggregation, and once the heterogeneous stem cell cluster reach a certain size they embark on re-differentiation. However, retrodifferentiation is capable of transdifferentiation and histogenesis, giving rise to cell lineages not related to the original tissue. More importantly, retrodifferentiation occurs *in vitro* and in response to MHC class II beta chain ligation, which appears to emulate the stress equivalent to limb amputation in newts [96]. The difference between dedifferentiation and retrodifferentiation is most probably semantic, since sequence of events involved in the reversion process to a heterogeneous intermediary (Fig. **2**) and stem cell states (blastema formation), is more difficult to elucidate in an *in vivo* complex situation. It should be stressed here that transdifferentiation of a mature adult cell refers to traversal of the differentiation barrier, adopting a completely different specialization fate. While redifferentiation refers to re-ontogony to the same specialization fate from where the reversion process started.

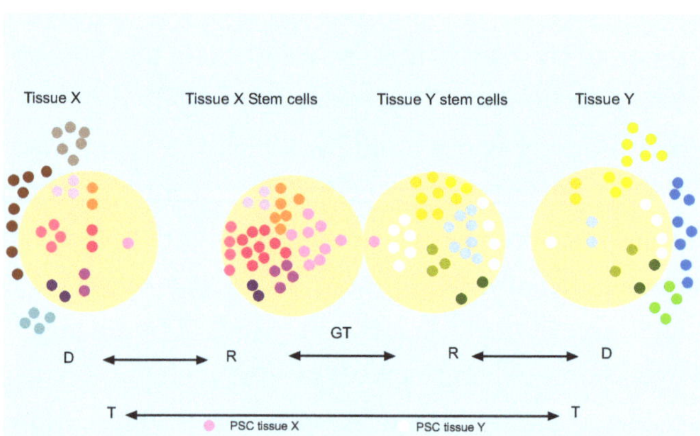

Fig. (3). The dynamic relationship between differentiation (**D**), retrodifferentiation (**R**), redifferentiation (**R** & **D**) and transdifferentiation (**T**). Large Light brown circle represents stem cell stage with hierarchical developmental potential, whereas more pluripotent (white and light pink circles labelled PSC) than others inside light brown circle. Cells occupying blue area are more committed specialized cells belonging to a specific tissue type depicted by coloured circles (tissue X) (tissue Y). Each circle represents a certain lineage in tissue X or Y. Arrow head depicts direction of conversion. Germ layer stem cells transition (GT) occurs once cells revert to a stem cell stage in response to specific culture condition leading to transdifferentiation and histogenesis.

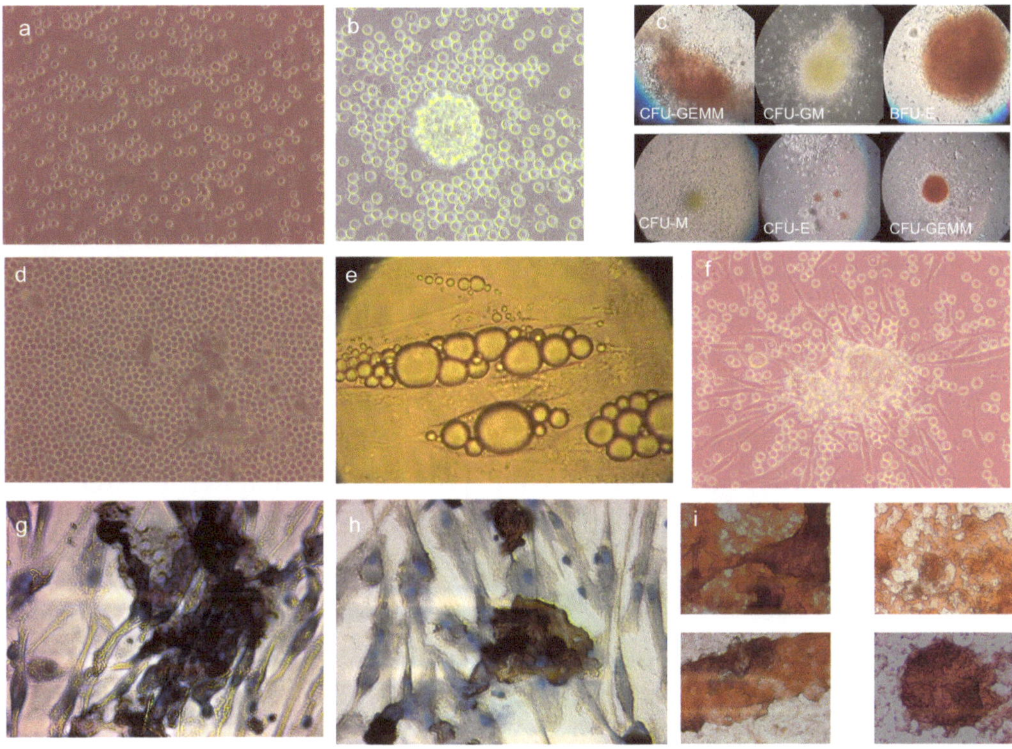

Fig. (4). Retrodifferentiation of human leukocytes (**a**) into the mesenchymal germ layer stem cells. Giving rise to the hematopoietic system (**b**) (redifferentiation) including cobble stone areas (**c**) with stromal and adipocytes (**d**) (under hematopoietic inductive culture condition). Alternatively leukocytes can be retrodifferentiated into osteogeneic progenitors, and upon feeding, transdifferentiate into bone nodules (**g**) producing cells staining red with alizarin red (**i**), with expression of osteopontin (**h**) and osteocalcin (**i**). Homocytic (**f**) aggregation is the hall mark of retrodifferentiation, and on transdifferentiation, cell colonies undergo re-morphogenesis (**b**) or trans-morphogenesis into another different tissue altogether (**e-l**).

PHENOTYPES OF THE RETRODIFFERENTIATED STEM CELL STATE

Ligation of monomorphic regions of the MHC class II antigen on human leukocytes with monoclonal antibody CR3/43, under hematopoietic-inducing culture conditions, promotes the loss of more mature hematopoietic markers such as the pan leukocyte antigens CD45 or the B cell specific markers CD19. Extinguishing of these lineage-associated antigens occurs during cell migration and homocytic aggregation in a cell culture dish. These changes have been observed using confocal microscopy and immunostaining using anti-human leukocyte markers such as CD 19 FITC or CD45 FITC with CD34 PE-CY5 conjugated [97]. Within minutes of inducing retrodifferentiation, leukocytes gradually extinguish lineage-associated markers CD19 or CD45 while homing to form clusters of cell colonies. This is always accompanied by upregulation of the hematopoietic marker CD34 by migratory cells and during homocytic aggregation. Immunophenotypic analysis using flow cytometry of mononuclear cells (MNC) before and 2 hrs and 24 hrs after induction of hematopoietic retrodifferentiation shows a significant increase in the relative number of cells expressing the hematopoietic stem cell marker CD34. The majority of CD34+ cells are CD45 low and are either positive or negative for the CD38 antigen, typical of committed and more primitive haematopoietic progenitor cells, respectively [98]. The latter type has been shown to possess more long-term SCID repopulating potential. A significant proportion of CD34+ cells co-express c-Kit or CD133 which increases by 24 hrs as they enter the cell cycle. Furthermore, an increase in the absolute number of myeloid and erythroid progenitors occurs indicating resumption of redifferentiation. The retrodifferentiated CD34 hematopoietic stem cells can be purified using anti-human CD34 magnetic beads with significant CD34 expression. Seeding of retrodifferentiated hematopoietic stem cells in methylcellulose containing human recombinant growth factors [99] results in redifferentiation to give rise to a variety of hematopoietic colonies (Fig. **4**) of different lineages such as colony-forming unit granulocyte, erythroid, monocyte, macrophage, megakaryocyte (CFU-GEMM), colony-forming unit granulocyte, monocyte, macrophage(CFU-GM), colony forming unit-monocyte, macrophage (CFU-M), blast forming unit-erythroid (BFU-E) and colony-forming unit erythroid (CFU-E). Interestingly, three hour retrodifferentiated hematopoietic stem cells derived from human male donors are capable of engraftment in minimally irradiated female non-obese diabetic/severe combined immunodeficient (NOD/SCID) mice. Human cell engraftment ensues in the bone marrow, thymus and spleen. This animal model is considered the gold standard of demonstrating potency of human hematopoietic stem cells [100]. Further characterization of engrafted cells demonstrate the integration of significant levels of human lymphohematopoietic cell lineages, comprised of T-lymphocytes, monocytes, erythrocytes and megakaryocytes, including platelets. In addition, significant levels of clonogenic human CD34+ cells differentiation when seeded in methylcellulose containing human recombinant factors. The thymi of engrafted animals contained maturating human thymocytes, while the spleen mainly consisted of T-lymphocytes. Fluorescence in situ hybridization (FISH) further identified the presence of human male X and Y chromosomes at engrafted sites, whilst the human origin of the cells was confirmed by a specific PCR assay for the human Cart-1 gene.

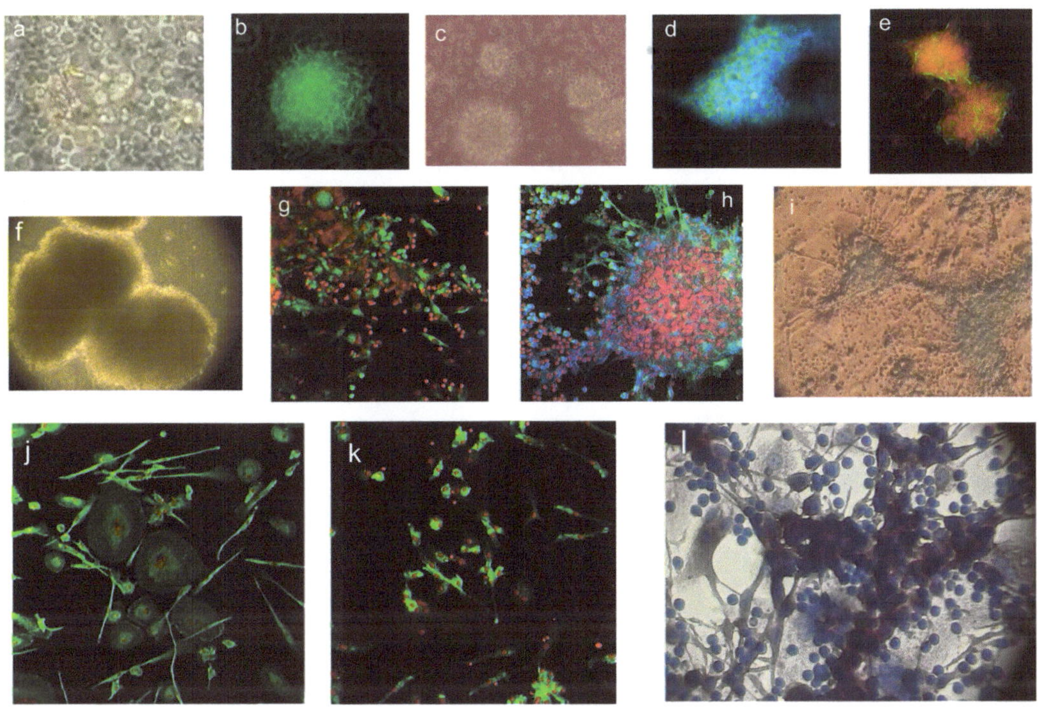

Fig. (5). Transient retrodifferentiation of human leukocytes (**a**) into OCT4 (**b** green) and nanog (**d** green) positive flat colonies upon induction in embryonic/neural culture condition. Purified undifferentiated cells capable of differentiation into embryoid body (**f**) when seeded in growth factor containing culture medium. Upon feeding, cell colonies transdifferentiate into nestin positive (**e** green) neurospheres (**c** phase contrast), which will extend neurites (**h**) and mature into neural axons (**i**) connecting with neighbouring neurospheres. Upon maturation, neurons will stain positive for TUBB3 (**K** green), Neurofilament (**j** green) and differentially for MAP2 (**I** brown and **h** green and blue) confined to cell body (blue) and tau (green) to axon. Glia stain with GFAP (**g** green).

Under neural/embryonic inductive culture conditions, and in response to ligation with CR3/43, reversion of leukocytes to colonies expressing OCT-4 and Nanog having flat morphology occur (Fig. **5**) [101, 102] by 24hrs in culture, as analyzed using immunohistochemistry. Furthermore, analysis using immunostaining and flow cytometry showed significant increase in the number of undifferentiated cells [103] that are both CD45 and CD34 negative. These cells can be purified or enriched by negative selection on a Midi Macs column using anti-human CD45 and glycophorin A-coated magnetic beads to exclude human leukocytes and red blood cells. The unbound cells were negative for hematopoietic-associated markers that are typically expressed by the erythroid, lymphoid and myeloid cell lineages including MHC class I and II antigens when analyzed using flow cytometry and immunostaining. In addition, such purified cells transcribe high levels of the embryonic stem cell marker OCT-4, and, to a lesser extent, nestin and CD34 and stain positive with antihuman OCT-4, using immunohistochemistry. These cells formed or differentiated into embryoid bodies when cultured in hanging drops or seeded in growth factors-containing methylcellulose culture media (Fig. **5**), respectively. Immunohistochemical analysis demonstrates expression of three germ layer markers: myocardial specific actin, alpha-fetoprotein, cytokertin 7 &20, desmin and s100.

Furthermore, induction of transdifferentiation on day 2 into neural cells occurs spontaneously on feeding with neural/embryonic inductive culture medium (without adding CR3/43 monoclonal antibody while feeding) (Fig. **5**). This is marked by expression of nestin by cell colonies, accompanied by down regulation of embryonic stem cell markers OCT-4 and NANOG as determined by immunohistochemistry. Upon nestin expression, colonies start to lose their round shape and project neurites which continue to extend progressively with time and subsequently reconnect to each other. Upon maturation into neural axons, they connect with other axons emanating from nearby colonies or neurospheres. Dual immunohistochemical staining of cultures at defined time points using antibodies to MAP-2 (microtubule associated protein 2) and Tau (T proteins), NF (neurofilamnets) and GFAP (glial fibrillary acidic protein) and oligodendrocyte and CD45, analyzed by confocal microscopy, showed differential expression of neural, glial and oligodendrocytic antigens [104]. For example, by 48 hrs, cellular spheres were co-stained positive for MAP-2 and Tau, indicative of neurospheres containing immature neurons. One and 2 weeks later, the spheres differentiated into mature neurons, with MAP-2 being confined to cell bodies and Tau to the axons (Fig. **5**). On the other hand, spheres that were analyzsed for NF and GFAP showed the presence of neural and glial precursors. However, 1 and 2 weeks later, such spheres differentiated into neurons expressing NF alone without GFAP, or astrocytic-like cells expressing GFAP without NF, the latter typical of glial cells. In contrast, co-staining for the pan-leukocyte marker CD45 and oligodendrocyte indicated the absence of CD45 expression at all-time points post retrodifferentiation, while oligodendrocyte staining was confined to the periphery of the spheres and persisted in 1 and 2 week old cultures. Moreover, time course analysis of neurotransmitters released by these neurons in response to depolarization revealed that, as they matured, the neurons produced increasing amounts of glutamate, GABA, tyrosine, dopamine and serotonin and uptake of taurine (unpublished data).

In contrast, 1 week post induction of human leukocyte retrodifferentiation and transdifferentiation under cardiopoietic [105] culture conditions grown on gelatinized six-well plates or chamber slides, cells transcribed the cardiac transcription factor GATA-4, human atrial natriuretic peptide (hANP) and cTnT. Dual immunohistochemical analysis of 1 week old cultures revealed the expression of myocardial-specific actin and cardiac-specific troponin I without CD45 expression. Three-day and 7-day human male leukocytes cultured under cardiac inductive conditions were capable of differentiating into fully mature cardiomyocytes when infused into non-irradiated Rnu/Rnu nude rats with needle stick damaged myocardia, and were integrated seamlessly into the area of damage. This was detected by XY chromosome *in situ* hybridization (FISH) and immunostaining for cardiac specific actin and HLA class I antigens (unpublished data).

Induction of human leukocyte retrodifferentiation under mesenchymal stem cell/ osteogeneic culture conditions (unpublished data) gave rise to colonies expressing the mesenchymal stem cell markers CD105, CD73 and CD90 after 24 hrs culture with a concomitant decrease in hematopoietic lineage specific markers CD45, CD3, CD 33 and CD19. After a week in culture (excluding CR3/43 antibody), cell colonies transdifferentiated into bone nodules, forming spheres which stained red with alizarin dye, and when immunostained with osteopontin and osteocalcin showed marked expression [106] (Fig. **4**).

Induction of human leukocyte retrodifferentiation under endoderm-hepatogeneic culture conditions [107] gave rise to cell colonies expressing CD34 with CD45 after 24 hours in culture, which upon feeding transdifferentiated into hepatocytes (Fig. **6**), expressing albumin and cytokeratin 18 after a week in culture (unpublished observation).

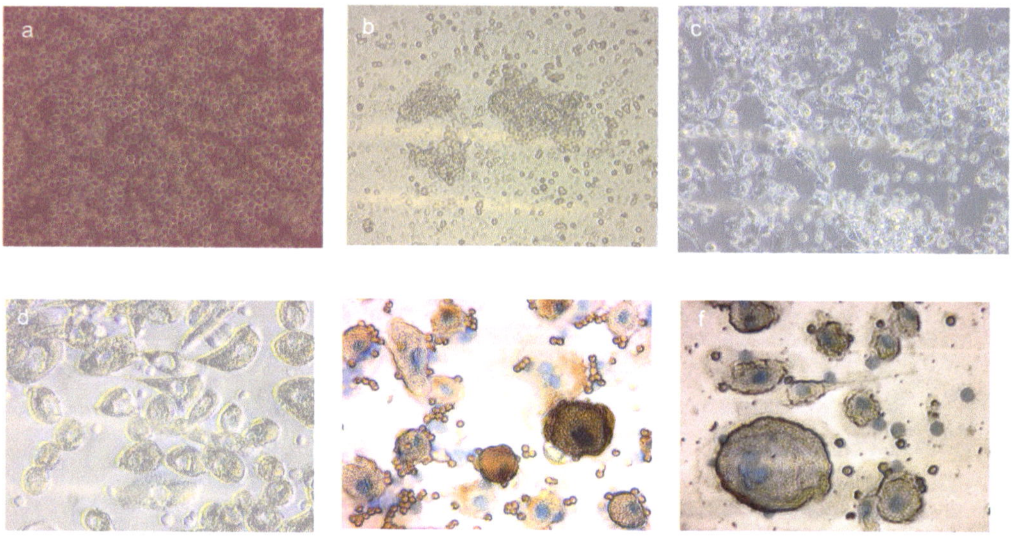

Cytokeratin 18 Human Albumin

Fig. (6). Leukocytes (**a**) retrodifferentiation and subsequent transdifferentiation into hepatocytes (**c-f**) staining positive cytokeratin 18 (**e** brown stain) and albumin (**f** brown stain blue stain show binucleated cells). Homocytic aggregation (**b**) always precedes transdifferentiation.

CLINICAL APPLICATION OF RETRODIFFERENTIATED STEM CELLS

Proofs of principle studies were carried out in two hematological conditions: acquired aplastic anemia and beta thalassemia major, following extensive analysis of the stability and safety of the retrodifferentiated hematopoietic stem cells (including karyotyping, g-banding and intradermal injection into mice). These studies showed no tumor development, and maintenance of normal karyotype even after 1 month in culture and 3 months in mice. Acquired Aplastic anaemia, is a life threatening rare bone marrow failure disorder. In this condition, depletion of the stem cell compartment, which is believed to be autoimmune in nature, gives rise to acellular bone marrow leading to hemorrhage, infection and severe anemia. The treatment of choice for this hematological condition is allogeneic bone marrow transplantation from fully HLA-matched sibling. Though this procedure is curative in the majority of young patients with aplastic anemia, extending this benefit to older patients or those lacking a family donor remains a major challenge. A proof of principle clinical study assessing the safety and efficacy of infusing autologous retrodifferentiated hematopoietic stem cells (RHSC) into four patients with aplastic anemia without the use of any pre- or post-conditioning regimen (including immunosuppression) was carried out. Without mobilization with growth factors, mononuclear cells were harvested from four patients by aphaeresis. These cells were induced to retrodifferentiate under hematopoietic culture conditions for 3 hrs. Autologous RHSC were washed and infused. Following single infusion, two patients became transfusion independent and have remained so for more than 10 years. Karyotyping and G-banding analysis pre and post-procedure in all patients remained the same. This pilot study was the first to demonstrate the functional utility of reprogrammed fully-differentiated adult cells in a clinical setting.

Beta-thalassemia [108] is a genetic red blood cell disorder affecting the beta-globin chain of the adult hemoglobin gene. This results in excess accumulation of unpaired alpha-chain gene products, leading to reduced red blood cell life span and the development of severe anemia. Current treatment of this disease involves regular blood transfusion and chelation therapy to lower blood transfusion–induced iron overload. Fetal haemoglobin (HbF) switching agents have been proposed to treat genetic blood disorders, such as sickle cell anaemia and betathalassemia, in an effort to compensate for the dysfunctional form of the beta-globin chain in adult haemoglobin (HbA). The rationale behind this approach is to pair the excess normal alpha-globin chain with the alternative fetal gamma-chain to promote red blood cell survival and ameliorate the anemia.

Infusion of autologous 3 hrs retrodifferentiated hematopoietic stem cells (RHSC) into 21 patients with beta-thalassemia was carried out. Follow up over 6 months showed a reduced mean blood transfusion requirement, increased mean fetal hemoglobin synthesis, and significantly lowered mean serum ferritin. This was always accompanied by an increase in mean corpuscular volume (MCV), mean corpuscular haemoglobin (MCH), and mean corpuscular hemoglobin concentration (MCHC). No adverse side effects in response to the infusion of autologous RHSC were noted. Since it was autologous, the procedure may have ameliorated the course of this genetic disorder through transient expression of HbF which was not maintained by 12 months post-treatment (unpublished data) due to reversion to adult hemoglobin synthesis. HbF to HbA switching occurs by 6 months in infants suffering from beta thalassemia [109].

More recently, a proof of principle clinical trial was conducted on patients with stage 4 liver cirrhosis having HCV infection. Though the results were preliminary, infusion of stem cells created under hepatogenic/endoderm culture conditions in these end stage patients significantly improved liver enzymes, such as albumin, ALT (alanine aminotransferase) and AST (aspartate aminotransferase), accompanied by increased % prothrombin concentration, alpha-fetoprotein and platelets counts. Up to six months post-treatment, fibroscan demonstrates amelioration of cirrhosis, changing from F4 to F3 in some patients. More importantly, the majority of patients did not require exogenous administration albumin due to non-occurrence of ascite (unpublished data).

CONCLUDING REMARKS

The ability of a group of specialized cells to reverse differentiate and give rise to blastema with the ability to redifferentiate and embark on remorphogenesis is tremendous. Equally significant is the ability of a heterogeneous population of human leukocytes to retrodifferentiate and form a heterogeneous population of stem cells with varied development potential. Moreover, pivotal to both mechanisms is the formation of cluster/s of cells capable of developing a precise three-dimensional configuration to offer the promise to reconstruct complex tissue or body parts. Retrodifferentiation enables the formation of human tissue *ex-vivo,* since it leads to transdifferentiation in response to altering culture media and conditions. Retrodifferentiation is a *bona fide* phenomenon, since rearranged immunoglobulin genes in B-cells and T-cell receptors revert to germ-line configurations in leukaemic cells and lymphoid cell lines [110]. This latter finding may be contentious, but evidence does exist. Preliminarily data involving sequential analysis of the transcriptional profiling of retrodifferentiated leukocytes transdifferentiated into neural tissue showed marked changes within 3 hrs of inducing retrodifferentiation. There were substantial changes in gene expression, up to 100-fold, associated with retrodifferentiation, which were visible as quickly as 3 hrs, with many sustained for 24 hrs. The transcripts changing during retrodifferentiation have previously been categorized, generically, in the 'response to stress' of the immune system, wounding, Wnt signaling pathway and pluripotency, senescence and autophagy, Matrix Metalloproteinases and pluripotency genes.

More specifically, pathways involved in integrin and focal adhesion, interferon signaling, interleukin signaling, and chemokines are the most strongly affected.

The conversion to the adherent state into nervous tissue is associated with more profound changes in gene expression, with at least 15 transcripts activated more than 100-fold. Conversion to the adherent state is associated with activation of extracellular matrix and matrix-binding proteins, and striking changes in chemokines, including expression of neural antigens and neurotransmitters including neurotransmission signaling pathways. Interestingly, transcripts such as matrix metalloproteinase (which is believed to prevent scar formation), immune response to stress and wounding, autophagy and senescence, and transcripts indicative of re-capitulating development have also been observed during epiomorphogenic regeneration.

More recently, emulation of blastema formation in aged human mesenchymal/ stromal cells by organizing cells in a 3-dimensional condensate [111] showed reversion of the differentiated state to an earlier stage of mesenchymal cell development as evidenced by switching from a CXCL12High CXCR4– to a CXCL12Low CXCR4+ Brachyury+ phenotype. An optimal autophagic response was required during dedifferentiation which promoted cytoplasmic remodeling, mitochondrial regression, and a bioenergetic shift from oxidative phosphorylation to anaerobic metabolism. Importantly, the dedifferentiated MSCs were capable of generating histologically-organized mesodermal structures when implanted *in vivo*; a property lacking in the parent, non-dedifferentiated MSC population grown in standard 2D conditions.

Finally, the ease by which human leukocytes can be harvested from peripheral blood using apheresis and the robustness of retrodifferentiation / transdifferentiation processes has allowed the design of single use kits for the production of a variety of stem cell types (manuscript in preparation). The automation of stem cells production for the regeneration of any tissue type with 100% efficiency, specificity and sterility will facilitate the wider application of stem cells therapy for many hematological and degenerative diseases for all age groups, thereby reducing waiting lists and the time to receiving stem cell therapy.

The ultimate goal of regenerative medicine is to regenerate without scars, reconstructing much needed tissue and complex organs to replace diseased or aged ones which can be achieved more readily and safely [112] by harnessing dedifferentiation or retrodifferentiation. For example, the combination of retrodifferentiation and transdifferentiation with bioprinting devices [113] will allow reconstruction of organs using a variety of scaffold materials which will propel regenerative medicine into human epi-morphogenesis with tremendous benefits and applications both *ex vivo* and *in vivo*.

CONFLICT OF INTEREST

The author confirms that the author has no conflict of interest to declare for this publication.

ACKNOWLEDGEMENTS

I would like to thank Dr. Keith Redpath for proof reading and editing of this chapter.

REFERENCES

[1] Abuljadayel IS. Induction of stem cell-like plasticity in mononuclear cells derived from unmobilised adult human peripheral blood. Curr Med Res Opin 2003; 19(5): 355-75. [http://dx.doi.org/10.1185/030079903125001901] [PMID: 13678473]

[2] Chaar ZY, Tsilfidis C. Newt opportunities for understanding the dedifferentiation process. ScientificWorldJournal 2006; 6 (Suppl. 1): 55-64. [http://dx.doi.org/10.1100/tsw.2006.327] [PMID: 17205187]

[3] Smith ME, Holgate CS, Williamson JM, Grigor I, Quirke P, Bird CC. Major histocompatibility complex class II antigen expression in B and T cell non-Hodgkins lymphoma. J Clin Pathol 1987; 40(1): 34-41.

[http://dx.doi.org/10.1136/jcp.40.1.34] [PMID: 3546388]

[4] Abuljadayel IS, Afghan RK, McCaffrey TA, *et al.* SCID repopulating cells derived from unmobilised adult human peripheral blood. Curr Med Res Opin 2004; 20(1): 87-100. [http://dx.doi.org/10.1185/030079903125002766] [PMID: 14741077]

[5] Abuljadayel IS, Mohanty D. Indian J Med Res 2012; 135: 106-25.

[6] Abuljadayel IS, Quereshi H, Ahsan T, *et al.* Infusion of autologous retrodifferentiated stem cells into patients with beta-thalassemia. ScientificWorldJournal 2006; 6: 1278-97. [http://dx.doi.org/10.1100/tsw.2006.229] [PMID: 17041717]

[7] Abuljadayel IS. Harnessing pluripotency from differentiated cells: a regenerative source for tissue-specific stem cell therapies. Curr Stem Cell Res Ther 2006; 1(3): 325-31. [http://dx.doi.org/10.2174/157488806778226812] [PMID: 18220877]

[8] Kumar V, Abbas AK, Fausto N, Aster JC. Tissue renewal, regeneration and repair. Pathologic Basis of Disease. Eighth ed. Philadelphia: Elsevier 2010; pp. 79-110.

[9] Krafts KP. Tissue repair: The hidden drama. Organogenesis 2010; 6(4): 225-33. [http://dx.doi.org/10.4161/org.6.4.12555] [PMID: 21220961]

[10] Mendelson A, Frenette PS. Hematopoietic stem cell niche maintenance during homeostasis and regeneration. Nat Med 2014; 20(8): 833-46. [http://dx.doi.org/10.1038/nm.3647] [PMID: 25100529]

[11] Lucas D, Scheiermann C, Chow A, *et al.* Chemotherapy-induced bone marrow nerve injury impairs hematopoietic regeneration. Nat Med 2013; 19(6): 695-703. [http://dx.doi.org/10.1038/nm.3155] [PMID: 23644514]

[12] Pereira C, Clarke E, Damen J. Hematopoietic colony-forming cell assays. Methods Mol Biol 2007; 407: 177-208. [http://dx.doi.org/10.1007/978-1-59745-536-7_14] [PMID: 18453257]

[13] von Vietinghoff S, Ley K. Homeostatic regulation of blood neutrophil counts. J Immunol 2008; 181(8): 5183-8. [http://dx.doi.org/10.4049/jimmunol.181.8.5183] [PMID: 18832668]

[14] Alenzi FQ, Alenazi BQ, Ahmad SY, *et al.* The haemopoietic stem cell: between apoptosis and self renewal. Yale J Biol Med 2009; 82(1): 7-18. [PMID: 19325941]

[15] Liu Z, Yang A, Wang Z, *et al.* Multiple apoptotic defects in hematopoietic cells from mice lacking lipocalin 24p3. J Biol Chem 2011; 286(23): 20606-14. [http://dx.doi.org/10.1074/jbc.M110.216549] [PMID: 21507940]

[16] Pietras EM1, Lakshminarasimhan R, Techner JM, *et al.* Re-entry into quiescence protects hematopoietic stem cells from the killing effect of chronic exposure to type I interferons. J Exp Med 2014; 211(2): 245-62.

[17] Baldridge MT, King KY, Goodell MA. Inflammatory signals regulate hematopoietic stem cells. Trends Immunol 2011; 32(2): 57-65. [http://dx.doi.org/10.1016/j.it.2010.12.003] [PMID: 21233016]

[18] Scheinberg P, Young NS. How I treat acquired aplastic anemia. Blood 2012; 120(6): 1185-96.

[19] Zeng Y, Katsanis E. The complex pathophysiology of acquired aplastic anaemia. Clin Exp Immunol 2015; 180(3): 361-70.
[http://dx.doi.org/10.1111/cei.12605] [PMID: 25683099]

[20] Bochtler T, Fröhling S, Krämer A. Role of chromosomal aberrations in clonal diversity and progression of acute myeloid leukemia. Leukemia 2015; 29(6): 1243-52.
[http://dx.doi.org/10.1038/leu.2015.32] [PMID: 25673237]

[21] Zhang S, Kipps TJ. The pathogenesis of chronic lymphocytic leukemia. Annu Rev Pathol 2014; 9: 103-18.
[http://dx.doi.org/10.1146/annurev-pathol-020712-163955] [PMID: 23987584]

[22] Burt RK, Barr W, Oyama Y, Traynor A, Slavin S. Future strategies in hematopoietic stem cell transplantation for rheumatoid arthritis. J Rheumatol Suppl 2001; 64(1): 42-8.
[PMID: 11642504]

[23] Tiniakos DG, Kandilis A, Geller SA. Tityus: a forgotten myth of liver regeneration. J Hepatol 2010; 53(2): 357-61.
[http://dx.doi.org/10.1016/j.jhep.2010.02.032] [PMID: 20472318]

[24] Michalopoulos GK. Liver regeneration after partial hepatectomy: critical analysis of mechanistic dilemmas. Am J Pathol 2010; 176(1): 2-13.
[http://dx.doi.org/10.2353/ajpath.2010.090675] [PMID: 20019184]

[25] Bataller R, Brenner DA. Liver fibrosis. J Clin Invest 2005; 115(2): 209-18.
[http://dx.doi.org/10.1172/JCI24282] [PMID: 15690074]

[26] Berenguer M, García-Eliz M, Baiguera C, Puoti M. Beyond cure: preventing and managing the complications of end-stage liver disease. Curr Opin HIV AIDS 2015; 10(5): 355-60.
[http://dx.doi.org/10.1097/COH.0000000000000184] [PMID: 26248123]

[27] Laflamme MA, Murry CE. Heart regeneration. Nature 2011; 473(7347): 326-35.
[http://dx.doi.org/10.1038/nature10147] [PMID: 21593865]

[28] Fan D, Takawale A, Lee J, Kassiri Z. Cardiac fibroblasts, fibrosis and extracellular matrix remodeling in heart disease. Fibrogenesis Tissue Repair 2012; 5(1): 15.
[http://dx.doi.org/10.1186/1755-1536-5-15] [PMID: 22943504]

[29] Nguyen TP, Qu Z, Weiss JN. Cardiac fibrosis and arrhythmogenesis: the road to repair is paved with perils. J Mol Cell Cardiol 2014; 70: 83-91.
[http://dx.doi.org/10.1016/j.yjmcc.2013.10.018] [PMID: 24184999]

[30] Kumta S, Ritz M, Hurley JV, Crowe D, Romeo R, OBrien BM. Acute inflammation in foetal and adult sheep: the response to subcutaneous injection of turpentine and carrageenan. Br J Plast Surg 1994; 47(5): 360-8.
[http://dx.doi.org/10.1016/0007-1226(94)90096-5] [PMID: 8087375]

[31] Kishi K, Okabe K, Shimizu R, Kubota Y. Fetal skin possesses the ability to regenerate completely: complete regeneration of skin. Keio J Med 2012; 61(4): 101-8.
[http://dx.doi.org/10.2302/kjm.2011-0002-IR] [PMID: 23324304]

[32] Ferguson MW, Duncan J, Bond J, *et al.* Prophylactic administration of avotermin for improvement of skin scarring: three double-blind, placebo-controlled, phase I/II studies. Lancet 2009; 373(9671): 1264-74.
[http://dx.doi.org/10.1016/S0140-6736(09)60322-6] [PMID: 19362676]

[33] Spallanzani L. An Essay on Animal Reproductions. London: Becket T 1769.

[34] Repesh LA, Oberpriller JC. Ultrastructural studies on migrating epidermal cells during the wound healing stage of regeneration in the adult newt, Notophthalmus viridescens. Am J Anat 1980; 159(2): 187-208.
[http://dx.doi.org/10.1002/aja.1001590207] [PMID: 6160761]

[35] Repesh LA, Oberpriller JC. Scanning electron microscopy of epidermal cell migration in wound healing during limb regeneration in the adult newt, Notophthalmus viridescens. Am J Anat 1978; 151(4): 539-55.
[http://dx.doi.org/10.1002/aja.1001510408] [PMID: 645617]

[36] Christensen RN, Tassava RA. Apical epithelial cap morphology and fibronectin gene expression in regenerating axolotl limbs. Dev Dyn 2000; 217(2): 216-24.
[http://dx.doi.org/10.1002/(SICI)1097-0177(200002)217:2<216::AID-DVDY8>3.0.CO;2-8] [PMID: 10706145]

[37] Butler EG. The effects of x-radiation on the regeneration of the forelimb of Amblystoma larvae. J Exp Zool 1933; 65: 271-315.
[http://dx.doi.org/10.1002/jez.1400650302]

[38] Butler EG. Studies on limb regeneration in x-rayed Amblystoma larvae. Anat Rec 1935; 62: 295-307.
[http://dx.doi.org/10.1002/ar.1090620308]

[39] Hay ED, Fischman DA. Origin of the blastema in regenerating limbs of the newt Triturus viridescens. An autoradiographic study using tritiated thymidine to follow cell proliferation and migration. Dev Biol 1961; 3: 26-59.
[http://dx.doi.org/10.1016/0012-1606(61)90009-4] [PMID: 13712434]

[40] Hay ED. Electron microscopic observations of muscle dedifferentiation in regenerating Amblystoma limbs. Dev Biol 1959; 1: 555-85.
[http://dx.doi.org/10.1016/0012-1606(59)90018-1]

[41] Brockes JP, Lo DC. Reversibility of the mononucleate-to-multinucleate myogenic transition during amphibian limb regeneration. Eye (Lond) 1994; 8(Pt 2): 151-4.
[http://dx.doi.org/10.1038/eye.1994.38] [PMID: 7958014]

[42] Casimir CM, Gates PB, Patient RK, Brockes JP. Evidence for dedifferentiation and metaplasia in amphibian limb regeneration from inheritance of DNA methylation. Development 1988; 104(4): 657-68.
[PMID: 3268408]

[43] Echeverri K, Clarke JD, Tanaka EM. *In vivo* imaging indicates muscle fiber dedifferentiation is a major contributor to the regenerating tail blastema. Dev Biol 2001; 236(1): 151-64.
[http://dx.doi.org/10.1006/dbio.2001.0312] [PMID: 11456451]

[44] Echeverri K, Tanaka EM. Ectoderm to mesoderm lineage switching during axolotl tail regeneration.

Science 2002; 298(5600): 1993-6.
[http://dx.doi.org/10.1126/science.1077804] [PMID: 12471259]

[45] Lo DC, Allen F, Brockes JP. Reversal of muscle differentiation during urodele limb regeneration. Proc Natl Acad Sci USA 1993; 90(15): 7230-4.
[http://dx.doi.org/10.1073/pnas.90.15.7230] [PMID: 8346239]

[46] Steen TP. Stability of chondrocyte differentiation and contribution of muscle to cartilage during limb regeneration in the axolotl *(Siredon mexicanum)*. J Exp Zool 1968; 167(1): 49-78.
[http://dx.doi.org/10.1002/jez.1401670105] [PMID: 5646635]

[47] Steen TP. Origin and differentiative capacities of cells in the blastema of the regenerating salamander limb. Am Zool 1970; 10(2): 119-32.
[http://dx.doi.org/10.1093/icb/10.2.119] [PMID: 5426254]

[48] Wallace H. The components of regrowing nerves which support the regeneration of irradiated salamander limbs. J Embryol Exp Morphol 1972; 28(2): 419-35.
[PMID: 4642994]

[49] Wallace H, Maden M, Wallace BM. Participation of cartilage grafts in amphibian limb regeneration. J Embryol Exp Morphol 1974; 32(2): 391-404.
[PMID: 4463210]

[50] Globus M, Vethamany-Globus S, Lee YC. Effect of apical epidermal cap on mitotic cycle and cartilage differentiation in regeneration blastemata in the newt, *Notophthalmus viridescens*. Dev Biol 1980; 75(2): 358-72.
[http://dx.doi.org/10.1016/0012-1606(80)90169-4] [PMID: 7372003]

[51] Mescher AL. Effects on adult newt limb regeneration of partial and complete skin flaps over the amputation surface. J Exp Zool 1976; 195(1): 117-28.
[http://dx.doi.org/10.1002/jez.1401950111] [PMID: 1255117]

[52] Tassava RA, Garling DJ. Regenerative responses in larval axolotl limbs with skin grafts over the amputation surface. J Exp Zool 1979; 208(1): 97-110.
[http://dx.doi.org/10.1002/jez.1402080111] [PMID: 381569]

[53] Singer M. The influence of the nerve in regeneration of the amphibian extremity. Q Rev Biol 1952; 27(2): 169-200.
[http://dx.doi.org/10.1086/398873] [PMID: 14949331]

[54] Brockes JP. Mitogenic growth factors and nerve dependence of limb regeneration. Science 1984; 225(4668): 1280-7.
[http://dx.doi.org/10.1126/science.6474177] [PMID: 6474177]

[55] Rose SM. The role of nerves in amphibian limb regeneration. Ann N Y Acad Sci 1948; 49(Art 5): 818-33.
[http://dx.doi.org/10.1111/j.1749-6632.1948.tb30966.x] [PMID: 18882684]

[56] Thornton CS. Amphibian limb regeneration and its relation to nerves. Am Zool 1970; 10(2): 113-8.
[http://dx.doi.org/10.1093/icb/10.2.113] [PMID: 5426253]

[57] Wallace H, Watson A, Egar M. Regeneration of subnormally innervated axolotl arms. J Embryol Exp Morphol 1981; 62: 1-11.

[PMID: 7276803]

[58] Wallace H. Vertebrate Limb Regeneration. Toronto: John Wiley & Sons 1981.

[59] Jarvik E. Basic Structure and Evolution of Vertebrates. London: Academic Press 1980; 1.

[60] Vlaskalin T, Wong CJ, Tsilfidis C. Growth and apoptosis during larval forelimb development and adult forelimb regeneration in the newt *(Notophthalmus viridescens)*. Dev Genes Evol 2004; 214(9): 423-31.
[http://dx.doi.org/10.1007/s00427-004-0417-1] [PMID: 15322877]

[61] McCusker CD, Gardiner DM. Understanding positional cues in salamander limb regeneration: implications for optimizing cell-based regenerative therapies. Dis Model Mech 2014; 7(6): 593-9.
[http://dx.doi.org/10.1242/dmm.013359] [PMID: 24872456]

[62] Bryant SV, French V, Bryant PJ. Distal regeneration and symmetry. Science 1981; 212(4498): 993-1002.
[http://dx.doi.org/10.1126/science.212.4498.993] [PMID: 17779956]

[63] French V, Bryant PJ, Bryant SV. Pattern regulation in epimorphic fields. Science 1976; 193(4257): 969-81.
[http://dx.doi.org/10.1126/science.948762] [PMID: 948762]

[64] Crawford K, Stocum DL. Retinoic acid coordinately proximalizes regenerate pattern and blastema differential affinity in axolotl limbs. Development 1988; 102(4): 687-98.
[PMID: 3168786]

[65] Endo T, Bryant SV, Gardiner DM. A stepwise model system for limb regeneration. Dev Biol 2004; 270(1): 135-45.
[http://dx.doi.org/10.1016/j.ydbio.2004.02.016] [PMID: 15136146]

[66] Satoh A, Gardiner DM, Bryant SV, Endo T. Nerve-induced ectopic limb blastemas in the Axolotl are equivalent to amputation-induced blastemas. Dev Biol 2007; 312(1): 231-44.
[http://dx.doi.org/10.1016/j.ydbio.2007.09.021] [PMID: 17959163]

[67] Tank PW. The effect of nonlimb tissues on forelimb regeneration in the axolotl, *Ambystoma-mexicanum*. J Exp Zool 1987; 244: 409-23.
[http://dx.doi.org/10.1002/jez.1402440308]

[68] Gardiner DM, Blumberg B, Komine Y, Bryant SV. Regulation of HoxA expression in developing and regenerating axolotl limbs. Development 1995; 121(6): 1731-41.
[PMID: 7600989]

[69] Maden M. Retinoids as endogenous components of the regenerating limb and tail. Wound Repair Regen 1998; 6(4): 358-65.
[http://dx.doi.org/10.1046/j.1524-475X.1998.60411.x] [PMID: 9824554]

[70] Niazi IA, Saxena S. Abnormal hind limb regeneration in tadpoles of the toad, Bufo andersoni, exposed to excess vitamin A. Folia Biol (Krakow) 1978; 26(1): 3-8.
[PMID: 648678]

[71] Riddle RD, Johnson RL, Laufer E, Tabin C. Sonic hedgehog mediates the polarizing activity of the ZPA. Cell 1993; 75(7): 1401-16.
[http://dx.doi.org/10.1016/0092-8674(93)90626-2] [PMID: 8269518]

[72] Simon HG, Kittappa R, Khan PA, Tsilfidis C, Liversage RA, Oppenheimer S. A novel family of T-box genes in urodele amphibian limb development and regeneration: candidate genes involved in vertebrate forelimb/hindlimb patterning. Development 1997; 124(7): 1355-66.
[PMID: 9118806]

[73] Gibson-Brown JJ, Agulnik SI, Silver LM, Niswander L, Papaioannou VE. Involvement of T-box genes Tbx2-Tbx5 in vertebrate limb specification and development. Development 1998; 125(13): 2499-509.
[PMID: 9609833]

[74] Boilly B, Cavanaugh KP, Thomas D, Hondermarck H, Bryant SV, Bradshaw RA. Acidic fibroblast growth factor is present in regenerating limb blastemas of axolotls and binds specifically to blastema tissues. Dev Biol 1991; 145(2): 302-10.
[http://dx.doi.org/10.1016/0012-1606(91)90128-P] [PMID: 2040374]

[75] Poulin ML, Patrie KM, Botelho MJ, Tassava RA, Chiu IM. Heterogeneity in the expression of fibroblast growth factor receptors during limb regeneration in newts *(Notophthalmus viridescens)*. Development 1993; 119(2): 353-61.
[PMID: 8287792]

[76] Mullen LM, Bryant SV, Torok MA, Blumberg B, Gardiner DM. Nerve dependency of regeneration: the role of Distal-less and FGF signaling in amphibian limb regeneration. Development 1996; 122(11): 3487-97.
[PMID: 8951064]

[77] DJamoos CA, McMahon G, Tsonis PA. Fibroblast growth factor receptors regulate the ability for hindlimb regeneration in Xenopus laevis. Wound Repair Regen 1998; 6(4): 388-97.
[PMID: 9824558]

[78] Martin GR. The roles of FGFs in the early development of vertebrate limbs. Genes Dev 1998; 12(11): 1571-86.
[http://dx.doi.org/10.1101/gad.12.11.1571] [PMID: 9620845]

[79] Yokoyama H, Yonei-Tamura S, Endo T, Izpisúa Belmonte JC, Tamura K, Ide H. Mesenchyme with fgf-10 expression is responsible for regenerative capacity in Xenopus limb buds. Dev Biol 2000; 219(1): 18-29.
[http://dx.doi.org/10.1006/dbio.1999.9587] [PMID: 10677252]

[80] Yokoyama H, Ide H, Tamura K. FGF-10 stimulates limb regeneration ability in *Xenopus laevis*. Dev Biol 2001; 233(1): 72-9.
[http://dx.doi.org/10.1006/dbio.2001.0180] [PMID: 11319858]

[81] Miyazaki K, Uchiyama K, Imokawa Y, Yoshizato K. Cloning and characterization of cDNAs for matrix metalloproteinases of regenerating newt limbs. Proc Natl Acad Sci USA 1996; 93(13): 6819-24.
[http://dx.doi.org/10.1073/pnas.93.13.6819] [PMID: 8692902]

[82] Vinarsky V, Atkinson DL, Stevenson TJ, Keating MT, Odelberg SJ. Normal newt limb regeneration requires matrix metalloproteinase function. Dev Biol 2005; 279(1): 86-98.
[http://dx.doi.org/10.1016/j.ydbio.2004.12.003] [PMID: 15708560]

[83] Tanaka EM, Brockes JP. A target of thrombin activation promotes cell cycle re-entry by urodele muscle cells. Wound Repair Regen 1998; 6(4): 371-81.
[http://dx.doi.org/10.1046/j.1524-475X.1998.60413.x] [PMID: 9824556]

[84] Tanaka EM, Drechsel DN, Brockes JP. Thrombin regulates S-phase re-entry by cultured newt myotubes. Curr Biol 1999; 9(15): 792-9.
[http://dx.doi.org/10.1016/S0960-9822(99)80362-5] [PMID: 10469562]

[85] Vascotto SG, Beug S, Liversage RA, Tsilfidis C. Expression profiles of elastase1 (NvElastaseI) and secretory leukocyte protease inhibitor (NvSLPI) during forelimb regeneration in adult *Notophthalmus viridescens* suggest a role in epithelial remodeling and delamination. Dev Genes Evol 2006; 216(9): 499-509.
[http://dx.doi.org/10.1007/s00427-006-0061-z] [PMID: 16508785]

[86] Odelberg SJ, Kollhoff A, Keating MT. Dedifferentiation of mammalian myotubes induced by msx1. Cell 2000; 103(7): 1099-109.
[http://dx.doi.org/10.1016/S0092-8674(00)00212-9] [PMID: 11163185]

[87] Simon HG, Nelson C, Goff D, Laufer E, Morgan BA, Tabin C. Differential expression of myogenic regulatory genes and Msx-1 during dedifferentiation and redifferentiation of regenerating amphibian limbs. Dev Dyn 1995; 202(1): 1-12.
[http://dx.doi.org/10.1002/aja.1002020102] [PMID: 7703517]

[88] Kumar A, Velloso CP, Imokawa Y, Brockes JP. The regenerative plasticity of isolated urodele myofibers and its dependence on MSX1. PLoS Biol 2004; 2(8): E218.
[http://dx.doi.org/10.1371/journal.pbio.0020218] [PMID: 15314647]

[89] Sandoval-Guzmán T, Wang H, Khattak S, *et al.* Fundamental differences in dedifferentiation and stem cell recruitment during skeletal muscle regeneration in two salamander species. Cell Stem Cell 2014; 14(2): 174-87.
[http://dx.doi.org/10.1016/j.stem.2013.11.007] [PMID: 24268695]

[90] Tata PR, Mou H, Pardo-Saganta A, *et al.* Dedifferentiation of committed epithelial cells into stem cells in vivo. Nature 2013; 503(7475): 218-23.
[PMID: 24196716]

[91] Gurdon JB, Uehlinger V. Fertile intestine nuclei. Nature 1966; 210(5042): 1240-1.
[http://dx.doi.org/10.1038/2101240a0] [PMID: 5967799]

[92] Campbell KH, McWhir J, Ritchie WA, Wilmut I. Sheep cloned by nuclear transfer from a cultured cell line. Nature 1996; 380(6569): 64-6.
[http://dx.doi.org/10.1038/380064a0] [PMID: 8598906]

[93] Takahashi K, Yamanaka S. Induction of pluripotent stem cells from mouse embryonic and adult fibroblast cultures by defined factors. Cell 2006; 126(4): 663-76.
[http://dx.doi.org/10.1016/j.cell.2006.07.024] [PMID: 16904174]

[94] Hu K. All roads lead to induced pluripotent stem cells: the technologies of iPSC generation. Stem Cells Dev 2014; 23(12): 1285-300.
[http://dx.doi.org/10.1089/scd.2013.0620] [PMID: 24524728]

[95] Maki N, Suetsugu-Maki R, Tarui H, Agata K, Del Rio-Tsonis K, Tsonis PA. Expression of stem cell

pluripotency factors during regeneration in newts. Dev Dyn 2009; 238(6): 1613-6. [http://dx.doi.org/10.1002/dvdy.21959] [PMID: 19384853]

[96] Mantegazza AR, Magalhaes JG, Amigorena S, Marks MS. Presentation of phagocytosed antigens by MHC class I and II. Traffic 2013; 14(2): 135-52. [http://dx.doi.org/10.1111/tra.12026] [PMID: 23127154]

[97] Live imaging of human leukocytes retrodifferentiation into stem cells. Available at: https://www.youtube.com/watch?v=5Gazt295VjM.

[98] Bhatia M, Wang JC, Kapp U, Bonnet D, Dick JE. Purification of primitive human hematopoietic cells capable of repopulating immune-deficient mice. Proc Natl Acad Sci USA 1997; 94(10): 5320-5. [http://dx.doi.org/10.1073/pnas.94.10.5320] [PMID: 9144235]

[99] de Wynter E, Ploemacher RE. Assays for the assessment of human hematopoietic stem cells. J Biol Regul Homeost Agents 2001; 15(1): 23-7. [PMID: 11388741]

[100] van der Loo JC, Hanenberg H, Cooper RJ, Luo FY, Lazaridis EN, Williams DA. Nonobese diabetic/severe combined immunodeficiency (NOD/SCID) mouse as a model system to study the engraftment and mobilization of human peripheral blood stem cells. Blood 1998; 92(7): 2556-70. [PMID: 9746798]

[101] Niwa H, Miyazaki J, Smith AG. Quantitative expression of Oct-3/4 defines differentiation, dedifferentiation or self-renewal of ES cells. Nat Genet 2000; 24(4): 372-6. [http://dx.doi.org/10.1038/74199] [PMID: 10742100]

[102] Chambers I, Silva J, Colby D, *et al.* Nanog safeguards pluripotency and mediates germline development. Nature 2007; 450(7173): 1230-4. [http://dx.doi.org/10.1038/nature06403] [PMID: 18097409]

[103] Thomson JA, Itskovitz-Eldor J, Shapiro SS, *et al.* Embryonic stem cell lines derived from human blastocysts. Science 1998; 282(5391): 1145-7. [http://dx.doi.org/10.1126/science.282.5391.1145] [PMID: 9804556]

[104] Palmer TD, Markakis EA, Willhoite AR, Safar F, Gage FH. Fibroblast growth factor-2 activates a latent neurogenic program in neural stem cells from diverse regions of the adult CNS. J Neurosci 1999; 19(19): 8487-97. [PMID: 10493749]

[105] Kehat I, Kenyagin-Karsenti D, Snir M, *et al.* Human embryonic stem cells can differentiate into myocytes with structural and functional properties of cardiomyocytes. J Clin Invest 2001; 108(3): 407-14. [http://dx.doi.org/10.1172/JCI200112131] [PMID: 11489934]

[106] Chaudhry GR, Yao D, Smith A, Hussain A. Osteogenic cells derived from embryonic stem cells produced bone nodules in three-dimensional scaffolds. J Biomed Biotechnol 2004; 2004(4): 203-10. [http://dx.doi.org/10.1155/S111072430431003X] [PMID: 15467160]

[107] Hannan NR, Segeritz CP, Touboul T, Vallier L. Production of hepatocyte-like cells from human pluripotent stem cells. Nat Protoc 2013; 8(2): 430-7. [http://dx.doi.org/10.1038/nprot.2012.153] [PMID: 23424751]

[108] Bank A. Understanding globin regulation in β-thalassemia: its as simple as α, β, γ, δ. J Clin Invest 2005; 115(6): 1470-3.
[http://dx.doi.org/10.1172/JCI25398] [PMID: 15931385]

[109] Miller JL. Signaled expression of fetal hemoglobin during development. Transfusion 2005; 45(7): 1229-32.
[http://dx.doi.org/10.1111/j.1537-2995.2005.00182.x] [PMID: 15987371]

[110] Abuljadayel IS. Memorandum: Retrodifferentiation of B lymphocytes into haematopoietic 'stem cells' induced by ligation of the MHC class II B chain. Stem Cell Research: Report from the Select Committee.

[111] Pennock R, Bray E, Pryor P, *et al.* Human cell dedifferentiation in mesenchymal condensates through controlled autophagy. Sci Rep 2015; 5(20): 13113.
[http://dx.doi.org/10.1038/srep13113] [PMID: 26290392]

[112] Fior J. Salamander regeneration as a model for developing novel regenerative and anticancer therapies. J Cancer 2014; 5(8): 715-9.
[http://dx.doi.org/10.7150/jca.9971] [PMID: 25258653]

[113] Gao G, Cui X. Three-dimensional bioprinting in tissue engineering and regenerative medicine. Biotechnol Lett 2015; 1-9.
[PMID: 26466597]

CHAPTER 2

The Mesendoderm: A Wellspring of Cell Lineages for Regenerative Medicine

Badwi B. Boumelhem[1], **Stephen J. Assinder**[1], **Christoph Hammans**[1], **Marietta P. Tanudisastro**[1], **Damien T. M. Le**[1], **Kurt W. L. Brigden**[1] and **Stuart T. Fraser**[1,2,*]

[1] *Discipline of Physiology, School of Medical Science, University of Sydney, Camperdown, New South Wales, Australia*

[2] *Discipline of Anatomy & Histology, School of Medical Science, University of Sydney, Camperdown, New South Wales, Australia*

Abstract: Regenerative medicine is centred around the premise that progenitor populations can be engineered to give rise to mature cell lineages forming a complex tissue architecture which in turn produces functional organs. The potency of the starting progenitor population is therefore a critical consideration. The mesendoderm is a rare population of cells present in the embryo only at gastrulation. This bipotent population gives rise to the mesoderm and the definitive endoderm and all mature cell types derived from these germ layers. Mesodermal progenitors generate cardiac, smooth and skeletal muscle, as well as the blood and vascular lineages, bone and connective tissue cells. The endoderm is the source of numerous cell lineages with potential utility for regenerative medicine including hepatocytes, pancreatic lineages and the epithelial cells of the respiratory, gastrointestinal and reproductive tracts. The development of numerous organs is dependent upon mesoderm-derived lineages interacting with endodermal-derived cell types. The kidney, adrenal gland, pancreas and genito-urinary tract development all require interactions between mesodermal and endodermal derivative cell types. Here, we describe the unique genetic programmes that lead to mesendoderm formation, the pathways leading to mesoderm and endoderm specification and examples where mature cell types from both germ layers interact to support their mutual development. We will also show how these programmes are being

[*] **Corresponding author Stuart T. Fraser:** Discipline of Anatomy & Histology, School of Medical Science, University of Sydney, Camperdown, New South Wales, Australia; Tel: +612 9036 3313; Fax: +612 9036 3316; E-mail: stuart.fraser@sydney.edu.au

Atta-ur-Rahman & Shazia Anjum (Eds.)

harnessed to direct the differentiation of pluripotent cells *in vitro* into mesendoderm-derived cells and tissues which can be used to improve the quality of human life. Finally, we will discuss considerations for combining stem cell differentiation with tissue engineering through 3D bioprinting modalities.

Keywords: 3D bioprinting, Embryonic development, Embryonic stem cells (ESC), Endoderm, Germ layer specification, Induced pluripotent stem cells (iPSC), Mesendoderm, Mesoderm, Regenerative medicine, Tissue engineering.

INTRODUCTION

The complex adult mammalian body is derived from three simple structures early in embryogenesis termed the germ layers. These are: the ectoderm, which forms the skin and central nervous system; the endoderm which gives rise to the epithelial tissues of the viscera such as the respiratory, gastrointestinal and genito-urinary tracts; and the mesoderm, which forms all of the connective tissue, blood, vessels and muscle tissues. These three germ layers, can be first distinguished at the developmental stage termed gastrulation. However, it has been proposed for some time that the mesoderm and endoderm arise from a single cell type with the potential to form both lineages. This cell type, termed the mesendoderm, can be identified in simpler animal models such as the frog embryo, and an equivalent cell type can be generated from mammalian pluripotent stem cells in culture. The mesendoderm is, in the end, responsible for the formation of essentially vast amounts of the body except for the brain, skin (derived from the ectoderm) and other tissues derived from a structure arising later in embryogenesis termed the neural crest.

All of the epithelial tissues contain mesoderm-derived lineages. For example, the gastro-intestinal epithelial tissue has a mesoderm-derived connective tissue component essential in maintaining structural integrity. Indeed, it is now clear that, in many organs, extensive cross-talk must take place between endoderm-derived tissues and mesodermal-derived structures during embryogenesis of organogenesis to proceed. The developing pancreas and liver (endoderm) require signals from underlying blood vessels (mesoderm) to form [1, 2]. Removal of the blood vessels leads to a loss of appropriate signals and failure of pancreatic development. In contrast, most mesoderm-derived organs lack epithelial structures. For example, the heart, skeletal muscle and bone marrow do not contain endoderm-derived cell types (Table **1**).

Table 1. Organs and cells derived from the mesendoderm.

Organ	Endoderm-derived Lineages	Mesoderm-derived Lineages
Salivary gland	Serous, mucosal and seromucosal epithelial cells	Connective tissue fibroblasts, macrophages endothelial cells, adipose cells
Trachea	Goblet cells columnar epithelial cells	Tracheal cartilage, lymphoid cells, macrophages
Lungs	Type I and II alveolar cells Clara/club cells, Goblet cells	Endothelial cells alveolar macrophages
Stomach	Surface mucous cells mucous neck cells entero-endocrine cells chief cells, parietal cells	Smooth muscle, adipocytes, endothelial cells fibroblasts, lymphoid cells
Liver	Hepatocyte cholangiocyte	Sinusoidal endothelial cell Kupffer cell, hepatic stellate cell
Gastrointestinal tract	Gastric epithelium glands (pyloric, cardiac, fundus)	Mesentric connective tissue fibroblasts smooth muscle, endothelial cells, lymphoid cells
Pancreas	Acinar cells, centroacinar cells pancreatic α, β and δ cells	Capillary endothelial cells connective tissue fibroblasts
Kidney	Tubular epithelial cells	Glomerular endothelial cells, mesangial cells podocytes, capsular stromal fibroblast
Prostate	Cuboidal epithelial cells columnar epithelial cells	Prostatic stroma fibroblasts, smooth muscle connective tissue
Urinary bladder	Urothelium	Smooth muscle, hematopoietic cells, endothelial cells
Bone		Osteocytes, osteoblasts, chondrocytes, endothelial cells, hematopoietic cells, adipocytes
Thymus	Cortical thymic epithelial cells medullary thymic epithelial cells	Thymocytes (developing T lymphocytes) dendritic cells, thymic macrophages
Spleen		Capsule stromal fibroblasts, erythrocytes lymphocytes, sinusoidal endothelium macrophages
Gonads		Leydig cells, Sertoli cells, follicular cells thecal cells

Understanding the biology underpinning mesendoderm development offers us insight in the formation of a vast amount of the human body. Defining the molecular processes that control mesendoderm formation and differentiation, as well as the formation of other similar bi-potent populations in the embryo, can be of great benefit to regenerative medicine. In particular, the *ex vivo* generation of functional mini-organs or organoids exhibiting appropriate morphology and tissue architecture and derived from a patient's own stem cells offers enormous potential for individuals with many pathological conditions. Successful 3D bioprinting and differentiation of stem cells into functional tissue equivalents as insulin-producing pancreatic islets, clusters of functional hepatocytes, lung or mini-kidney tissue will have a profound effect on a vast cohort of patients, most of whom are waiting for matching organ donors. However, an understanding of the embryological processes required in the formation of these tissues is required before we can proceed to generating new tissues *in vitro*. This chapter aims to inform the reader of the developmental processes taking place to form and differentiate the mesendoderm and to demonstrate how this basic biological knowledge can be applied to regenerative medicine *via* induced pluripotent stem cells, tissue engineering and 3D bioprinting.

The Germ Layers

In contrast to the formation of bi-laminar organisms such as jellyfish, mammalian embryogenesis requires the formation of three distinct three germ layers; ectoderm, endoderm and mesoderm. It is these three layers of cells that give rise to all of the different cell lineages in the adult human body. The ectoderm gives rise to the external layer of the skin and the central nervous system. The endoderm forms all of the major epithelial cell lineages giving rise to the different tracts of the body such as the respiratory, gastrointestinal and genitourinary systems. All of these systems must form tube-like structures with tight junctions forming between cells to seal the external environment from the internal body and yet still allow transport (of oxygen and nutrients from food) or secretion and excretion (of carbon dioxide, urine, faeces, sperm *etc.*). In complex animals, such as mammals, a third germ layer exists which forms all of the supportive structures between the external skin, central nervous systems and internal epithelial organs. Supportive structures such as the skeletal system, the integumentary system, blood and lymphatic vessels, heart, smooth and skeletal muscle and blood are all derived form the mesoderm.

The Mesendoderm

Investigations into the origins of the endoderm and mesoderm led to the concept that a bi-potent progenitor cell exists early in development capable of giving rise to each germ layer. This cell type has been termed “the mesendoderm”. Evidence for the existence of the mesendoderm has been collecting for decades [3]. When individual cells in early embryos where labelled and the fate of these cells explored, it was observed that some labelled single cells gave rise to cell types arising from both the mesoderm and the endoderm. In the developing nematode *C.elegans* at the 4 cell stage of development there is a single mesendodermal blastomere that, during the subsequent cleavage stage, will give rise to two daughter blastomeres: one a mesodermal progenitor and the other an endodermal progenitor [4]. Origins of the mesendoderm population were identified by removing cells fated to become ectoderm or endoderm, and then culturing both tissues in close proximity, resulting in the formation of cells of a mesodermal origin [5]. Studies on the nematodes and zebrafish have revealed a bi-potent cell type that is able to generate both mesoderm and endoderm. The tissues that are generated from the mesendoderm are shown in Fig. (**1**).

The existence of a bi-potent mesendodermal progenitor was supported by studies showing shared gene expression profiles in candidate mesendodermal cells across multiple species. The soluble factor Nodal (and related factors) was consistently identified as a key regulator of mesendoderm development in species as far ranging as the African clawed frog *Xenopus*, zebrafish and the mouse [6]. Binding of Nodal to its receptor activated phosphorylation of the Smad pathway inducing gene expression. Amongst the downstream targets of Nodal signalling is the transcriptional regulator Eomesodermin (*Eomes*) [7]. Eomesodermin then functions in multiple ways to focus differentiation along the mesendoderm lineages. Eomesodermin inhibits ectoderm differentiation [8]. Eomesodermin also induces expression of a range of transcription factors critical to mesendoderm development including Brachyury (Bry/*T*), Mix-like 1 (*Mixl1*) and Goosecoid (*Gsc*). These form transcriptional complexes that regulate mesendoderm differentiation. Lhx1, another downstream target of Eomesodermin also partners with other transciptional regulators that play critical roles in mesendoderm development [9]. A grossly simplified model of mesendoderm formation is presented in Fig. (**2**).

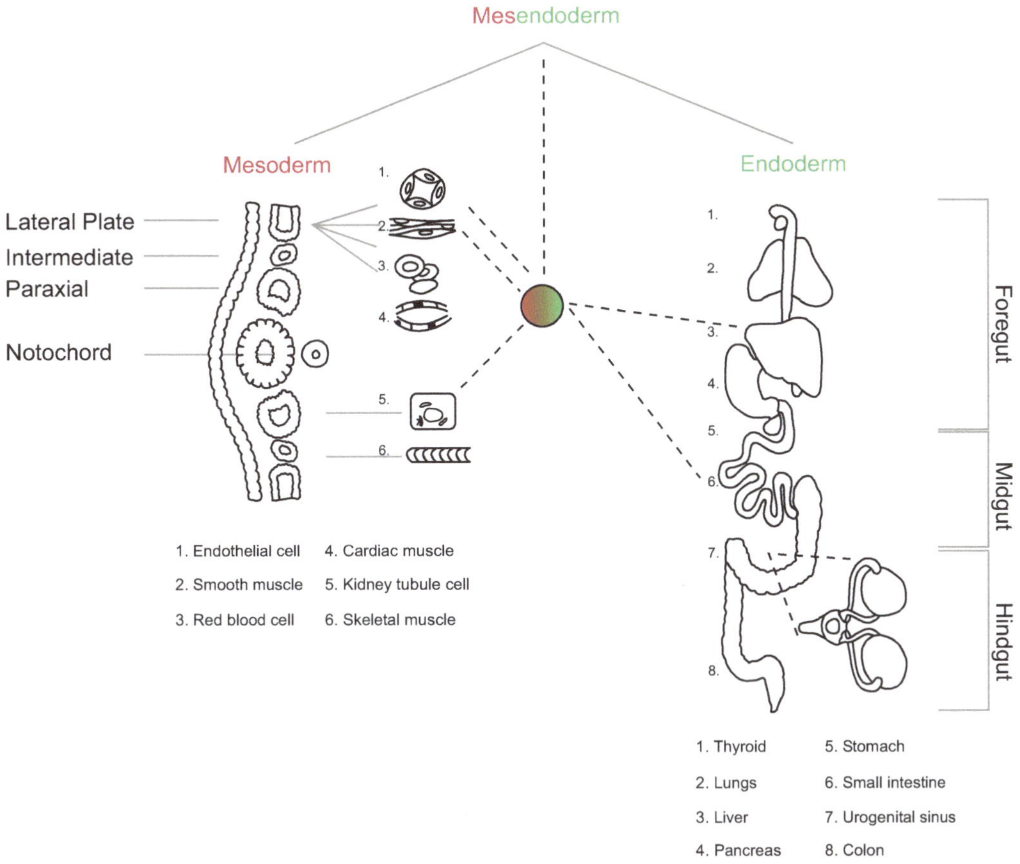

Fig. (1). Derivatives of the mesendoderm. The mesendoderm gives rise to cell types from both the mesoderm (red) and endoderm (green) germ layers. Mesoderm derivatives include: endothelial cells, smooth muscle cells, red blood cells and cardiac cells from the lateral plate mesoderm; kidney tubule cells from intermediate mesoderm and skeletal muscle from the paraxial mesoderm. Endoderm derivatives include: the thyroid, lungs, liver and pancreas from the foregut endoderm; stomach and small intestine from the midgut endoderm and urogenital sinus and colon from the hindgut endoderm.

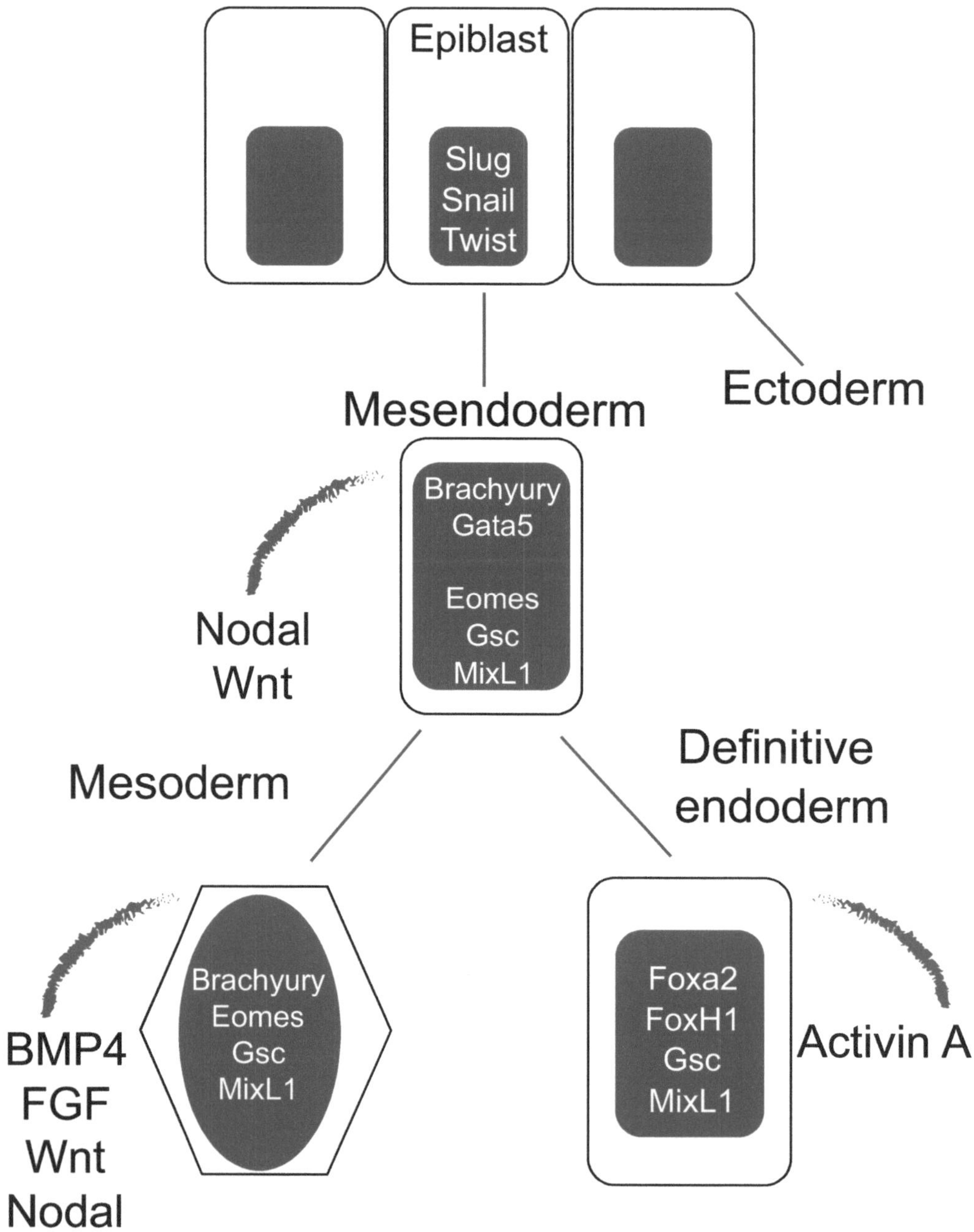

Fig. (2). A simplified model of mesendoderm formation. Rounded rectangular cells represent E-cadherin$^+$ epithelial cells. Gradients of critical soluble morphogenic and growth factors are shown next to each developmental stage. Transcription factors playing critical roles in each lineage are shown within the schematic nucleus.

Mesoderm Differentiation

Eomesodermin can induce expression of genes which play critical roles in mesoderm development [7]. In particular, cardiac and head mesoderm is derived from *Eomes*-expressing progenitors [10]. Other transcriptional regulators expressed in the mesendoderm develop more restricted roles in the mesoderm including Brachyury that becomes restricted to the axial mesoderm responsible for notochord formation. The soluble factor Bone morphogenetic factor-4 (Bmp4) is an essential requirement for mesoderm development across many species [11]. Genetic ablation of the gene encoding Bmp4 in the mouse leads to early embryonic lethality with many embryos failing to proceed through gastrulation or form mesodermal cells [12]. A combination of soluble factor signalling through Bmp4, and transcriptional regulator complexes (including Brachyury, Mixl1, eomesodermin and goosecoid) lead to expression of mesoderm-specific transcription factors such as Mesp1 and commit the progenitor cells which have received these signals to become mesodermal cells. Specification of mesoderm into more committed progenitors requires a range of different soluble factors generating a gradient of cell signalling according to cellular location within the developing embryo. This gradient of signalling in turn initiates differentiation transcriptional programmes leading the specification of distinct types of mesoderm progenitors. This is a gross simplification of a truly complex system, however it serves to offer us a model for mimicking early embryological events *in vitro* to generate cell types that may be of use for regenerative medicine.

Endoderm Specification

The endoderm gives rise to the epithelial lining of the three major internal tubes: the respiratory, gastrointestinal and genitourinary tracts, as well as the organs that branch off them (Fig. **1**). Development of definitive endoderm (DE) in mammals begins at gastrulation, when endoderm precursors migrate from the epiblast to the anterior primitive streak followed by implantation into the visceral endoderm (VE). This movement is facilitated by an epithelial-to-mesenchymal transition (EMT) which is dependent upon the expression of genes *Snail and Slug*, both of which momentarily suppress E-cadherin [13]. In *Xenopus* and zebrafish, migration of endodermal precursors is dependent upon mesoderm-derived

Sdfl/Cxcl12b, which acts on endoderm expressing Cxcr4 [14]. Observations in Zebrafish models suggest that there is a feedback check on differentiation of cells and its neighbouring environment [15]. Inhibition of the Activin A receptor and Nodal-related proteins in *Xenopus* embryo provided the first insight into the underlying mechanisms of endoderm formation in vertebrates [13]. Activin and nodal are part of the TGF-β family. Findings from explant studies suggested that the inhibition of TGF-β signalling in the endoderm lead to increased expression of mesodermal and ectodermal markers at the cost of endoderm marker expression. The separation of the three embryonic germ layers is in part, due to TGF-β signalling [16]. Nodal signalling, a subclass of the TGF-β superfamily has been shown to induce formation of mesoderm and endoderm in vertebrates [17]. Ligands that belong to the TGF-β family (including Activin A and Nodal) operate through the homologous intracellular mediators Smad2 and Smad3 [18]. Mutations of the *Nodal* allele led to the loss of endoderm differentiation in embryos due to reductions in gene expression of *Smad2* and *Smad3* [18]. Overexpression of Activin A or Nodal-related factors in *Xenopus* and zebrafish however, confirmed mesodermal and endodermal cell fates required a gradient of nodal and activin-related signals [6, 19]. Cells furthest from the source of Nodal express mesodermal genes, due to the exposure of low concentration of Nodal [20, 21]. The canonical Wnt signalling pathway activates Nodal expression during embryogenesis [13]. Specification of DE requires both Wnt signalling and Nodal expression as mouse embryos lacking *Nodal* fail to form a primitive streak [22]. *Wnt3*$^{-/-}$ mice do not form a primitive streak or mesoderm highlighting its importance in germ layer formation [23]. *Wnt3* expression during gastrulation is found in the posterior visceral endoderm shortly before mesendodermal cells begin migrating into the primitive streak [24]. Wnt3 has been shown to induce Nodal (and *vice versa*) through feedback signalling which demonstrates the ability of the canonical Wnt pathway to directly enhance endoderm generation in mammals [23].

In the vast network of endodermal transcription factors, there are two key families: the GATA family and the FoxA family of Forkhead transcription factors. Members of both gene families are downstream targets of Nodal, however, the direct mechanisms regulating them are still poorly understood. *Gata4*, *Gata5* and

Gata6 play roles in endoderm development [25]. *Gata4*$^{-/-}$ and *Gata6*$^{-/-}$ embryos exhibit reduced development of both the visceral endoderm and the definitive endoderm however *Gata5*$^{-/-}$ mice did not show any endoderm defects [26 - 28]. Inactivation of members of the FoxA family of Forkhead genes results in a reduction but not a cessation of endoderm generation [29]. *FoxA2* is required for the formation of the foregut, but not the mid- or hindgut [30]. Members of the GATA and Forkhead families are proposed to synergistically form a complex stimulating endoderm gene transcription [13, 31, 32]. Downstream factors stemming from the Nodal pathway Sox17 and MixL1 have also been shown to play a part in endoderm generation [33]. In the developing mouse embryo, *Sox17* expression appears in the DE between 7.5-8.5 days post coitus (dpc). *Sox17*$^{-/-}$ mice show a reduction in DE formation with a failure of midgut expansion and elevated levels of apoptosis in the foregut [34]. These findings suggest that Sox17 plays a role in maintaining endoderm integrity. *Mixer* has been shown to regulate *Sox17* and other endodermal genes through feedback loops involving members of the GATA family [35]. *Mixl1*$^{-/-}$ mouse embryos exhibit reduced DE generation and deformities in the hindgut development [36]. *Mixl1*-overexpression in *Xenopus* induced surplus formation of the endoderm with an increase in Nodal expression [37]. These finding suggest that *Mixl1* is part of a feedback loop regulating Nodal.

In vitro Models of Mesendoderm Formation

Dissecting the molecular pathways regulating the formation of the cells, tissues and organs derived from the mesendoderm has been conducted in a number of animal models. Simpler organisms such the nematode *C.elegans* have proven useful. Externally fertilising species such as the African clawed frog *Xenopus* and zebrafish have also been useful as individual cells in the early embryo can be labelled and their derivative cell types assessed for retention of the label. The development of null mutant mouse strains proved a boon for understanding the specific roles of genes in mesendoderm development. In many cases, the genes ablated have proven essential for embryonic survival and the null mutant embryos have died *in utero*. The intra-uterine development of the mouse embryo has also proven to be a hurdle for analyzing developmental processes.

Embryonic Stem (ES) Cell Differentiation as in vitro Model of Mesendoderm Development

A number of research groups have capitalized on the ability of embryonic stem (ES) cells to remain pluripotent under controlled cell culture conditions to recapitulate developmental processes and events *in vitro* [38]. For mouse ES cells this typically requires administration of leukemia inhibitory factor (LIF) to culture medium. Once LIF is removed, mouse ES cells will spontaneously differentiate according to the signals the ES cells receive from the cell culture environment. Embryonic stem (ES) cells are derived from the inner cell mass of the blastocyst and are extremely useful pluripotent, undifferentiated cells that have the capability to generate any cell type in the body. ES cells can also be used as an *in vitro* model of embryogenesis. This is particularly useful as mammalian development is challenging to study due to the *in utero* gestation and small size of early embryos. ES cell cultures can be scaled up, genetically modified, exposed to chemically defined media and directed along specific differentiation pathways. Some strategies used for generating and isolating mesendoderm, endoderm and mesoderm progenitors during pluripotent stem cell *in vitro* differentiation are presented in Fig. (**3**).

ES cells derived from the mouse or humans have been used to identify critical stages and processes in mesendoderm development. Upon withdrawal of any agents maintaining self-renewal, ES cells will begin to spontaneously differentiate *in vitro*. The direction of differentiation will depend upon the signals present in the cell culture system. Soluble growth factors, morphogens, cytokines and chemokines can all induce differentiation. These factors, as well as components of the media, exogenously added additives (vitamins, salts, lipids) or inhibitors can all be included to drive differentiation along specific lineages, or to improve the yield of specific cell types. The surface on which the ES cells differentiate can also influence differentiation. Different extracellular matrix components such as fibronectin, laminin or vitronectin activate distinct integrin proteins and transduce different signalling into the ES cell cultures. Stiffness of the matrix on which cells are growing also influences differentiation [39]. The addition of fetal calf serum drives ES differentiation away from ectodermal lineages towards mesendoderm. In serum-free, chemically-defined systems, the addition of soluble Nodal directs

ES cell differentiation towards mesendoderm [40]. Once ES cells have been differentiated out of their self-renewing state, genes specific for the mesendoderm lineage will be expressed. These genes can be used as indicators of mesendoderm differentiation either in mRNA readout systems or as tags which can be utilized to enrich and purify mesendoderm from all other cells in the differentiation culture.

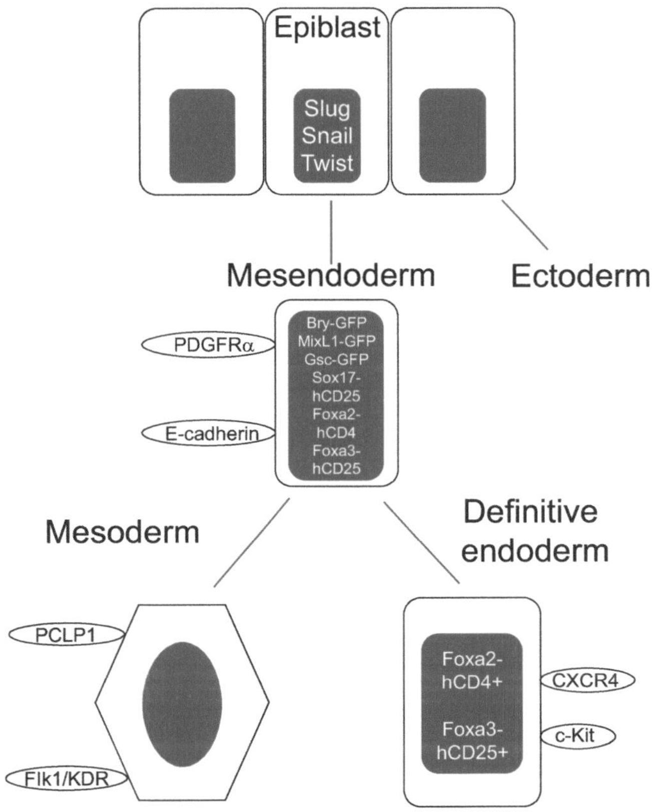

Fig. (3). Differentiating cultures of pluripotent stem cells such as ES and iPS cells can be purified at distinct steps of differentiation using a broad range of methodologies. Lineage- or stage-dependent surface markers are indicated here including PDGFRα and E-cadherin for mesendoderm. Reporter transgenes (represented within the nucleus of each cell type) can be used if coupled to a fluorescent protein such as GFP (green fluorescent protein) or used to drive a surface protein such as human CD4 (hCD4) or human CD25 (hCD25) on the cell surface. Fluorescently conjugated monoclonal antibodies against these human cell surface proteins can then be used to distinguish mouse cells at different stages of differentiation allowing for purification, transcriptome analysis or re-culture.

A number of systems have been developed using the promoter regions of mesendoderm-specific genes such as Brachyury (*T*), goosecoid (*Gsc*) or Mixl1 to drive fluorescent protein expression thus indicating successful differentiation into the mesendoderm. The regulatory elements driving T, Gsc and Mixl1 have all be used to drive green fluorescent protein (GFP) expression in ES differentiation cultures. When cells expressing GFP are purified using fluorescence-activated cell sorting (FACS) they were found to express high levels of mesendoderm genes (Fig. **3**). Combining this with cell surface protein expression can help improve the purity of mesendoderm differentiation. Tada and colleagues found that Gsc-GFP$^+$ cells express the epithelial (E)-cadherin and the platelet-derived growth factor receptor alpha chain and that FACS can be used to purify this population [41]. Similar approaches can be used in the differentiation of human ES (hES) cell cultures to obtain useful quantities of purified mesendodermal cells which can then be used in tissue engineering or 3D printing. Other groups have used mesendodermal regulatory elements to drive expression of human surface proteins on mouse cells. Species-specific fluorescent antibodies against these non-functional human surface proteins (CD4 or CD25) can then detect surface expression of these molecules and be used to monitor mesendoderm development or purify these cells by FACS [42].

Mesendodermal cells derived *in vitro* from ES cultures can then be differentiated into mesodermal cells or endoderm-derived lineages. For example, Bry-GFP$^+$ cells purified by FACS have been used to differentiate mesodermal lineages such as hematopoietic and endothelial lineages as well as endodermal lineages including pancreatic and hepatic lineages [43 - 45]. Modulating mesendoderm regulators in ES differentiation cultures can improve the yield of desired cell types. Activation of inducible *Mixl1* gene accelerated generation of mesodermal and endodermal cell types [46].

Following differentiation from mesendoderm, cell-surface markers can identify mesodermal progenitors. Mesodermal cells lack epithelial surface proteins during epithelial-mesenchymal transition. The vascular endothelial growth factor-2, also known as Flk1 or KDR (in humans) is expressed on the surface of lateral plate mesodermal cells fated to become blood or blood vessel progenitors [47]. Therefore mesodermal differentiation to the E-cadherin$^-$ Flk1$^+$ stage allows for enriched of lateral plate mesoderm. The hematopoietic committed Flk1$^+$ population could be further enriched according to α4-integrin or podocalyxin expression [48, 49]. Vascular and cardiac mesodermal progenitors derived from ES differentiation cultures can also be purified using a different combination of surface markers. Endodermal progenitors can also be identified from ES differentiation cultures. A combination of surface of expression of CXCR4 and c-Kit has proven to be an effective means for assessing endodermal differentiation *in vitro*. In the context of embryonic stem (ES) and induced pluripotent stem (iPS) cell cultures, feedback from the differentiating mesoderm influences the formation and expansion of endodermal precursors. Depending on the feedback provided by the mesoderm, endodermal precursors may differentiate within a heterogeneous culture or die *in vitro* [13]. Together, this highlights the importance of mesoderm-endoderm interactions and the development of the later during gastrulation. This will be expanded upon in this chapter.

Induced Pluripotent Stem Cells as a Novel Source of Mesendodermal Progenitors for Regenerative Medicine

While ES cells have been used extensively since the early 1980s, their use for regenerative medicine was limited to helping define appropriate differentiation conditions to generate cell types useful for regenerative therapies. The utility of pluripotent stem cells for regenerative medicine changed dramatically when, in 2006, Shinya Yamanaka and colleagues reported the generation of induced pluripotent stem (iPS) cells from terminally differentiated adult fibroblasts. This revolution led to the widespread and routine production of iPS cell lines from a range of species including humans [50]. Induced pluripotent stem cells exhibit very similar properties to ES cells including pluripotency, teratoma formation and the ability to differentiate into a broad range of cell types upon receiving the appropriate signals *in vitro*. Similar to ES cells, iPS can also be potentially used for therapies, disease models or drug screening. One advantage they have over ES is that they can be made readily from adult cells, avoiding the ethically challenging aspects of ES cell biology. Patient-specific iPS have been generated in which the genetic lesion underpinning the pathology the patients experience has then been repaired [51 - 53]. Since the discovery of iPS cells, groups have made great strides in publishing cellular models outlining a vast multitude of conditions and potential therapies. Here, we will show the current standings of differentiation of mature cell types and how these can be implemented for regenerative medicine.

Modelling Interactions Between Endoderm- and Mesoderm-derived Cell Types Using Pluripotent Stem Cells

As shown in Table **1**, many endoderm-derived tissues and organs have a significant mesoderm-derived component. To form physiologically and medically relevant cell types, tissues or organs for regenerative medicine, an understanding of how the mesendoderm gives rise to these complex biological structures is required. Here, we will describe medically important organ systems that are being mimicked *in vitro* to generate useful 3D structures for regenerative medical interventions.

Differentiation of Pancreatic β-cells as a Treatment for Diabetes

The pancreas is often described as two organs with distinct endocrine and exocrine components [54]. The exocrine acinar cells are responsible for the production of digestive enzymes whereas the endocrine cells of the islets of Langerhans regulate blood sugar levels [54]. The insulin-producing cell type of the islet of Langerhans is the β-cell. Autoimmune destruction of the β-cell leads to type-1 diabetes [55]. Type-2 diabetes manifests as resistance to insulin in the peripheral tissues [55]. Collectively, diabetes affects 300 million people worldwide [56]. One of the biggest limitations in curing type-1 diabetes is finding a renewable source of functional β-cells. Generation of mature, functional β-cells from embryonic and induced pluripotent stem cells has raised the exciting possibility for a near-limitless source of β-cells for transplantation [38].

The pancreas is an endodermal organ at the posterior foregut level [57]. The sites of pancreatic induction occur precisely where the endodermal epithelium contacts the mesodermal endothelium of major blood vessels [1]. Pancreatic progenitors are found at E8.5 in the mouse with expression of homeodomain transcription factor *Pdx1* [54]. Pancreatic buds will then form from $Pdx1^+$ cells [58]. Culturing endodermal cells with the dorsal aortae induces expression of insulin and *Pdx1*, where expression was found adjacent to the endothelium, highlighting the importance of endothelial cells in pancreatic cell induction [1]. While *Pdx1* is the first indicator of pancreatic generation, it is not solely limited to the pancreas [38]. It is also found in other foregut endodermal tissues such as the stomach and duodenum [38]. Commitment of pancreatic cell differentiation occurs with co-expression of *Pdx1* with transcription factor *Ptf1a* [59]. Secretion of Fibroblast growth factor 10 (FGF10) from the mesenchyme expands *Pdx1*$^+$/*Ptf1a*$^+$ cell number [38]. Genetic lineage tracing studies have found that *Pdx1*$^+$/*Ptf1a*$^+$ progenitor cells give rise to all mature pancreatic cells [59, 60]. Furthermore, ablation or mutation of *Pdx1* and/or *Ptf1a* has resulted in blunted pancreatic growth and differentiation [59]. There have been few studies that examine human embryonic pancreas development, however, the similarities between rodent and human pancreatic organogenesis is evidently clear [61]. Understanding the embryology of pancreatic development has paved the way for differentiation of pancreatic cells from stem cells.

Equipped with this knowledge, researchers have been able to generate mature, functional β-cells *in vitro* from ES and iPS cells. The protocol established by D'Amour and colleagues has been the most routinely used, however, modifications have been made. Briefly, DE was induced by Activin-A. Pancreatic fate was then specified through retinoic acid and FGF10 signalling [62]. Maturation of β-cells included supplementation with IGF-1, HGF, and extendin-1 [62]. Confirmation of the differentiated cell types was sequential, with induction of FOXA2$^+$, CXCR4$^+$, SOX17$^+$ endoderm, generation of PDX1$^+$/PDF1a$^+$ pancreatic progenitors and development of NGN3$^+$/NKX2.2$^+$ endocrine cells [38]. Production of insulin secreting cells was determined by the expression of C-peptide, which is released when proinsulin is converted to insulin [38].

Transplantation of β-cells for the purpose of curative purposes has had mixed results. One of the biggest hurdles is the maturation of β-cells to produce insulin prior to or following transplantation. In streptozocin-induced diabetic mice, Jiang and colleagues transplanted glucose-responsive stem cell derived β-cells that reversed the hyperglycemic state of 30% of mice. Graft analysis showed the recipient mice expressed human C-peptide and PDX1 [63]. Addressing the issue of the amount of β-cells derived from stem cells, Pagliuca and colleagues were able to generate hundreds of millions of insulin-producing β-cells *in vitro*. Following transplantation, the differentiated β-cells were able to secrete human insulin in a glucose-regulated manner [56].

Developing New Systems for Regenerative Medical Intervention During Liver Disease

The liver is a multi-purpose organ that serves roles in metabolic, endocrine and exocrine functions [64]. These functions largely serve to restore homeostasis through detoxification, metabolism of dietary compounds, regulation of glucose and control of blood homeostasis [64]. The primary cell type of the liver is the hepatocyte [65]. Hepatocytes and biliary epithelial cells (cholangyocytes) are endodermal derivatives supported by mesodermal-derived stromal cells, endothelial cells, Kuppfer cells and blood vessels [65].

The embryological origin of the liver begins at day 8 of gestation (E8.0), located in the foregut endoderm [65]. At E9.0, the embryonic liver forms a thickened epithelium, making it morphologically distinguishable [66]. The next stage of hepatogenesis is the formation of the liver bud. This occurs at E9.5 when endodermal hepatoblasts (hepatocyte progenitors) shed from the epithelium and occupy the mesodermal septum transversum mesenchyme (STM) [66]. The STM gives rise to the mesodermal cell types of the liver including: macrophages, stromal fibroblasts, sinusoidal endothelial cells and stellate cells [67]. At E10.5-E11.0 of development, haematopoietic progenitors migrate from the yolk sac, aorta-gonad-mesonephros and placenta to the foetal liver, which becomes one of the major sites of blood production [68]. This is supplemented as the liver becomes vascularised and populated by haematopoietic cells between E10.0-E15.0 [64].

Endothelial cells which reside between the epithelium and STM aid in the migration of hepatoblasts into the stroma [69]. Hepatoblasts are bi-potential as they can differentiate into cholangyocytes and hepatocytes depending on their location in the developing liver. Hepatoblasts in the portal vein will differentiate into cholangyocytes that will line the lumen whereas hepatoblasts in the parenchyma will differentiate into hepatocytes [64]. Maturation and function of these cells begins at E13.0 and continues after birth [69]. In all, hepatogenesis demonstrates the sequence of tissue interactions between the endoderm and mesoderm.

Hepatoblast specification and regulation of liver morphogenesis are dependent upon vascular endothelial growth factor receptor 2 (KDR) [70]. Prior to hepatic specification, KDR+ cells are located in the endoderm at E8.0 with expression waning throughout hepatogenesis [70]. $KDR^{-/-}$ mouse embryos lacked the development of endothelial cells, which in turn, instruct liver development [2]. Endothelial cells have also been shown to aid in hepatocyte and cholangiocyte specification through Wnt and Notch pathways [71]. While KDR expressing endothelial cells were believed to be restricted to the mesodermal lineage, Goldman and colleagues revealed the bi-potency of KDR^{+} cells using human and mouse embryonic stem cells (2). Endodermal cells lack KDR expression, however, when cultured in media supporting differentiation of hepatic cells, two populations arise; KDR^{+} hepatic progenitors and KDR^{-} hepatic cells [2]. The KDR^{+} hepatic progenitors serve to stimulate differentiation and maturation of the KDR^{-} hepatic cells, which further enforces the interaction between the mesoderm and endoderm [70].

Understanding the development of the hepatic niche *in vivo* paved the way for researchers to recapitulate the development of the liver *in vitro* using stem cells. Differentiation of hepatocytes from stem cells begins at the endoderm. Generating DE from stem cells is routine with the addition of Activin-A and Wnt3A [72]. Next, supplementation with FGF-2 and BMP-4 initiates hepatic induction [73]. Hepatoblast maturation and expansion requires the addition of hepatic growth factor, FGF4, Oncostatin M and Dexamethasone in cultures [64]. Differentiated hepatic-like cells look and function like hepatocytes. These functions include: secretion of albumin, storage of glycogen, expression of hepatic enzymes and metabolism of drugs [74].

Chronic liver diseases such as cirrhosis and hepatitis affect millions worldwide [75]. Transplantation is the only curative means for chronic liver diseases [75]. Other means of restoration are needed. One such possibility is the transplantation of hepatocytes, which could delay the progression of the disease and act as a conduit for patients awaiting a liver transplant [76]. Despite the success of hepatic-like cell differentiation from stem cells, their implantation and integration into mice with various liver injury models have shown only a modest rescue of function. Their inability to repopulate diseased livers *in vivo* may be attributed to stem cells that have not fully maturated into hepatocytes [70]. Success was seen in hepatectomised mouse given a bioartifical liver coupled with mES cell-derived hepatocytes [77]. Liver- function and overall survival improved compared to mice that did not receive the mES-derived hepatocytes [77]. While the use of stem cell derived hepatocytes for regenerative purposes is still being refined, they have been a useful source for modelling liver diseases. Yoshida and colleagues demonstrated the prowess of induced pluripotent stem cells at modelling the hepatitis C virus, which can be expanded upon for potential drug treatments [78].

Endodermal and Mesodermal Derivative Lineages in Treatment for Disease of the Genitourinary Tract

Interactions between the mesoderm and endoderm-derivative cell lineages can be observed in the genitourinary tract. Derivatives that comprise the mesodermal germ layer here include: the nephron, ureter, renal pelvis, glomerular endothelial and mesangial cells, renal interstitial cells, adventitial fibroblasts and ureteral and bladder smooth muscle [78]. Endodermal derivatives include the luminal and basal epithelial cells of the bladder, urethra and prostate [78]. These mesodermal cell types originate from a narrow strip of tissue located bilateral of the embryonic midline termed the intermediate mesoderm [79]. Preceding the formation of the kidney and ureter is the mesoderm-derived nephric duct. During the development of the nephric duct, the surrounding intermediate mesoderm undergoes a mesenchymal-to-epithelial (MET) transformation [80]. This inductive process results in the generation of mature nephrons.

Development of the Bi-potential Gonad and Reproductive Tracts

The intermediate mesoderm gives rise to the urogenital system (kidneys, gonads and associated ducts). The reproductive tract develops as a sexually bi-potential structure, in close association with the urinary tract, for the first 7 weeks of human development [81] (Fig. **1**). The intermediate mesoderm extends to form bilateral nephrogenic chords during the second week of development in the human. These chords will eventually become bilateral urogenital ridges around 5 weeks when development of gonadal tissue starts with the indifferent gonadal phase. At this stage the mesothelium toward the middle of the mesonephros thickens to form the gonadal urogenital ridges. Primary sex chords (epithelial digitations) form and extend into the mesenchyme and remain as undifferentiated gonadal ridges that are invaded by migrating primordial germ cells during week 6 of gestation (see later). Three weeks of gestation heralds the formation of the pronephros, a primitive kidney, which is continuous caudally with the mesonephros (Wolffian duct). Around 4 weeks this caudal end meets and fuses with an endoderm lined sinus at the caudal end of the gut tube called the cloaca. The region anterior to the fusion point will become the bladder and pelvic urethra of the neonate. The posterior portion of the cloaca forms the urogenital sinus and the anorectal canal. The cranial region of the Wolffian duct regresses to form a lumened duct that will persist in the male and form the efferent ducts, epididymides and vasa deferentia.

At around 6 weeks of gestation, focal infolding of the intermediate mesoderm-derived coelomic epithelium, lateral to the mesonephric ducts occurs to generate the paramesonephric ducts [82]. These ducts are the precursors of the uterus, uterine ducts, cervix and upper portion of the vagina in the female. Hence by the end of week six of gestation the foetus possesses two genital ridges, each on the medial surface of a pair of patent mesonephric (Wolfian) ducts and collectively called the urogenital ridges. Lateral to each urogenital ridge is a paramesonephric (Müllerian) duct. The metanephros, mesonephric ducts and paramesonephric ducts all join to the bi-potential urogenital sinus. Up to this point, both male and female foetuses are indistinguishable. Sex determination and sexual differentiation occurs after seven weeks of gestation, and once the gonadal ridges have been populated with migrating primordial germ cells [83]. This chapter will not deal with sexual differentiation, the reader is directed to recent reviews [84, 85].

Development of Kidney Organoids from Human Embryonic and Induced Pluripotent Stem Cells as a Model of Kidney Development

Kidney transplants are the most common form of organ transplantation worldwide. There is a clear need for new medical interventions for the treatment of renal disease. The formation of the kidney is a highly complex process however; recapitulation of renal organogenesis *in vitro* offers great hope to alleviate the challenges experienced by patients with renal failure. Development of the mammalian kidney is a product of the reciprocal interactions between the endodermal uretic bud and mesodermal mesenchyme [80]. Derivatives of the uretic bud give rise to the renal collecting ducts, epithelial lining of the renal calyces and the urothelial lining of the ureter [80]. Conversely, the mesenchyme undergoes a mesenchyme-to-epithelial transition and leads to the differentiation of the nephron segments, including: the glomerulus, proximal and distal tubules and the loops of Henle [80]. Differentiation of embryonic and induced pluripotent stem cells into renal lineages has been successful [86 - 89]. Vigneau and colleagues utilized the Bry-GFP mouse ES cell line to obtain renal progenitor cells [91]. ES cells were differentiated to generate Bry-GFP+ cells that were then purified by fluorescence-activated cell sorting (FACS). Bry-GFP+ progenitors could form renal progenitors expressing marker genes such Cadherin-11 and Wt1. Upon transplantation into live neonatal kidneys, these cells could integrate into the renal tubules [90].

Takasato and colleagues have recently reported the successful induction of both the uretic epithelium and metanephric mesenchyme *in vitro* [92]. Their system has also been extended for induction of kidney formation in organoid cultures [91]. Briefly, treatment of human ES and iPS cells with FGF9 and retinoic acid promotes differentiation of the uretic epithelium with exposed Wnt signalling aiding in the differentiation of the metanephric mesenchyme [91]. Indeed, differentiation of both cells types have been made possible by replicating the signalling and temporo-spatial development of the kidney *in vivo*. Generation of kidney organoids have given rise to the exciting possibility of testing models of kidney diseases and the ability to screen drugs for toxicity for cellular therapy.

The Urogenital Sinus and Accompanying Organs

The urogenital sinus originates initially as a sub-compartment of the cloaca; a transient endoderm-derived cavity, located at the posterior end of the gut [92]. Cross talk between the mesenchyme and epithelium are critical for the development of urogenital tissues. Secretion of *Sonic hedgehog* (Shh) from the endoderm is key to patterning and separation of the cloaca [92]. Failure of the cloaca to separate during foetal development has severe clinical consequences and is thought to underlie the pathogenesis of a wide range of urogenital malformations [80].

The genitourinary tract (GUT) encompasses the entirety of the genital ridge as well as the organs that filter waste and fluid from the blood: the kidneys, ureters, bladder and the urethra. The kidneys also control blood pressure through the secretion of hormones, angiotensin and aldosterone [93]. The passage of urine from its production by the nephron and elimination by the urethra highlights the most critical function of the GUT [94]. The tract is susceptible to malfunctions at any point, which may impair the complete emptying of the bladder. Issues involving the bladder are currently remedied by surgically replacing the bladder wall with either intestinal or stomach tissue [95]. However, the transplantation leads to eventual salt imbalances, infection, stones and in some cases, cancer [96]. Ideally, the replacement of the bladder wall should be comprised of a bioengineered scaffold augmented with stem cells that promote cell growth, proliferation and survival.

Kidney diseases result in the degradation of the microstructural components, which lead to an obstruction in the passage of urine [94]. The accumulation of urine along the tract leads to infection and scarring and eventual failure [97]. The diseases that fall under this category include: urinary tract infections, glomerulonephritis, polycystic kidney disease, haemolytic uremic syndrome and hydronephrosis [97]. The male genital tract comprises of the bladder, urethra, prostate, seminal vesicles, gonads, penis and testes [94]. Physiologically, they produce and transport sperm, secrete sex hormones and facilitate erections. Complications in the tract may lead to infertility and erectile dysfunction [95]. Correlations of the genital tract with the urinary tract lead to clashes in diagnosis, which also complicates matters. Here, stem cells can be used to model the development of the male genital tract, including the prostate. This would provide insight into the mechanisms that underpin normal growth which are lacking, and applied into a cell/scaffold graft to alleviate impairments of the genital tract.

Differentiation of Urothelium from Mouse, Human and Adult Stem Cells

The urothelium of the bladder is unique as it lines the luminal surface and acts as a near impermeable membrane [98]. This tissue is composed of basal, intermediate and superficial cells, which express unique structural proteins called uroplakins (UPK) [99]. Assembly of the four subtypes of UPK (UPK1a, UPK1b, UPK2 and UPK3) results in plaque formation on the luminal surface [98]. Presence of UPK is a marker of urothelial differentiation. Of all the cell types that have been differentiated from ES and iPS cells, studies looking into the differentiation of urothelial cells are limited. Differentiation of urothelium from human ES and iPS cells as well as mouse ES cells has been reported [98, 100, 101]. In mouse ES cells, GATA4 signalling has been shown to drive urothelial differentiation, however, in human ES cells; GATA4 signalling was shown to be minimal [98, 101].

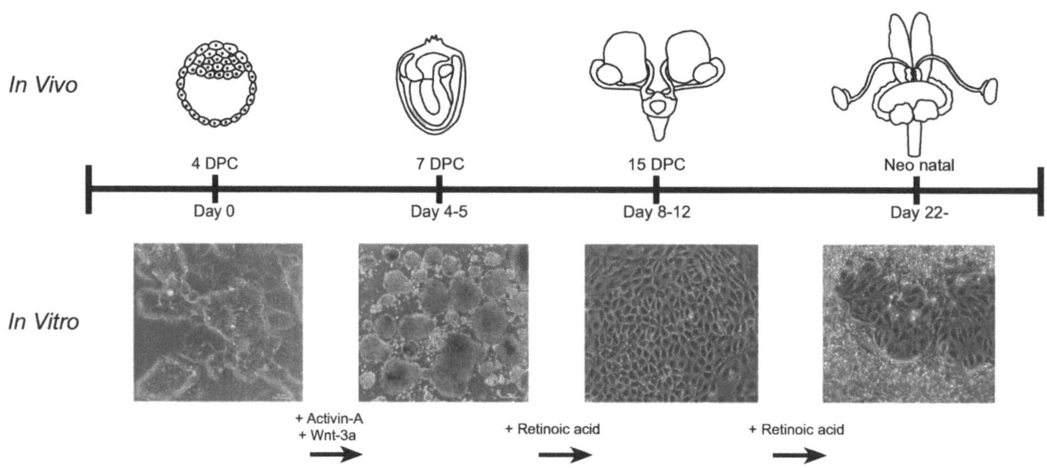

Fig. (4). Timeline of *in vivo* development of the urogenital sinus and *in vitro* differentiation of urothelium mouse embryonic stem cells. The inner cell mast of the blastocyst (*in vivo*, 4 days post coitus (DPC)) gives rise to all three germ layers: the ectoderm, the mesoderm and the endoderm (*in vivo*, 7 DPC, green). Mimicking this *in vitro* with mouse embryonic stem cells, cultures treated with Activin-A and Wnt-3a lead to definitive endoderm induction (*in vitro* Day 4-5). Development of the urogenital sinus continues from 15 DPC (*in vivo*) and matures after birth (neonatal). Differentiation of definitive endoderm into urothelium requires treatment with retinoic acid (*in vitro*, D8-12, D22-onwards).

As the bladder is of endodermal origin, the first step in the generation of urothelium (and all endodermal cell types) is the temporal differentiation of ES and iPS cells into DE with Activin A [72]. Following induction of DE, the above groups supplemented their cultures with retinoic acid and urothelial-specific medium for differentiation of urothelium in mouse and human ES and iPS cells respectively (Fig. **4**) [98, 101]. Confirmation of urothelium was determined by expression of the UPK proteins *via* immunohistochemical, real time RT-PCR and flow cytometric analyses [98, 101]. Despite the limited number of studies aiming to differentiate urothelium from ES and iPS cells, the above methodologies have shown that is possible to do so without the need for matrices or co-cultures. For tissue regenerative purposes, the next step would be to augment scaffolds with differentiated urothelial cells to repair damaged tissue.

Primordial Germ Cell Development

Many of the processes of early specification and migration in the human are assumed similar to those in other mammalian models including the mouse [102]. Primordial germ cells (PGC) in mice originate from proximal epiblast cells to form a cluster of precursor PGCs in the incipient allantois in extra-embryonic mesoderm at approximately E7.25 [103, 104]. Following specification, their numbers expand as they start to migrate to and through the developing gut endoderm, into and through the mesentery to colonise the gonadal ridges [102]. Precursor PGCs are specified in the epiblast prior to gastrulation through bone morphogenic protein (BMP) signalling between the extra-embryonic ectoderm and visceral endoderm [83]. Mice lacking BMP2, BMP4, BMP8b or SMAD mediators of BMP receptor signalling all exhibit reduced induction and proliferation of early PGCs along with defective mesoderm formation and allantois development [105, 106]. Evidences show that it is extra-embryonic ectoderm derived BMPs acting on BMP receptors expressed by visceral endoderm that results in the formation of PGCs in a dose dependent manner [107]. In addition to inducing a founding population of PGCs in the mouse, BMPs are also important for PGC migration to, and proliferation at the genital ridge [108]. As well as gradients of stimulatory BMPs, the induction of epiblast cells to PGCs is also dependent on gradients of inhibitory factors [109]. BMPs subsequently induce expression of *Prdm1* that encodes the protein BLIMP1, a key regulator of PGC specification [110]. BLIMP1 acts to repress progression to the mesodermal somatic lineage [110, 111] and to promote epigenetic reprogramming of early PGCs [109].

In vitro Generation of PGCs and Regenerative Medicine

Early PGCs can undergo de-differentiation into pluripotent germ cells when treated with FGF and LIF in culture due to suppression of *Prdm1* expression and BLIMP1 synthesis [112, 113]. This knowledge of BLIPM1 and its interactions with other key factors, PRDM14 and AP2g, that act to suppress mesodermal genes [109], have helped establish that the majority of epiblast cells from whole explants are all able to be directed to a PGC fate in the presence of BMP4 *in vitro* [114]. These PGCs can further differentiate to gametes that are able to produce viable offspring [115]. Hayashi and colleagues also demonstrated that PGCs could be derived from embryonic stem cells under specific culture regimen. Furthermore, iPSCs derived from somatic cells can be induced to form PGCs provides an exciting prospect in regenerative medicine [109].

As well as providing a model for the study of the molecular regulation of germ cell differentiation, it also has potential for a source of unlimited numbers of gametes for treatment of infertility, including transplantation. This is particularly pertinent to induced infertility in cancer patients following certain treatments (*e.g.* gametotoxic chemotherapies, whole body and gonad radiation therapies). Advancements in treatment protocols has resulted in increasing numbers of cancer survivors and hence significant subsequent morbidities due to resulting infertility [116]. This has led to clinical recommendations that sperm and egg banking to preserve fertility be considered prior to treatment [117]. Whilst this is an option for adolescent and adult patients, it is not an option for pre-pubertal patients. One potential future is the transplantation of germ stem cells isolated from the pre-pubertal gonad or of PGCs/germ stem cells derived in culture from iPSCs [115].

In the male, spermatogonial stem cells collected prior to chemotherapy might be transplanted back to the testis at a later appropriate date. Whilst this has been trialled in humans, no follow-up reports of long-term maintenance, which includes live births and fertility in these subjects is apparent [118, 119]. However, such successful outcome measures have been achieved in other mammals, including non-human primates [120, 121].

Adipogenesis and Differentiation of Adipocytes from Stem Cells.

Adipose tissue arises concurrently with the developing kidneys, urogenital sinus and primordial germ cells.

Mammals have two types of adipose tissue, white (WAT) and brown (BAT). Both WAT and BAT serve distinct roles in energy metabolism with WAT storing energy in the form of triglycerides while BAT expends energy as heat [122]. Lineage tracing studies have shown that BAT originates from myogenic factor 5 (Myf5)-expressing progenitors, compared to the non-Myf5 lineage progenitors that give rise to WAT [123]. WAT can be further subdivided into subcutaneous and visceral WAT, with visceral shown to have implications in metabolic dysfunction such as obesity [124].

Adipose tissue is currently thought to be derived from the mesoderm germ layer [122]. Characterisation of adipocyte differentiation has been described in two phases, determination and terminal phase [125]. The determination phase results in the conversion of a stem cell into a pre-adipocyte, the details of which remain largely unexplored [126]. The terminal phase of differentiation describes the maturation of pre-adipocytes using adipocyte cell lines 3T3-L1 and 3T3-F44a [125, 127]. These cells lines are useful as they are restricted into differentiating into other cells types and culturing primary adipocytes are difficult to maintain [125, 128].

Another model that can be utilised to detail adipogenesis is adipocytes derived from ESC and iPSCs. While ESCs and iPSCs recreate the key steps that occur during development *in vivo*, replicating the signalling events required for adipocyte differentiation *in vitro* is difficult [128]. It is useful then, to initiate differentiation into the adipocytes using growth factors known to terminally differentiate adipocytes [128]. Transcriptional regulation of adipocyte differentiation is largely due to peroxisome proliferator-activated receptor γ2 (PPARγ2) [128]. Therefore, differentiation of WAT from ESCs and iPSCs routinely use PPARγ2 agonist such as rosiglitazone along with adipogenic factors insulin and dexamethasone [128, 126]. Further inclusion of all-trans retinoic acid aids in the differentiation of adipocytes *in vitro* [129, 126].

The above process describes the differentiation of adipocytes that resemble WAT phenotypically and functionally. Conversely, differentiation of brown-like adipocytes from ESCs and iPSCs has only recently been achieved [130 - 132, 128]. For differentiation of BAT, a combination of PPARγ2, CCAAT/enhancer-binding protein β (CEPBβ) and PR domain containing 16 (PRDM16) can be used [128]. Lee and Cowan (2014) describe the addition of lentiviruses of the above transcription factors for a high yield of brown adipocyte differentiation from human pluripotent stem cells [128]. Alternatively, addition of reported brown adipocyte inducer BMP7 can be added in cultures along with KIT ligand, tyrosine kinase 3 ligand, interleukin-6 and vascular endothelial growth factor [132, 133]. Confirmation of brown adipocyte differentiation from ESCs and iPSCs is expression of BAT specific genes *UCP1* and *PRDM16* and functional assays inducing thermogenesis from β-adrenergic receptor stimulation [132].

Mesendoderm and Tissue Engineering for Regenerative Medicine

We have described the developmental processes regulating the formation of organs derived from the mesendoderm. While pluripotent stem cells offer tremendous hope for regenerative medicine, many of the differentiation protocols developed thus far have been lacking the architectural and three-dimensional (3D) signals that are essential for proper organogenesis *in vivo*. A revolution in engineering, 3D printing, is now being applied to modelling embryonic processes from stem cells to obtain material for transplantation that is more appropriate in tissue and organ architecture than previously considered feasible. Here, we will discuss the application of 3D printing to the formation of mesendoderm-derived tissues with some insight from our own experiences.

Introduction to 3D Printing

There are few technological innovations that have been received with as much as enthusiasm and excitement as 3D printing. The prospect of printing into the third dimension has captivated the imaginations of scientists and laypeople alike. The rapid development and decreasing cost of this technology has meant that the promise this technology has offered is now becoming a reality. 3D printing is the process of manufacturing a customised three-dimensional object. 3D printing, otherwise known as additive manufacturing, was first developed by the engineer and scientist, Charles Hull. In the 1980s, Hull sought to improve the lengthy and error-prone process of producing plastic prototypes from photopolymers. In 1983, Hull produced the first 3D-printed object, a small plastic cup [134]. In 1986, Hull obtained the patent for *stereolithography* (one 3D printing technique which is outlined below), followed by a number of others on 3D printing technology [135]. 3D printing has since undergone extensive development and expansion so that a number of variations of the technology are now available for applications that range from commercial prototyping to medical implants, the food industry and weapon production in the defence forces [135].

To understand the digital workflow which gives rise to a 3D printed object, the different stages of the printing process have been detailed below and are represented in Fig. (**5**). The process of printing an object begins with a CAD (computer aided design) file. CAD is a commonplace method of developing a digital representation of a 3D objects and can be readily performed by a number of different available software [135]. Once a CAD file has been developed, the CAD model is converted into another file type known as .STL (standard tessellation language). STL files store information about each surface of the object in the form of vertices' coordinates. The .STL file is interpreted by the printer, which then performs another conversion to a G-file. The G-file is a deconstruction of the object into two-dimensional cross-sections which are then printed layer by layer on top of each other [135].

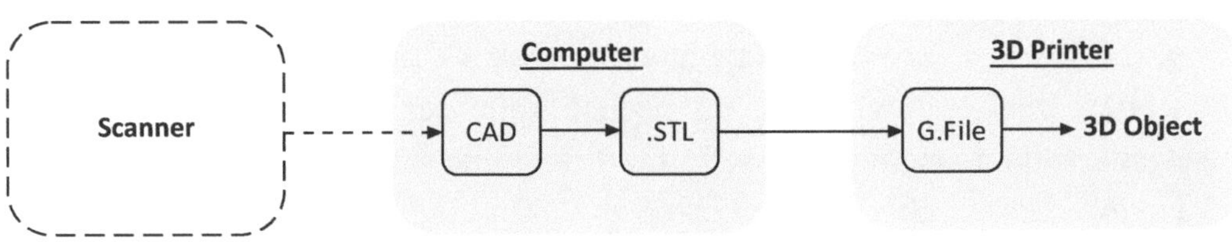

Fig. (5). Illustration showing the digital files and conversions involved in producing a 3D printed object. Note that the 3D CAD object may be based on a newly developed design or a scan of an existing object.

3D Printing Technologies

There is currently a number of available 3D printing techniques, six of which have been briefly explained below. It should be noted that while the following techniques use quite different technical approaches, the underlying principle of 3D printing (or additive manufacturing) remains the same. Each printer produces a specific 2D cross-sectional shape before shifting the existing layers and adding a subsequent 2D shape on top or below. Repeated additions of cross-sections finally result in the 3D object.

Stereolithography involves the hardening of a photopolymerizing liquid resin as UV light is beamed at a bath of resin. The beam is projected onto the base of the bath in the shape of a desired 2D cross-section. After the photopolymer is cross-linked, the cross-section is lifted by a stage and the next 2D layer can be solidified at the interface between the resin and UV light. This was the technique first described and patented by Charles Hull in 1986 [135]. Inkjet printing refers to the traditional method of printing ink onto paper in 2 dimensions. Two different techniques can be used for 3D inkjet printing as outlined include "drop on demand" and "continuous flow" printing. The principle of a drop on demand (DOD) inkjet printer is that pulses of voltage and pressure at the paper and ink interface direct the ink droplet to the desired place in the 2D plane. However, through reasonably simple modifications, this technique has been adapted to print in 3D. The adapted method works by printing 'binding' liquids in a way that binds a layer of powder (substrate particles 50 - 100 μm diameter) into a specific 2D shape. A stage is then lowered after printing each layer, a further layer of particles is added on, and the next 2D cross-section can be added. This method of printing is not confined to use with photopolymerizing materials, so allows a greater diversity of substrates that can be bound by the printed liquid [135]. Continuous inkjet printing uses electrically charged plates as well as patterns of pressure waves in the printer head to direct droplets of ink to the desired site on the 2D paper. Once again, with reasonably simple modifications, the printer can also position droplets of liquids other than ink onto the 2D plane, which can then be lowered to print subsequent 2D cross sections from the STL file. Successive stage lowering and addition of cross-sections then produces the desired 3D construct. The benefit of this technique is that any liquid with appropriate viscosity can be printed.

Selective laser sintering operates under the same basic process as 3D Inkjet printing. Layers of particles are hardened into 2D cross sections before lowering a stage. However rather than using a binding liquid to bind the particles, the substrate particles are melted by a CO_2 laser before amalgamating and solidifying in the shape of the desired cross-section [135]. Fused Deposition Modelling (FDM) is perhaps the most widely used 3D printing technique when considering personal use and commercial prototype production. Fused deposition modelling involves depositing a melted thermoplastic filament onto a 2D plane over a stage. The plastic filament is rolled into a heating unit and then extruded through a nozzle. The printer head is then guided in two dimensions by belts to produce the specified 2D cross-section. Upon completion of this cross-section, the stage is lowered and a further cross-section is deposited based on the STL file. A limitation of this method is that it must use thermoplastic materials such as polycarbonate and polystyrene. Furthermore, heterogeneity in the plastic substrate and the way that the STL file is sliced into a printer path introduces defects in the model [135]. Pressure Based Extrusion (PBE) functions similarly to fused deposition modelling (FDM). A printer head can be manipulated in two dimensions to extrude 2D cross-sectional shapes, while a platform may be raised and lowered in order to print successive cross-sections. However PBE varies from FDM in that the extruded material originates from a syringe or reservoir that is emptied by creating a pressure gradient. This gives the printer a wider range of available printing substrates, as any fluid with appropriate viscosity can be extruded. One such printer is the Fab@Home Model 3.0 which is described in the section below [135]. Laminated Object Manufacturing of 3D printing cuts 2D layers out of sheets of materials such as paper, plastic or metal. Each cross-section is cut using a laser, knife or welder and fused onto the sections below by heating before the stage is lowered and the next sheet of material is added over the top [135].

Note that all of these 3D printing techniques have specific advantages and limitations that make them suitable for use in different fields and applications. Considerations which must be undertaken before choosing a printing method in a research setting include: the ability to print with different materials (discussed in the next chapter), the resolution, shapes and sizes that can be achieved by each printer as well as the cost, which may range from hundreds of dollars for a desktop printer, to a cost in excess of US$250,000 for industrial type printers [135].

3D Printing of Biomaterials in Tissue Engineering

Biomaterials were mainly used as coatings to support the transplantation of medical devices into patients. However, in recent times, the focus has shifted to using biomaterials as the main components of implantable scaffolds for regenerative medicine applications [136]. This is due the ability of biomaterials to promote the viability, proliferation and differentiation of cells. The increased interest in creating biomaterial scaffolds appears to coincide with the development of 3D printing [137]. As described above, various 3D printing processes have been developed and refined over the years. Consequently, due to advances in cell biology, engineering and biomaterials science, more biomaterials with different properties have also been fabricated [138]. This has allowed for the production of tissue-engineered scaffolds for a wide range of applications including restoration of anatomical defects, reconstruction of complex organs and tissues, as well as scaffolds for stem cell differentiation [138]. However, challenges still remain in perfecting the structure of tissue-engineered scaffolds so that they are able to reflect the intricate architecture of human extracellular matrices as this will optimise and guide the proliferation, differentiation and function of cells towards specific tissue types [137]. This is important because the ultimate goal of 3D bioprinting is to produce scaffolds with anatomically and physiologically similar properties to specific tissues and organs for use in regenerative medicine [137].

Although the printing processes are important in developing and manipulating the architecture of these scaffolds, factors such as the biomaterial and cell source also have to be considered. Therefore, this section will provide an overview of the general criteria that can be used to select biomaterials and cells for 3D bioprinting, as well as the various modifications that can be applied to aid in the generation of specific tissue-engineered scaffolds. Furthermore, this section will also discuss the tests that can be employed to analyse the properties of the scaffold and observe the progressive development of the included cells to determine the overall success of the final product.

Selecting the Biomaterial for 3D Bioprinting

3D bioprinting offers the potential of a tissue engineering approach thats even more comprehensive and widely relevant than that of scaffolding [139]. Bioprinting involves the printing of 'bioinks', which comprise live cells with a fluid that acts as an extracellular matrix, most commonly a hydrogel and may also include biochemical factors, proteins or drugs [140]. The rationale of this technique is that by bioprinting cells as part of an architecture, they will readily differentiate to the range of cells which make up a functional tissue [139]. The biological nature of the bioinks imposes a number of limitations on the printing process. So far, 3D inkjet continuous printing and pressure-based extrusion (PBE) have been most successfully applied to this printing approach. This technique may be particularly useful in soft tissues such as the cardiovascular system. This approach has had limited clinical relevance but recent research has revealed that this is a promising field. Challenges which need to be overcome include low cell survival after printing, a lack of well-directed cell differentiation protocols, insufficient vascularisation and the creation of matrices through which metabolites will diffuse [140]. When selecting suitable biomaterials for 3D bioprinting, various factors have to be considered, such as compatibility with biological products as well as the printing process being utilised [137]. With more biomaterials being engineered specifically for 3D bioprinting, the criteria for suitable characteristics have become more detailed and precise. Therefore, some of the more essential qualities that are now expected in 3D bioprinting biomaterials will be discussed.

Printability: Printability is an important factor in determining the suitability of biomaterials for 3D bioprinting, as it affects the printers ability to accurately deposit biomaterials when producing the desired structure [137]. When determining the printability of a biomaterial, the type of printer being used must first be taken into consideration as the processing parameters and components will vary [138]. For example, continuous inkjet (CIJ) printing involves forcing a liquid through a small diameter orifice so that it produces a stream of smaller droplets that can be guided accurately to print a precise structure [141]. Therefore, bioinks are often used in CIJ printing, as they are predominantly liquid/gel solutions [142]. Another example is fused deposition modelling (FDM), which involves the extrusion of thermoplastic polymer filaments through a nozzle onto a platform in a layer-by-layer process [142]. Thus, in FMD, thermoplastic polymers such as poly (ε-caprolactone) are usually used for printing. Other factors affecting printability such as viscosity, gelling methods and melting temperature, also need to be considered to allow for optimal manipulation of the biomaterial during printing [137]. Bioinks must be gelled almost immediately after printing by polymer crosslinkers, photoactivation or thermal activation [143]. Thus, since this limitation does not allow the production of self-supporting structures, bioinks are usually more viscous to prevent the structure from collapsing as the gel is polymerised [144]. However, in FDM, it is more appropriate to consider the thermal properties of the biomaterial such as melting temperature or glass transition temperature as they determine the viscosity of the extruded biomaterial [145]. This is because viscosity decreases as temperature increases, and this will most likely affect the ability to control the deposition of the biomaterial.

Mechanical and structural properties: The mechanical and structural properties of the biomaterial are dependent on the tissue type it is mimicking such as skin, bone and cardiac tissue [137]. For example, elasticity is one of the major mechanical characteristics of soft tissues such as skin, blood vessels, cardiac and skeletal muscle, so elastomeric biomaterials such as α-elastin hydrogels have been developed to mimic this property [146]. However, it has become increasingly difficult to engineer biomaterials that possess the desired mechanical and structural properties without compromising its other characteristics such as cell viability [144]. More recently, methods have been devised to modify biomaterials to attain particular properties while maintaining other important characteristics. For example, a recent study by Rutz and colleagues (2015) has attempted to devise a bioink method that produces bioinks from a variety of synthetic and natural biomaterials [144]. This has allowed them to modify the composition, degree of cross linking and polymer concentration to optimise the mechanical and structural properties of the bioink, while maintaining printability.

Degradability: In tissue engineering, it is necessary to use biomaterials that degrade at a rate that is slower or equal to the cells ability to produce its own extracellular matrix (ECM) [147]. This is particularly important as 3D bioprinted structures are intended to be temporary and only used to guide the proliferation and differentiation of cells that will eventually replace damaged tissue. As such, hydrogels are often used because along with other biological, structural and mechanical properties, they are biodegradable [143]. However, when comparing naturally derived hydrogels (*e.g.* alginate, gelatin and elastin) and synthetic hydrogels (*e.g.* poly (ethylene glycol) (PEG) and poly (vinyl alcohol) (PVA)), both types of hydrogels are biodegradable but vary in their ability to control their rate of degradation [146]. Thus, it may be more suitable to use a combination of both depending on the application. Furthermore, linear aliphatic polyesters such as poly(lactic acid) (PLA), poly (glycolic acid) (PGA), and their copolymers poly(lactic acid-co-glycolic acid) (PLGA) are also used in printing scaffolds because they are both biodegradable and biocompatible [148]. When considering the degradability of the biomaterial, the products must be non-toxic so that they can be readily metabolised and cleared from the body [137]. Li and colleagues (2015) demonstrated that the DNA-based combination hydrogel utilised was biodegradable as it could be degraded by proteases (endoproteinase Glu-C) and nucleases (*Eco*RI and *Bam*HI).

Biocompatibility: The selected biomaterials should not be immunogenic and be able to coexist within the host since 3D-bioprinted structures are intended to be alternative transplant options for patients [149]. Generally, natural biomaterials such as alginate, collagen and fibrin are biocompatible due to their naturally derived origin [136]. However, synthetic polymers such as poly(lactic-co-glycolic acid) (PGA) and hydroxyapatite can also be made biocompatible by modifying particular properties such as composition, morphology and degradation [150].

Cell viability: Cell viability should also be considered when selecting biomaterials, especially those that incorporate cells before printing such as hydrogels [151]. Certain properties are desired when maximising cell viability in 3D-printed scaffolds. Lower-viscosity biomaterials are generally preferred as they provide a suitable environment for maintaining cell viability and function [137]. This is in contrast with high-viscosity biomaterials as they have been shown to decrease cell viability due to the increased shear stresses experienced during printing. Furthermore, biomaterials with low thermal conductivity and the ability to protect cells during printing may also increase viability and function after printing [152].

Biomimicry: The selected biomaterial should be able to support the proliferation, differentiation and function of cells by providing the appropriate signals *via* cell-material interactions or soluble factors [153]. As such, growth factors can be added to scaffolds *via* various methods to promote the differentiation of cells into tissue-specific cell lineages [148]. This will be discussed in greater detail in the next section. Alternatively, modifying the architecture of the scaffold can also enhance cellular signalling and interaction [154]. Due to advances in engineering and material science, modifications can now be achieved at the macro (*e.g.* shape), micro (*e.g.* porosity) and nano (*e.g.* surface modifications for cellular adhesion) levels [138]. For example, changing the porosity of the scaffold has been shown to enhance the biomimetic properties of scaffolds as it increases oxygenation throughout the scaffold to promote the viability and differentiation of multipotent stromal cells [155]. Porosity also enhances the delivery of osteogenic and angiogenic agents in bone tissue-engineered scaffolds [156]. However, porosity reduces mechanical properties such as compressive strength and increases the complexity of printing [156], so the degree of porosity that can be achieved is often limited.

Surface nanoarchitecture such as ridges, steps and grooves can also be included to increase biomolecule attachment for cellular adhesion, proliferation, and differentiation [154]. This is because the components of tissues and organs that cells interact with (*e.g.* extracellular matrices (ECM) are nanometres in dimension so mimicking this characteristic in scaffolds will enhance and guide tissue growth [157]. For example, Elias and colleagues demonstrated that osteoblast function was increased in alumina and carbon fibre biomaterials possessing nanometre surface dimensions that approximated those of hydroxyapatite found in bone [148]. More recently, there have been attempts to develop biomaterials with enhanced biomimetic qualities through an understanding of the composition and arrangement of extracellular matrices (ECM) in the targeted tissues [137]. This has been achieved through tissue decellularisation protocols, which produce intact ECM scaffolds for structural and physiological analysis. Tissue decellularisation involves the lysis of cells in tissues and organs using decellularisation agents as well as the removal of cell remnants by rinsing [158]. However, there are certain limitations to consider when using this method including possible toxicity from retention of decellularisation agents and difficulties in maintaining vascular structures [137]. Despite this, the possibility of incorporating ECM components into scaffolds using 3D bioprinting technology would be immensely beneficial to tissue engineering and regenerative medicine, as it would optimise biodegradability, biocompatibility and biomimicry.

Scaffolding and 3D-Bioprinting: 3D printing may currently be incorporated into tissue engineering in two different ways, both of which will be outlined in the section that follows. The first of these is the use of 3D printed scaffolds, which enable the fabrication of an extracellular architecture into which stem cells may be seeded, or the cells surrounding an implant may migrate into [159]. This approach has been particularly successful in the engineering of bone and tracheal tissues [160]. The second method of incorporating 3D printing into tissue engineering has been through 3D bioprinting. Bioprinting refers to the fabrication of 3D structures which incorporate living cells and bioactive moieties in a defined spatial distribution [161]. This technique also lends itself to stem cell research, where differentiation processes under different cellular architectures can be explored. An example of such research operating in a lab at the University of Sydney is given later in this section. The following sections will explore and discuss the current standing and the ongoing research with regard to 3D printing approaches in the engineering of various mesendodermal tissues.

Once the desired biomaterial has been selected, cells then need to be combined with the biomaterial to ultimately produce scaffolds that could potentially be used as replacements for disease or damaged tissues [136]. When determining which cell types to use for printing, it is firstly important to choose a cell type that is able to rapidly expand into the required numbers for printing [137]. If there are too little cells, this may result in reduced cell viability, but if there are too many cells, this may lead to hyperplasia or apoptosis [137]. Tissues and organs are complex structures composed of multiple cell types with specific biological functions. In tissue engineering, the aim is to replicate the structures of certain tissues or organs, so it is important that the engineered scaffolds contain cells that mimic the physiological properties of the targeted tissues. Thus, multiple cell types or stem cells (which can differentiate and proliferate into the required cell types) should be used [162].

Since the 3D bioprinted structure will ultimately be transplanted into patients, the structure must be biocompatible and resist rejection by the host immune system. This can be achieved by using an autologous cell source or tolerance-induction strategies [163, 164]. Furthermore, the cells must be robust enough to survive the bioprinting processes such as shear stress and pressure, and physiological stresses after transplantation such as enzymes and toxins [137]. Stem cells are the preferred choice of cells for 3D bioprinting as they are able to proliferate (*via* self-renewal) to generate large numbers of undifferentiated and multipotent cells that are able to transform into multiple functional tissue-specific cell types [137]. Stem cells can be produced autologously and therefore should not be rejected after transplantation [163]. Since ES cells and iPS cells clearly demonstrate these properties along with other cell-specific advantages, they are now being used for 3D bioprinting [151]. More recently, mesenchymal stromal cells have also been successfully isolated, expanded and differentiated for clinical trials [137]. This highlights that research into advancing cell-culture techniques could make more stem cell populations viable as cell sources, thus increasing the possible applications of 3D bioprinting. Cells can either be suspended in the biomaterial for printing or adhered to the printed scaffold. This depends on the biomaterial and printing processes being utilised [136]. When using softer biomaterials such as hydrogels, the cells can be suspended in the biomaterial before printing [140]. Cells are able to retain cell viability and function when suspended in a polymerised gel. However, when using more rigid biomaterials such as thermoplastic polymers, the polymerisation conditions are often too harsh for cells to survive printing [165]. Therefore, cells are often cultured separately and adhered to the surface instead, allowing them to proliferate and differentiate to ultimately produce an extracellular matrix and functional proteins and saccharides that compose the tissue [141].

Biomimetic products: Biomaterial scaffolds often require the delivery of soluble factors to provide additional signals for cellular proliferation, differentiation and function as cell-cell and cell-material interactions may not be sufficient enough to achieve this [153, 148]. Thus, various methods have been developed to deliver biological factors such as growth factors to aid in the development of tissue-engineered scaffolds. Growth factors can also be directly added to printing solutions before constructing scaffolds [166]. TGF-β, BMP and VEGF integrated into scaffolds promote the recruitment and differentiation of osteoprogenitor cells leading to regulation of osteogenesis, tissue regeneration and ECM production [156, 167].

Gelatin methacrylate microspheres have been used to delivery various biological factors for tissue engineering applications including stem cell differentiation within scaffolds [168, 169]. However, when using microspheres, the movement and amalgamation of spheres often results in the inconsistent diffusion of growth factors throughout the scaffold. Therefore, microspheres should be used in conjunction with macro-porous and nano-fibrous scaffolds that enhance the ability to spatially control the release of biological factors [148]. Coating techniques can also be used to enhance biomimicry as growth factors can be added to the surface of scaffolds [170]. This was highlighted by Bose and colleagues, where a layer of hydroxyl carbonate apatite (HCA) on the surface of a 3D bioglass scaffold was able to adsorb proteins and growth factors promoting new bone formation *in vivo*.

3D-printed scaffolds can be tested for cell viability, differentiation and growth as well as structural, mechanical and degradative properties to provide important information towards optimising scaffold design [171]. The biomaterial can be chosen to maximise cell viability in tissue-engineered scaffolds. However, the printing process may stress the cells, and cause injury or apoptosis [137]. In syringe-based printing, the nozzle diameter as well as the dispensing pressure have been found to affect the cell viability immediately after printing which can result in apoptosis and necrosis [172, 173]. An increase in dispensing pressure and decreasing nozzle diameter was found to decrease cell viability. Inkjet bioprinting usually records cell viabilities greater than 85%, whereas microextrusion printing reports viabilities range from 40–80% and laser-assisted bioprinting reports viabilities in excess of 90% [174, 175]. Cell viability should be monitored after printing to ensure that sufficient number of cells remain in the 3D printed scaffolds for proliferation and differentiation into the specific tissues or organs. Live/dead assays involve the use of different stains for both live and dead cells (*e.g.* Calcein acetoxymethylester stains live cells bright green and ethidiumhomodimer-2 stains dead cells red) and are often conducted at different periods throughout the experiment to observe changes in cell viability over time [176].

3D-bioprinted stem cells must proliferate and differentiate into the appropriate lineages. It is predominantly the 3D environment (which consists of the selected biomaterial with or without biomimetic products) that affects cell differentiation in tissue-engineered scaffolds [153]. Therefore, to determine if the 3D environment is promoting or inhibiting the proliferation and differentiation of cells, gene or protein expression that are characteristic to specific cell types should be assessed [177]. Differentiation of human satellite cells into multinucleated myotubes was assessed by qRT-PCR to analyse the expression of myogenic transcription factors as well as immunocytochemistry using monoclonal antibodies that targeted muscle-specific proteins [177]. Tissue-engineered scaffolds may interfere with these analytical methods [176]. Structural, mechanical and degradative properties are often tested to determine the quality of the scaffold design in mimicking specific tissues [178]. Jeong & Hollister demonstrated that the porosity of scaffolds could affect its mechanical and degradative properties. Therefore, similar tests can be conducted on the 3D-printed scaffolds to assess its mechanical and degradative properties [179].

Scaffolding: Scaffolding results in a 3D structure that can mimic the extracellular matrix in which cells migrate or are seeded into the structure after printing. The scaffold then provides support and encourages the proliferation and differentiation of tissue specific cells [139]. More traditional techniques exist for the fabrication of scaffolds, however these are limited by a lack of control over the structure's shape and porosity [159]. 3D printing overcomes these limitations by custom manufacturing the scaffold in a layer by layer manner to optimise both the shape and porosity for potential implantation [180]. The manufacture of 3D printed scaffolds has been incorporated into the tissue engineering of a number of different mesendodermal-derived tissues and organs. The following sections will describe how this has been achieved in bone and tracheal tissues.

Applications of Scaffolding in Mesendodermal Tissues

Stem Cells, 3D Printing and Bone Healing

The tissue engineering of bone structures is perhaps the area in which 3D printing has had the greatest tangible impact to date. Bone, as a tissue, possesses unique properties that have been harnessed in tissue engineering approaches. Bone tissue consists of cancellous and cortical bone. Together, these produce complex structures which have porosities from less than 10% in cortical bone, to 50-90% porosity in cancellous bone [156]. At a cellular level, bone is a dynamic structure. Osteoblast cells continuously deposit new bone matrix while old bone is resorbed by osteoclasts. The collection of these processes is known as bone remodelling and gives bone tissue self-healing properties. Despite this self-healing, severe fractures often require external intervention [156]. Amongst these interventions to heal the bone, tissue engineering approaches are now becoming more common. Some earlier approaches include bone autografts and allografts, both of which seek to incorporate an exogenous structure into the bone in an attempt to strengthen and improve healing *in vivo*. These grafting solutions present problems including graft availability, infection, donor site morbidity and the need for additional operations [140]. Biocompatible scaffolds have become viable alternatives to these approaches. The scaffolds attempt to mimic properties of the extracellular matrix and have traditionally been produced by methods which are limited in their ability to specify scaffold shape and pore size [156]. The incorporation of 3D printing into the bone tissue engineering approach has enabled the fabrication of scaffolds without these shortcomings [159].

A number of techniques have now been used to produce high precision, custom defined, 3D scaffolds. The power of this approach lies in the ability to tailor the 3D structure to the patient, as well as the ability to produce interconnected porous structures that become vascularised and populated by osteoblasts and parenchymal cells. Of the available scaffold printing techniques, Drop on Demand 3D Inkjet Printing (described above) has been most commonly adopted for the manufacture of scaffolds [156]. A number of powder and binding liquid combinations have been used in this approach, but the availability of a strong biocompatible complex remains an issue and a limitation. VEGF, FGFs and

BMPs are examples of growth factors that may have the effect of encouraging and accelerating the incorporation of desirable cells and subsequent healing of bone. These growth factors may be used together with stem cells to engineer a tissue that comes closer in structure and function to the implantation site *in vivo*. Muscle-derived stem cells have been cultured and bioprinted onto a printed scaffold containing BMP-2 resulting in more functional bone tissue [156]. Current limitations to this approach include the viability of the cells after bioprinting as well as that of the incorporated growth factor [156].

Stem Cells and 3D Bioprinting of the Trachea

The trachea is an organ in which 3D scaffold engineering by 3D printers has already had a major impact. The application of 3D printed tracheas has already been demonstrated in animal models clinically to correct tracheal stenosis [181, 182]. As scaffolding techniques become elaborated, the tissues printed match those of the native tissues more closely. Park and colleagues have produced a tissue-engineered trachea that has mechanical properties very similar to those of the native organ by adapting a cylindrical scaffold to the shape of a bellow [183]. A sacrificial scaffold was engineered using a micro-stereolithography technique with an alkali soluble photopolymer [183]. Each layer is printed to form a negative mold of the trachea, with the uncured photopolymer dissolved with isopropyl alcohol. Sustained release of TGFβ1, through a functionalised gelatin sponge, induces the growth of tracheal cartilage [183]. Alternating layers of the gelatin sponge, one with TGFβ1 and one seeded with the chondrocytes, were laid into the sacrificial scaffold. These structures were implanted subcutaneously into nude mice and the results show that the bellow structure were less resistant to three-point bending and more resistant to radial compression than the cylindrical structure [183]. This mechanical outcome is comparable to that of the biological trachea and can be further developed as an engineered substitute in the case of tracheal stenosis or other diseases affecting the airways. In 2014, Zopf and colleagues successfully produced and implanted a custom, 3D printed airway splint into a critical, paediatric tracheobronchomalacia patient under FDA emergency-use exemption [184]. The procedure restored patency of the bronchus and essentially reversed the respiratory distress of the patient. This case demonstrates the potential which tissue engineering holds and paves the way for more extensive applications in the future.

3D Bioprinting of Stem Cells for the Treatment of Cardiac Disease

Cardiovascular structures are an obvious candidate for tissue engineering approaches with cardiovascular disease (CVD) accounting for approximately 30% of all deaths globally [185]. As treatment of myocardial infarction improves, increasing numbers of CVD patients now survive cardiovascular events and live with partially damaged heart muscle [186]. Bioprinting offers an avenue to manufacture tissues to replace damaged myocardium as well as other structures such as the heart valves and vessels. Currently engineered cardiovascular tissues are limited in their ability to support live cells, meaning that tissues have been limited to thicknesses of ~ 100 μm, mainly as a result of the diffusion limitation for oxygen [187]. 3D printing may be promising way to overcome this, as it is able to print thick tissues with uniformly dispensed cells, which are also porous enough to sustain nutrient and gaseous exchange by cells. Promising research by Gaetani and colleagues have used pressure base extrusion methods to print human cardiac-derived cardiomyocyte progenitor cells (hCMPCs) in a sodium alginate hydrogel [188]. The group successfully maintained viability after printing, and demonstrated promising differentiation of cells. hCMPCs have proved to be a promising substrate for cardiovascular bioprinting as these cells are committed to the cardiac lineage, can differentiate into cardiomyocytes and can successfully proliferate *in vitro*. When these cells were printed into a 3D construct, they responded positively in comparison to 2D cultures with increased expression of cardiomyocyte transcription factors [188]. While the sodium alginate supported the hCMPCs in this experiment, the use of a decellularised extracellular matrix in a heart-derived bioink has shown great promise. Experiments by Pati and colleagues show that it is able to increase the expression of cardiac-specific genes Myh6 and Actn1 when printed with human inferior turbinate-tissue derived mesenchymal stromal cells (hTMSCs) [189].

Valvular heart disease is expected to triple by 2050. Developing countries struggle with the medical needs of those afflicted with rheumatic fever and its implications on heart valves; 30% of those born with congenital heart defects do not survive [190]. Heart valve malfunction as seen in stenosis or regurgitation cannot self-heal and must be replaced [190, 191]. The successful engineering of functional bioprinted heart valves would have a similarly revolutionary impact for many patients requiring heart valve replacement. Patients currently seeking a heart valve replacement are limited to choosing an artificial valve which requires ongoing anticoagulant therapy, or a xenograft or allograft which requires replacement 25 years at most [192]. Specifically, children and younger adults require aortic valve replacements that can grow, regenerate, and remodel with their still growing bodies. Therefore, an option for a fully compatible and biological heart valve replacement would be vital to the improvement of quality of life of the patients and to avoid future replacement surgeries.

With 3D printing, significant progress has been made on the bioengineering of heart valves. Duan and colleagues have successfully cultured encapsulated root sinus smooth muscle cells (SMCs) (81.4 +/- 3.4% viability) and aortic valve leaflet interstitial cells (VICs) (83.2 +/- 4.0% viability) in the valve root and leaflets respectively, of alginate/gelatin valve conduits [191]. SMCs added to stiff matrix resulted in increased expression of alpha-smooth muscle actin while VICs in soft matrix resulted in increased vimentin expression. These results suggest that complex aortic valve hydrogel conduits are possible to construct, especially with further advances in 3D printing technology. Although the prospect of being able to custom print a heart valve is exciting, there are still limitations such as an unfit construct, long-term malformation, and loss of anisotropic function. A badly constructed structure would result in mechanical stress on the construct and ultimately, calcification, regurgitation and hemodynamic inefficiency of the heart. A long-term transplant malformation such as a uniform reduction or increase in thickness or remodelling could result in reduced function, regurgitation and calcification and will require further repair. Over time, anisotropic function may also cease and become isotropic, which may cause calcification [190].

Bioengineering Blood Vessels

Previous attempts have been made to produce vasculature by seeding 3D printed scaffolds. However this technique is limited by the fact that due to the printing conditions, cells cannot be printed into the scaffold directly, and the seeding of the scaffold at a later time poses a problem. Scaffold-free bioprinting methods have been investigated. Norotte and colleagues bioprinted blood vessels through the layering of various vascular cell types around agarose rods [193]. Layers of smooth muscle cells and fibroblasts formed vasculature of 0.9-2.5mm in diameter. The agarose rods were then pulled out of the construct leaving a cylindrical structure however this method possessed some limitations. Structurally, only open-ended vasculature may be constructed using this method and more complex structures may be harder to achieve. Further research into the use of thermoreverisble or photosensitive gels could improve the final steps of this process. Photolithography, a descendant of the stereolithography technique, has been investigated as another method of scaffold free bioprinting. This technique involves the exposure of photosensitive hydrogel to light through a photomask. The areas of the hydrogel exposed to the light polymerise and form a cross-linked material. Then, the rest of the uncrosslinked gel can be washed away leaving a polymerised structure designed by the shape of the light shown onto the photosensitive gel. This method is promising as the photosensitive hydrogel can be prepared to have specific porosity and mechanical properties, which affect the growth of the stem cells [194]. Bioprinted vascular structure made from GelMA hydrogel with endothelial progenitor cells (EPCs) and mesenchymal stem cells (MSCs) have been subcutaneously transplanted into immunodeficient mice. GelMA is a unique hydrogel in that it is a functionalised gelatin hydrogel with control over the degree of porosity and degradability for optimal cell viability post-printing. The results show that 7 days post-transplantation, lumen was present throughout the structure with murine erythrocytes flowing through the vasculature. This purports that the transplanted vessels had successfully fused with the host vasculature [195]. Using laser beams, the photolitography technique selectively polymerises the hydrogel used creating 3D structures.

3D Stem Cell Bioprinting in an Undergraduate Research Laboratory

To provide an example of a 3D Bioprinter in a research laboratory, the setup of an operational pressure based extrusion printer (PBE) has been described below. The printer is used for the purpose of stem cell research at the University of Sydney Charles Perkins Centre. This section will outline the general setup, operation and uses of this printer. The printer described above is a Fab@Home Model 3.0 PBE printer that is manufactured at Cornell University and is commercially available.

The chassis of the printer is produced from 3D printed acrylic panes. Five motors are then positioned within the chassis to drive either a belt, to manipulate the location of the printer head, or to drive one of two plungers that generate the pressure required for the extrusion of the syringe contents (Fig. **6**).

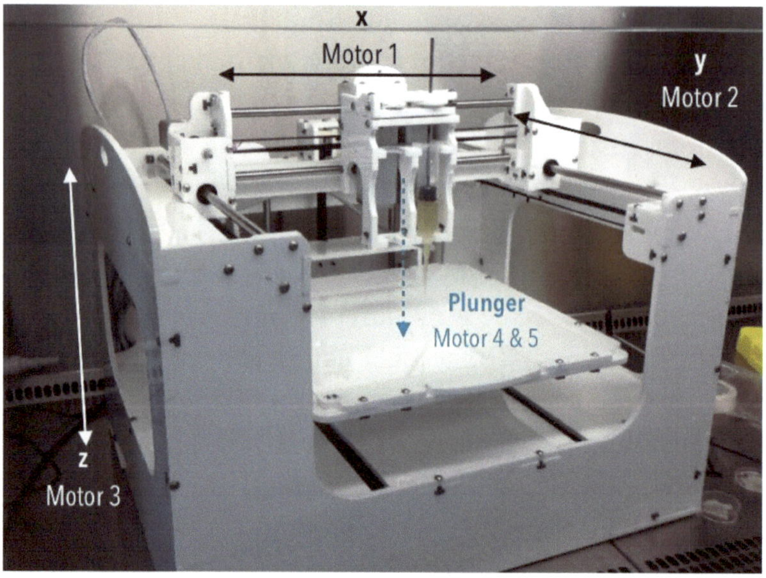

Fig. (6). Annotated set-up of the Fab@Home Model 3.0 Bioprinter in the X-lab undergraduate teaching facility of the Charles Perkins Centre. The printer is located in a biosafety cabinet (BSC) where it can be sterilized by UV light and used to print under cell culturing conditions. The annotations depict the printer axes of movement in 3D space.

The printer can be operated by software in one of two ways. Firstly, the motors may be driven by a basic programming language that sends movement trajectories directly to each motor. This gives the operator a high degree of control over the exact print path and extrusion speed but is limited to producing straight-line structures. Secondly, the printer may be driven by 'Fab Studio' and 'Fab Interpreter' software that allows the printing of objects from CAD (computer aided design) files. In this mode, the virtual 3D CAD object is converted automatically into a printer path that can then be executed.

The undergraduate research team operating this printer has been investigating differentiation of the cardiac lineage from embryonic stem cells (ESCs) in the context of bioprinted hydrogels. The group has successfully produced alginate bioinks containing ESCs, which can be printed into a calcium chloride ($CaCl_2$) bath to produce a dimerised hydrogel with precise architecture. Based on our own practical experience, we observed that the different viscosities of non-autoclaved and autoclaved alginate affected our ability to produce accurate and consistent structures using the Fab@Home printer. More specifically, the autoclaved alginate was considerably less viscous than the non-autoclaved alginate. Upon further analysis, we observed that this appeared to increase the time taken to polymerise autoclaved alginate with calcium chloride ($CaCl_2$) compared to non-autoclaved alginate. We believe that this was most likely due to the autoclaving process decreasing the degree of polymerisation in the alginate [196]. Therefore, this highlights how the factors affecting printability also need to be considered when selecting biomaterials for 3D bioprinting.

This hydrogel has successfully produced viable structures *in vitro*, including a sheet of communicating contractile cardiomyocytes. The hydrogel has also been used to produce a number of printed 3D architectures including sheets of hydrogel (Fig. 7). The pictured sheets can be combined with different growth factors with the rationale of differentiating different morphologies at different levels within the gel.

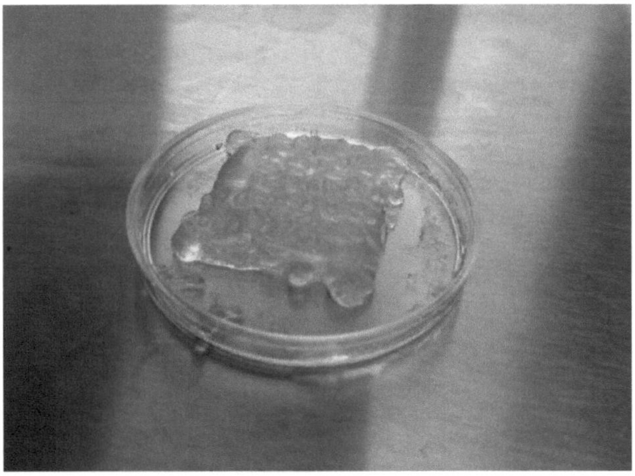

Fig. (7). This image shows an alginate-$CaCl_2$ hydrogel, which has been bioprinted in 2 layers which are combined with different growth factors for differentiation of mouse embryonic stem cells into cardiovascular lineages.

Future Directions in Tissue Engineering

3D bioprinting is the new frontier of clinical care, especially in the case of transplantation and cell therapy. It mitigates the risk of tissue rejection and the need for immunosuppressants. Organs and tissues can be customised to each patient and would eliminate the shortage of organ donors. At the moment, seeding the required stem cells onto a decellularised scaffold has been the closest to developing a whole functioning organ; however, the development of 3D printing technology allows control over the finer aspects of the process such as cell placement, cell concentration, and diameter of printed cells. 3D printing layers cells, growth factors, and the scaffold matrix (usually hydrogels), selectively to create tissue or whole organs [197]. These processes are outlined above. The clinical application of 3D bioprinted organs and tissues are still a couple of decades away, but there are a multitude of applications in the field emerging such as live implants of organs or parts of organs, *in situ* printing, and replacement of animal testing commercially and in drug discovery [137, 197].

3D bioprinting and tissue engineering have been used to successfully culture bone, vasculature, trachea, and heart tissue with ability to be transplanted. However, there are still issues with 3D bioprinting. Many of the whole organ bioengineering have only been produced in small scale and must be translated to human-sized organs. Specifically, the diffusion of oxygen between the host and transplanted tissue limits the engineered tissue width to under 200μm [197]. However, cutting edge vascular bed tissue engineering could solve this problem as new studies have shown successful perfusion of transplanted vasculature in murine models [198]. The integration of vasculature into current organ and tissue bioengineering would be vital, as thicker tissues will need a circulatory system to maintain cellular metabolic functions, mature, and fuse together. Tissues will grow in width as studies will continue to add various layers of multiple cell types to form a complete structure. A vascular network embedded onto the organ or tissue engineered would provide the cells with the oxygen, gas, nutrient, and waste exchange required by normal healthy living tissue [197].

Unfortunately, bifurcated and branched vessels have been challenging to grow with only a few studies investigating this area. A protocol for developing complex vascular network still needs to be engineered before integration with whole organs [199]. Kusuma and colleagues have successfully implanted a vascular network that fused with the host vasculature and established blood flow [200]. Their method involved the culturing of a bicellular vascular population from human pluripotent stem cells (hPSCs) in a synthetic matrix. These mature to early vascular cells (EVCs) and then to endothelial cells and pericytes, which are vital for stable vasculature. The use of 3D printing technology could hopefully further this research in that the development of the vascular bed could move out of the synthetic matrix and into an engineered whole organ.

In situ printing, the direct printing of live organs or tissue during open surgery is a leading procedure following the progress in 3D bioprinting technology. It has already been done with a skin lesion filled with keratinocytes and fibroblasts, but its full potential is to be able to print an organ straight onto the patient [199]. The development of whole organs from our skin to the bone means that we can test drugs and collect accurate results with respect to reaction, side effects, and benefits. Currently, the results of animal testing for anything from cosmetics to cutting edge cancer treatments must be translated to accurate human body responses. Although it is expensive and labour intensive, the results normally are unable to predict that of a normal human body response as animals and humans would have differences such as variable immunogenic make up and gene regulation [201]. Furthermore, human clinical trials must be thoroughly conducted before deciding whether the drug is effective, ineffective or too toxic [202]. Therefore, advances in 3D bioprinting of tissues and whole organs would greatly benefit research of how various drugs affect the corresponding organs. Parameters such as immunogenicity, effect of environmental factors, and pre-existing diseases can be strictly set to best observe the effect of the drug on the patient. For example, in the study to find a drug against Duchenne Muscular Dystrophy (DMD), an mdx murine model is being used. DMD is a genetic X-linked disorder where the dystrophin protein is absent causing skeletal muscle deterioration [203]. Instead of the murine model, skeletal muscle with DMD can be 3D printed, cultured and tested against the drug in question with the results already in context to a human response. Although human clinical trials should still be required, these results from the 3D printed tissue could be a more accurate prediction of the human outcome than would the results from an animal model.

CONCLUDING REMARKS

Revolutions in both biology and engineering have taken place in the past decade. The ability to induce pluripotent stem cells from adult cells has opened the field of regenerative medicine enormously as it may alleviate the difficulties associated with transplantation such as immune rejection. Advances in 3D bioprinting have expanded our possibilities of printing functional tissues and organs for tailor-made regenerative medical interventions. As many of the diseases currently afflicting patients worldwide involve either tissues or organs originating from the mesendoderm, understanding the underlying biological processes leading to the formation of these tissues is crucial. Combining our biological understanding of mesendoderm formation with patient-specific iPS cell generation and 3D bioprinting technologies holds great promise for the future of regenerative medicine.

CONFLICT OF INTEREST

The authors confirm that they have no conflict of interest to declare for this publication.

ACKNOWLEDGEMENTS

We apologise for the many hard working researchers whose work we have not included in this chapter. We also acknowledge the members of the Laboratory of Blood Cell Development and the Andrology Research Group, School of Medical Science, Faculty of Medicine, University of Sydney. The Laboratory of Blood Cell Development gratefully acknowledges past support from the Sydney Medical School, the NWG Macintosh Award, the National Breast Cancer Foundation and the Feline Health Research Fund. MPT was supported by the University of Sydney Outstanding Achievement Scholarship, James Aitken Scholarship, and the Fairfax prize (2013). DTML was supported by the University of Sydney Outstanding Achievement Scholarship, James Aitken Scholarship (2013) and the University of Sydney Academic Merit Prize (2013). CH, MPT and DTML were engaged in an undergraduate research group run by STF in senior Physiology units of study at the University of Sydney. We are grateful for the support of the Charles Perkins Centre in purchasing the 3D bioprinter described in this chapter. We also wish to acknowledge the help of Renee Barber, Denise McDonell, Virginia Klomp and Dr Soma Vignarajan from the Charles Perkins Centre in the purchase, establishment and operation of the 3D bioprinter described here. STF is the University of Sydney Sesquicentennial Senior Lecturer in Molecular Embryology. Finally, our thanks to Prof. Richard N. Mortie for all his wit and intelligence throughout this writing process-it has been a long time coming.

REFERENCES

[1] Lammert E, Cleaver O, Melton D. Induction of pancreatic differentiation by signals from blood vessels. Science 2001; 294(5542): 564-7.
[http://dx.doi.org/10.1126/science.1064344] [PMID: 11577200]

[2] Goldman O, Han S, Hamou W, *et al.* Stem Cell Reports. Authors 2014; 3(4): 556-65.

[3] Rodaway A, Patient R. Mesendoderm. an ancient germ layer? Cell 2001; 105(2): 169-72.
[http://dx.doi.org/10.1016/S0092-8674(01)00307-5] [PMID: 11336666]

[4] Maduro MF, Meneghini MD, Bowerman B, Broitman-Maduro G, Rothman JH. Restriction of mesendoderm to a single blastomere by the combined action of SKN-1 and a GSK-3beta homolog is mediated by MED-1 and -2 in *C. elegans*. Mol Cell 2001; 7(3): 475-85.
[http://dx.doi.org/10.1016/S1097-2765(01)00195-2] [PMID: 11463373]

[5] Nieuwkoop PD. Short historical survey of pattern formation in the endo-mesoderm and the neural anlage in the vertebrates: the role of vertical and planar inductive actions. Cell Mol Life Sci 1997; 53(4): 305-18.
[http://dx.doi.org/10.1007/PL00000608] [PMID: 9137623]

[6] Agius E, Oelgeschläger M, Wessely O, Kemp C, De Robertis EM. Endodermal Nodal-related signals and mesoderm induction in *Xenopus*. Development 2000; 127(6): 1173-83.
[PMID: 10683171]

[7] Ryan K, Garrett N, Mitchell A, Gurdon JB. Eomesodermin, a key early gene in *Xenopus* mesoderm differentiation. Cell 1996; 87(6): 989-1000.
[http://dx.doi.org/10.1016/S0092-8674(00)81794-8] [PMID: 8978604]

[8] Nelson AC, Cutty SJ, Niini M, *et al.* Global identification of Smad2 and Eomesodermin targets in zebrafish identifies a conserved transcriptional network in mesendoderm and a novel role for Eomesodermin in repression of ectodermal gene expression. BMC Biol 2014; 12(81): 81.
[http://dx.doi.org/10.1186/s12915-014-0081-5] [PMID: 25277163]

[9] Costello I, Nowotschin S, Sun X, *et al.* Lhx1 functions together with Otx2, Foxa2, and Ldb1 to govern anterior mesendoderm, node, and midline development. Genes Dev 2015; 29(20): 2108-22.
[http://dx.doi.org/10.1101/gad.268979.115] [PMID: 26494787]

[10] Costello I, Pimeisl I-M, Dräger S, Bikoff EK, Robertson EJ, Arnold SJ. The T-box transcription factor

Eomesodermin acts upstream of Mesp1 to specify cardiac mesoderm during mouse gastrulation. Nat Cell Biol 2011; 13(9): 1084-91.
[http://dx.doi.org/10.1038/ncb2304] [PMID: 21822279]

[11] Nikaido M, Tada M, Saji T, Ueno N. Conservation of BMP signaling in zebrafish mesoderm patterning. Mech Dev 1997; 61(1-2): 75-88.
[http://dx.doi.org/10.1016/S0925-4773(96)00625-9] [PMID: 9076679]

[12] Winnier G, Blessing M, Labosky PA, Hogan BL. Bone morphogenetic protein-4 is required for mesoderm formation and patterning in the mouse. Genes Dev 1995; 9(17): 2105-16.
[http://dx.doi.org/10.1101/gad.9.17.2105] [PMID: 7657163]

[13] Grapin-Botton A. Endoderm specification. StemBook. The Stem Cell Research Community 2008.
[http://dx.doi.org/10.3824/stembook.1.30.1]

[14] Fukui A, Goto T, Kitamoto J, Homma M, Asashima M. SDF-1 α regulates mesendodermal cell migration during frog gastrulation. Biochem Biophys Res Commun 2007; 354(2): 472-7.
[http://dx.doi.org/10.1016/j.bbrc.2007.01.007] [PMID: 17239342]

[15] Clements D, Woodland HR. Changes in embryonic cell fate produced by expression of an endodermal transcription factor, Xsox17. Mech Dev 2000; 99(1-2): 65-70.
[http://dx.doi.org/10.1016/S0925-4773(00)00476-7] [PMID: 11091074]

[16] Henry GL, Brivanlou IH, Kessler DS, Hemmati-Brivanlou A, Melton DA. TGF-beta signals and a pattern in Xenopus laevis endodermal development. Development 1996; 122(3): 1007-15.
[PMID: 8631246]

[17] Dougan ST, Warga RM, Kane DA, Schier AF, Talbot WS. The role of the zebrafish nodal-related genes squint and cyclops in patterning of mesendoderm. Development 2003; 130(9): 1837-51.
[http://dx.doi.org/10.1242/dev.00400] [PMID: 12642489]

[18] Liu Y, Festing M, Thompson JC, *et al.* Smad2 and Smad3 coordinately regulate craniofacial and endodermal development. Dev Biol 2004; 270(2): 411-26.
[http://dx.doi.org/10.1016/j.ydbio.2004.03.017] [PMID: 15183723]

[19] Thisse B, Wright CV, Thisse C. Activin- and Nodal-related factors control antero-posterior patterning of the zebrafish embryo. Nature 2000; 403(6768): 425-8.
[http://dx.doi.org/10.1038/35000200] [PMID: 10667793]

[20] Shen MM. Nodal signaling: developmental roles and regulation. Development 2007; 134(6): 1023-34.
[http://dx.doi.org/10.1242/dev.000166] [PMID: 17287255]

[21] Dubrulle J, Jordan BM, Akhmetova L, *et al.* Response to Nodal morphogen gradient is determined by the kinetics of target gene induction. eLife 2015; 4(4)
[http://dx.doi.org/10.7554/eLife.05042] [PMID: 25869585]

[22] Conlon FL, Lyons KM, Takaesu N, *et al.* A primary requirement for nodal in the formation and maintenance of the primitive streak in the mouse. Development 1994; 120(7): 1919-28.
[PMID: 7924997]

[23] Liu P, Wakamiya M, Shea MJ, Albrecht U, Behringer RR, Bradley A. Requirement for Wnt3 in vertebrate axis formation. Nat Genet 1999; 22(4): 361-5.
[http://dx.doi.org/10.1038/11932] [PMID: 10431240]

[24] Brennan J, Lu CC, Norris DP, Rodriguez TA, Beddington RS, Robertson EJ. Nodal signalling in the epiblast patterns the early mouse embryo. Nature 2001; 411(6840): 965-9.
[http://dx.doi.org/10.1038/35082103] [PMID: 11418863]

[25] Aronson BE, Stapleton KA, Krasinski SD. Role of GATA factors in development, differentiation, and homeostasis of the small intestinal epithelium. Am J Physiol Gastrointest Liver Physiol 2014; 306(6): G474-90.
[http://dx.doi.org/10.1152/ajpgi.00119.2013] [PMID: 24436352]

[26] Jacobsen CM, Narita N, Bielinska M, Syder AJ, Gordon JI, Wilson DB. Genetic mosaic analysis reveals that GATA-4 is required for proper differentiation of mouse gastric epithelium. Dev Biol 2002; 241(1): 34-46.
[http://dx.doi.org/10.1006/dbio.2001.0424] [PMID: 11784093]

[27] Koutsourakis M, Langeveld A, Patient R, Beddington R, Grosveld F. The transcription factor GATA6 is essential for early extraembryonic development. Development 1999; 126(9): 723-32.
[PMID: 10383242]

[28] Molkentin JD, Tymitz KM, Richardson JA, Olson EN. Abnormalities of the genitourinary tract in female mice lacking GATA5. Mol Cell Biol 2000; 20(14): 5256-60.
[http://dx.doi.org/10.1128/MCB.20.14.5256-5260.2000] [PMID: 10866681]

[29] Ang SL, Wierda A, Wong D, *et al.* The formation and maintenance of the definitive endoderm lineage in the mouse: involvement of HNF3/forkhead proteins. Development 1993; 119(4): 1301-15.
[PMID: 8306889]

[30] McKnight KD, Hou J, Hoodless PA. Foxh1 and Foxa2 are not required for formation of the midgut and hindgut definitive endoderm. Developmental Biology 2010; 337(2): 471-81.

[31] Bossard P, Zaret KS. GATA transcription factors as potentiators of gut endoderm differentiation. Development 1998; 125(24): 4909-17.
[PMID: 9811575]

[32] Cirillo LA, Lin FR, Cuesta I, Friedman D, Jarnik M, Zaret KS. Opening of compacted chromatin by early developmental transcription factors HNF3 (FoxA) and GATA-4. Mol Cell 2002; 9(2): 279-89.
[http://dx.doi.org/10.1016/S1097-2765(02)00459-8] [PMID: 11864602]

[33] Sakaguchi T, Kuroiwa A, Takeda H. A novel sox gene, 226D7, acts downstream of Nodal signaling to specify endoderm precursors in zebrafish. Mech Dev 2001; 107(1-2): 25-38.
[http://dx.doi.org/10.1016/S0925-4773(01)00453-1] [PMID: 11520661]

[34] Kanai-Azuma M, Kanai Y, Gad JM, *et al.* Depletion of definitive gut endoderm in Sox17-null mutant mice. Development 2002; 129(10): 2367-79.
[PMID: 11973269]

[35] Sinner D, Kirilenko P, Rankin S, *et al.* Global analysis of the transcriptional network controlling *Xenopus* endoderm formation. Development 2006; 133(10): 1955-66.
[http://dx.doi.org/10.1242/dev.02358] [PMID: 16651540]

[36] Tam PP, Khoo PL, Lewis SL, *et al.* Sequential allocation and global pattern of movement of the definitive endoderm in the mouse embryo during gastrulation. Development 2007; 134(2): 251-60.
[http://dx.doi.org/10.1242/dev.02724] [PMID: 17151016]

[37] Howell M, Itoh F, Pierreux CE, *et al. Xenopus* Smad4beta is the co-Smad component of developmentally regulated transcription factor complexes responsible for induction of early mesodermal genes. Dev Biol 1999; 214(2): 354-69.
[http://dx.doi.org/10.1006/dbio.1999.9430] [PMID: 10525340]

[38] Murry CE, Keller G. Differentiation of embryonic stem cells to clinically relevant populations: lessons from embryonic development. Cell 2008; 132(4): 661-80.
[http://dx.doi.org/10.1016/j.cell.2008.02.008] [PMID: 18295582]

[39] Engler AJ, Sen S, Sweeney HL, Discher DE. Matrix elasticity directs stem cell lineage specification. Cell 2006; 126(4): 677-89.
[http://dx.doi.org/10.1016/j.cell.2006.06.044] [PMID: 16923388]

[40] Warmflash A, Sorre B, Etoc F, Siggia ED, Brivanlou AH. A method to recapitulate early embryonic spatial patterning in human embryonic stem cells. Nat Methods 2014; 11(8): 847-54.
[http://dx.doi.org/10.1038/nmeth.3016] [PMID: 24973948]

[41] Tada S, Era T, Furusawa C, *et al.* Characterization of mesendoderm: a diverging point of the definitive endoderm and mesoderm in embryonic stem cell differentiation culture. Development 2005; 132(19): 4363-74.
[http://dx.doi.org/10.1242/dev.02005] [PMID: 16141227]

[42] Gadue P, Huber TL, Nostro MC, Kattman S, Keller GM. Germ layer induction from embryonic stem cells. Exp Hematol 2005; 33(9): 955-64.
[http://dx.doi.org/10.1016/j.exphem.2005.06.009] [PMID: 16140142]

[43] Fehling HJ, Lacaud G, Kubo A, *et al.* Tracking mesoderm induction and its specification to the hemangioblast during embryonic stem cell differentiation. Development 2003; 130(17): 4217-27.
[http://dx.doi.org/10.1242/dev.00589] [PMID: 12874139]

[44] Nostro MC, Cheng X, Keller GM, Gadue P. Wnt, activin, and BMP signaling regulate distinct stages in the developmental pathway from embryonic stem cells to blood. Cell Stem Cell 2008; 2(1): 60-71.
[http://dx.doi.org/10.1016/j.stem.2007.10.011] [PMID: 18371422]

[45] Kubo A, Shinozaki K, Shannon JM, *et al.* Development of definitive endoderm from embryonic stem cells in culture. Development 2004; 131(7): 1651-62.
[http://dx.doi.org/10.1242/dev.01044] [PMID: 14998924]

[46] Willey S, Ayuso-Sacido A, Zhang H, *et al.* Acceleration of mesoderm development and expansion of hematopoietic progenitors in differentiating ES cells by the mouse Mix-like homeodomain transcription factor. Blood 2006; 107(8): 3122-30.
[http://dx.doi.org/10.1182/blood-2005-10-4120] [PMID: 16403910]

[47] Nishikawa SI, Nishikawa S, Kawamoto H, *et al. In vitro* generation of lymphohematopoietic cells from endothelial cells purified from murine embryos. Immunity 1998; 8(6): 761-9.
[http://dx.doi.org/10.1016/S1074-7613(00)80581-6] [PMID: 9655490]

[48] Ogawa M, Kizumoto M, Nishikawa S, Fujimoto T, Kodama H, Nishikawa SI. Expression of α4-integrin defines the earliest precursor of hematopoietic cell lineage diverged from endothelial cells. Blood 1999; 93(4): 1168-77.
[PMID: 9949159]

[49] Zhang H, Nieves JL, Fraser ST, *et al.* Expression of podocalyxin separates the hematopoietic and vascular potentials of mouse embryonic stem cell-derived mesoderm. Stem Cells 2014; 32(1): 191-203.
[http://dx.doi.org/10.1002/stem.1536] [PMID: 24022884]

[50] Takahashi K, Yamanaka S. Induction of pluripotent stem cells from mouse embryonic and adult fibroblast cultures by defined factors. Cell 2006; 126(4): 663-76.
[http://dx.doi.org/10.1016/j.cell.2006.07.024] [PMID: 16904174]

[51] Phillips MJ, Perez ET, Martin JM, *et al.* Modeling human retinal development with patient-specific induced pluripotent stem cells reveals multiple roles for visual system homeobox 2. Stem Cells 2014; 32(6): 1480-92.
[http://dx.doi.org/10.1002/stem.1667] [PMID: 24532057]

[52] Menon T, Firth AL, Scripture-Adams DD, *et al.* Lymphoid regeneration from gene-corrected SCID-X1 subject-derived iPSCs. Cell Stem Cell 2015; 16(4): 367-72.
[http://dx.doi.org/10.1016/j.stem.2015.02.005] [PMID: 25772073]

[53] Hanna J, Wernig M, Markoulaki S, *et al.* Treatment of sickle cell anemia mouse model with iPS cells generated from autologous skin. Science 2007; 318(5858): 1920-3.
[http://dx.doi.org/10.1126/science.1152092] [PMID: 18063756]

[54] L Charles Murtaugh and Daniel Kopinke. Pancreatic stem cells. StemBook

[55] Madsen OD. Stem cells and diabetes treatment. APMIS 2005; 113(11-12): 858-75.
[http://dx.doi.org/10.1111/j.1600-0463.2005.apm_418.x] [PMID: 16480455]

[56] Pagliuca FW, Millman JR, Gürtler M, *et al.* Generation of functional human pancreatic β cells in vitro. Cell 2014; 159(2): 428-39.
[http://dx.doi.org/10.1016/j.cell.2014.09.040] [PMID: 25303535]

[57] Murtaugh LC. Pancreas and beta-cell development: from the actual to the possible. Development 2007; 134(3): 427-38.
[http://dx.doi.org/10.1242/dev.02770] [PMID: 17185316]

[58] Ohlsson H, Karlsson K, Edlund T. IPF1, a homeodomain-containing transactivator of the insulin gene. EMBO J 1993; 12(11): 4251-9.
[PMID: 7901001]

[59] Kawaguchi Y, Cooper B, Gannon M, Ray M, MacDonald RJ, Wright CV. The role of the transcriptional regulator Ptf1a in converting intestinal to pancreatic progenitors. Nat Genet 2002; 32(1): 128-34.
[http://dx.doi.org/10.1038/ng959] [PMID: 12185368]

[60] Gu G, Dubauskaite J, Melton DA. Direct evidence for the pancreatic lineage: NGN3+ cells are islet progenitors and are distinct from duct progenitors. Development 2002; 129(10): 2447-57.
[PMID: 11973276]

[61] Piper K, Brickwood S, Turnpenny LW, *et al.* Beta cell differentiation during early human pancreas development. J Endocrinol 2004; 181(1): 11-23.
[http://dx.doi.org/10.1677/joe.0.1810011] [PMID: 15072563]

[62] Shi Y, Hou L, Tang F, *et al.* Inducing embryonic stem cells to differentiate into pancreatic β cells by a novel three-step approach with activin A and all-trans retinoic acid. Stem Cells 2005; 23(5): 656-62. [http://dx.doi.org/10.1634/stemcells.2004-0241] [PMID: 15849173]

[63] Jiang J, Au M, Lu K, *et al.* Generation of insulin-producing islet-like clusters from human embryonic stem cells. Stem Cells 2007; 25(8): 1940-53. [http://dx.doi.org/10.1634/stemcells.2006-0761] [PMID: 17510217]

[64] Zorn. Liver Development. StemBook. The Stem Cell Research Community 2008.

[65] Zhao R, Duncan SA. Embryonic development of the liver. Hepatology 2005; 41(5): 956-67. [http://dx.doi.org/10.1002/hep.20691] [PMID: 15841465]

[66] Tremblay KD, Zaret KS. Distinct populations of endoderm cells converge to generate the embryonic liver bud and ventral foregut tissues. Dev Biol 2005; 280(1): 87-99. [http://dx.doi.org/10.1016/j.ydbio.2005.01.003] [PMID: 15766750]

[67] Asahina K, Zhou B, Pu WT, Tsukamoto H. Septum transversum-derived mesothelium gives rise to hepatic stellate cells and perivascular mesenchymal cells in developing mouse liver. Hepatology 2011; 53(3): 983-95. [http://dx.doi.org/10.1002/hep.24119] [PMID: 21294146]

[68] Al-Drees MA, Yeo JH, Boumelhem BB, *et al.* Making Blood: The Haematopoietic Niche throughout Ontogeny. Stem Cells Int 2015; 2015(10): 571893. [PMID: 26113865]

[69] Medlock ES, Haar JL. The liver hemopoietic environment: I. Developing hepatocytes and their role in fetal hemopoiesis. Anat Rec 1983; 207(1): 31-41. [http://dx.doi.org/10.1002/ar.1092070105] [PMID: 6638531]

[70] Goldman O, Han S, Sourisseau M, *et al.* KDR identifies a conserved human and murine hepatic progenitor and instructs early liver development. Cell Stem Cell 2013; 12(6): 748-60. [http://dx.doi.org/10.1016/j.stem.2013.04.026] [PMID: 23746980]

[71] Han S, Dziedzic N, Gadue P, Keller GM, Gouon-Evans V. An endothelial cell niche induces hepatic specification through dual repression of Wnt and Notch signaling. Stem Cells 2011; 29(2): 217-28. [http://dx.doi.org/10.1002/stem.576] [PMID: 21732480]

[72] D'Amour KA, Agulnick AD, Eliazer S, Kelly OG, Kroon E, Baetge EE. Efficient differentiation of human embryonic stem cells to definitive endoderm. Nat Biotechnol 2005; 23(12): 1534-41. [http://dx.doi.org/10.1038/nbt1163] [PMID: 16258519]

[73] Gouon-Evans V, Boussemart L, Gadue P, *et al.* BMP-4 is required for hepatic specification of mouse embryonic stem cell-derived definitive endoderm. Nat Biotechnol 2006; 24(11): 1402-11. [http://dx.doi.org/10.1038/nbt1258] [PMID: 17086172]

[74] Shiraki N, Umeda K, Sakashita N, Takeya M, Kume K, Kume S. Differentiation of mouse and human embryonic stem cells into hepatic lineages. Genes Cells 2008; 13(7): 731-46. [http://dx.doi.org/10.1111/j.1365-2443.2008.01201.x] [PMID: 18513331]

[75] Schuppan D, Afdhal NH. Liver cirrhosis. Lancet 2008; 371(9615): 838-51. [http://dx.doi.org/10.1016/S0140-6736(08)60383-9] [PMID: 18328931]

[76] Agarwal S, Holton KL, Lanza R. Efficient differentiation of functional hepatocytes from human embryonic stem cells. Stem Cells 2008; 26(5): 1117-27.
[http://dx.doi.org/10.1634/stemcells.2007-1102] [PMID: 18292207]

[77] Soto-Gutiérrez A, Kobayashi N, Rivas-Carrillo JD, *et al.* Reversal of mouse hepatic failure using an implanted liver-assist device containing ES cell-derived hepatocytes. Nat Biotechnol 2006; 24(11): 1412-9.
[http://dx.doi.org/10.1038/nbt1257] [PMID: 17086173]

[78] Rasouly HM, Lu W. Lower urinary tract development and disease. Wiley Interdiscip Rev Syst Biol Med 2013; 5(3): 307-42.
[http://dx.doi.org/10.1002/wsbm.1212] [PMID: 23408557]

[79] Tripathi P, Guo Q, Wang Y, *et al.* Midline signaling regulates kidney positioning but not nephrogenesis through Shh. Dev Biol 2010; 340(2): 518-27.
[http://dx.doi.org/10.1016/j.ydbio.2010.02.007] [PMID: 20152829]

[80] Santos dos JRP, Piscione TD Development of the kidney and lower urinary tract. Berlin, Heidelberg: Springer Berlin Heidelberg 2013; pp. 1-58.

[81] Rao PK, Burnett AL. Development of the male reproductive system. London: Springer London 2012; pp. 11-24.

[82] Healey A. Embryology of the female reproductive tract medical radiology. Berlin, Heidelberg: Springer Berlin Heidelberg 2012; pp. 21-30.

[83] De Felici M. Primordial germ cell biology at the beginning of the XXI century. Int J Dev Biol 2009; 53(7): 891-4.
[http://dx.doi.org/10.1387/ijdb.082815mf] [PMID: 19598110]

[84] Sajjad Y. Development of the genital ducts and external genitalia in the early human embryo. J Obstet Gynaecol Res 2010; 36(5): 929-37.
[http://dx.doi.org/10.1111/j.1447-0756.2010.01272.x] [PMID: 20846260]

[85] Tanaka SS, Nishinakamura R, Nishinakamura R, Nishinakamura R. Regulation of male sex determination: genital ridge formation and Sry activation in mice. Cell Mol Life Sci 2014; 71(24): 4781-802.
[http://dx.doi.org/10.1007/s00018-014-1703-3] [PMID: 25139092]

[86] Mae S, Shono A, Shiota F, *et al.* Monitoring and robust induction of nephrogenic intermediate mesoderm from human pluripotent stem cells. Nat Commun 2013; 4: 1367-7.
[http://dx.doi.org/10.1038/ncomms2378] [PMID: 23340407]

[87] Kang M, Han Y-M. Differentiation of human pluripotent stem cells into nephron progenitor cells in a serum and feeder free system. PLoS One 2014; 9(4): e94888.
[http://dx.doi.org/10.1371/journal.pone.0094888] [PMID: 24728509]

[88] Taguchi A, Kaku Y, Ohmori T, *et al.* Redefining the *in vivo* origin of metanephric nephron progenitors enables generation of complex kidney structures from pluripotent stem cells. Cell Stem Cell 2014; 14(1): 53-67.
[http://dx.doi.org/10.1016/j.stem.2013.11.010] [PMID: 24332837]

[89] Lam AQ, Freedman BS, Morizane R, Lerou PH, Valerius MT, Bonventre JV. Rapid and efficient differentiation of human pluripotent stem cells into intermediate mesoderm that forms tubules expressing kidney proximal tubular markers. J Am Soc Nephrol 2014; 25(6): 1211-25. [http://dx.doi.org/10.1681/ASN.2013080831] [PMID: 24357672]

[90] Vigneau C, Polgar K, Striker G, *et al.* Mouse embryonic stem cell-derived embryoid bodies generate progenitors that integrate long term into renal proximal tubules *in vivo*. J Am Soc Nephrol 2007; 18(6): 1709-20.
[http://dx.doi.org/10.1681/ASN.2006101078] [PMID: 17475814]

[91] Takasato M, Er PX, Chiu HS, *et al.* Kidney organoids from human iPS cells contain multiple lineages and model human nephrogenesis. Nature 2015; 526(7574): 564-8.
[http://dx.doi.org/10.1038/nature15695] [PMID: 26444236]

[92] Gupta A, Bischoff A, Peña A, Runck LA, Guasch G. The great divide: septation and malformation of the cloaca, and its implications for surgeons. Pediatr Surg Int 2014; 30(11): 1089-95.
[http://dx.doi.org/10.1007/s00383-014-3593-8] [PMID: 25217828]

[93] Hall JE. Control of blood pressure by the renin-angiotensin-aldosterone system. Clin Cardiol 1991; 14(8) (Suppl. 4): IV6-IV21.
[http://dx.doi.org/10.1002/clc.4960141802] [PMID: 1893644]

[94] Roberts CS. An overview of the genitourinary system. In: Walker HK, Hall WD, Hurst JW, White JM, Eds. Clinical Methods: The history, physical and laboratory examinations. 3rd ed., Boston: Butterworths 1990.

[95] Becker C, Jakse G. Stem cells for regeneration of urological structures. Eur Urol 2007; 51(5): 1217-28.
[http://dx.doi.org/10.1016/j.eururo.2007.01.029] [PMID: 17254699]

[96] Adamowicz J, Kowalczyk T, Drewa T. Tissue engineering of urinary bladder – current state of art and future perspectives. 2013; 66(2): 202-6.

[97] Basile DP, Anderson MD, Sutton TA. Pathophysiology of acute kidney injury. Compr Physiol 2012; 2(2): 1303-53.
[PMID: 23798302]

[98] Osborn SL, Thangappan R, Luria A, Lee JH, Nolta J, Kurzrock EA. Induction of human embryonic and induced pluripotent stem cells into urothelium. Stem Cells Transl Med 2014; 3(5): 610-9.
[http://dx.doi.org/10.5966/sctm.2013-0131] [PMID: 24657961]

[99] Wu X-R, Kong X-P, Pellicer A, Kreibich G, Sun T-T. Uroplakins in urothelial biology, function, and disease. Kidney Int 2009; 75(11): 1153-65.
[http://dx.doi.org/10.1038/ki.2009.73] [PMID: 19340092]

[100] Oottamasathien S, Wang Y, Williams K, *et al.* Directed differentiation of embryonic stem cells into bladder tissue. Dev Biol 2007; 304(2): 556-66.
[http://dx.doi.org/10.1016/j.ydbio.2007.01.010] [PMID: 17289017]

[101] Mauney JR, Ramachandran A, Yu RN, Daley GQ, Adam RM, Estrada CR. All-trans retinoic acid directs urothelial specification of murine embryonic stem cells *via* GATA4/6 signaling mechanisms. PLoS One 2010; 5(7): e11513.
[http://dx.doi.org/10.1371/journal.pone.0011513] [PMID: 20644631]

[102] De Felici M. Origin, Migration, and Proliferation of Human Primordial Germ Cells. London: Springer London 2012; pp. 19-37.

[103] Ginsburg M, Snow MH, McLaren A. Primordial germ cells in the mouse embryo during gastrulation. Development 1990; 110(2): 521-8.
[PMID: 2133553]

[104] Lawson KA, Hage WJ. Clonal analysis of the origin of primordial germ cells in the mouse. Ciba Found Symp 1994; 182: 68-84.

[105] Ying Y, Zhao G-Q. Cooperation of endoderm-derived BMP2 and extraembryonic ectoderm-derived BMP4 in primordial germ cell generation in the mouse. Dev Biol 2001; 232(2): 484-92.
[http://dx.doi.org/10.1006/dbio.2001.0173] [PMID: 11401407]

[106] Tremblay KD, Dunn NR, Robertson EJ. Mouse embryos lacking Smad1 signals display defects in extra-embryonic tissues and germ cell formation. 2001; 128(18): 3609-21.

[107] Lawson KA, Dunn NR, Roelen BA, *et al.* Bmp4 is required for the generation of primordial germ cells in the mouse embryo. Genes Dev 1999; 13(4): 424-36.
[http://dx.doi.org/10.1101/gad.13.4.424] [PMID: 10049358]

[108] Dudley BM, Runyan C, Takeuchi Y, Schaible K, Molyneaux K. BMP signaling regulates PGC numbers and motility in organ culture. Mech Dev 2007; 124(1): 68-77.
[http://dx.doi.org/10.1016/j.mod.2006.09.005] [PMID: 17112707]

[109] Magnúsdóttir E, Surani MA. How to make a primordial germ cell. Development 2014; 141(2): 245-52.
[http://dx.doi.org/10.1242/dev.098269] [PMID: 24381195]

[110] Kurimoto K, Yabuta Y, Ohinata Y, Shigeta M, Yamanaka K, Saitou M. Complex genome-wide transcription dynamics orchestrated by Blimp1 for the specification of the germ cell lineage in mice. Genes Dev 2008; 22(12): 1617-35.
[http://dx.doi.org/10.1101/gad.1649908] [PMID: 18559478]

[111] Ohinata Y, Payer B, O'Carroll D, *et al.* Blimp1 is a critical determinant of the germ cell lineage in mice. Nature 2005; 436(7048): 207-13.
[http://dx.doi.org/10.1038/nature03813] [PMID: 15937476]

[112] Durcova-Hills G, Tang F, Doody G, Tooze R, Surani MA. Reprogramming primordial germ cells into pluripotent stem cells. PLoS One 2008; 3(10): e3531.
[http://dx.doi.org/10.1371/journal.pone.0003531] [PMID: 18953407]

[113] Leitch HG, Blair K, Mansfield W, *et al.* Embryonic germ cells from mice and rats exhibit properties consistent with a generic pluripotent ground state. Development 2010; 137(14): 2279-87.
[http://dx.doi.org/10.1242/dev.050427] [PMID: 20519324]

[114] Ohinata Y, Ohta H, Shigeta M, Yamanaka K, Wakayama T, Saitou M. A signaling principle for the specification of the germ cell lineage in mice. Cell 2009; 137(3): 571-84.
[http://dx.doi.org/10.1016/j.cell.2009.03.014] [PMID: 19410550]

[115] Hayashi K, Ohta H, Kurimoto K, Aramaki S, Saitou M. Reconstitution of the mouse germ cell specification pathway in culture by pluripotent stem cells. Cell 2011; 146(4): 519-32.
[http://dx.doi.org/10.1016/j.cell.2011.06.052] [PMID: 21820164]

[116] Schover LR. Patient attitudes toward fertility preservation. Pediatr Blood Cancer 2009; 53(2): 281-4. [http://dx.doi.org/10.1002/pbc.22001] [PMID: 19301387]

[117] Lee SJ, Schover LR, Partridge AH, *et al.* American Society of Clinical Oncology recommendations on fertility preservation in cancer patients. J Clin Oncol 2006; 24(18): 2917-31. [http://dx.doi.org/10.1200/JCO.2006.06.5888] [PMID: 16651642]

[118] Radford J, Shalet S, Lieberman B. Fertility after treatment for cancer. Questions remain over ways of preserving ovarian and testicular tissue. BMJ 1999; 319(7215): 935-6. [http://dx.doi.org/10.1136/bmj.319.7215.935] [PMID: 10514140]

[119] Radford J. Restoration of fertility after treatment for cancer. Horm Res 2003; 59 (Suppl. 1): 21-3. [http://dx.doi.org/10.1159/000067840] [PMID: 12566716]

[120] Hermann BP, Sukhwani M, Winkler F, *et al.* Spermatogonial stem cell transplantation into rhesus testes regenerates spermatogenesis producing functional sperm. Cell Stem Cell 2012; 11(5): 715-26. [http://dx.doi.org/10.1016/j.stem.2012.07.017] [PMID: 23122294]

[121] Shetty G, Uthamanthil RK, Zhou W, *et al.* Hormone suppression with GnRH antagonist promotes spermatogenic recovery from transplanted spermatogonial stem cells in irradiated cynomolgus monkeys. Andrology 2013; 1(6): 886-98. [http://dx.doi.org/10.1111/j.2047-2927.2013.00126.x] [PMID: 24124124]

[122] Berry DC, Stenesen D, Zeve D, Graff JM. The developmental origins of adipose tissue. Development 2013; 140(19): 3939-49. [http://dx.doi.org/10.1242/dev.080549] [PMID: 24046315]

[123] Shan T, Liang X, Bi P, Zhang P, Liu W, Kuang S. Distinct populations of adipogenic and myogenic Myf5-lineage progenitors in white adipose tissues. J Lipid Res 2013; 54(8): 2214-24. [http://dx.doi.org/10.1194/jlr.M038711] [PMID: 23740968]

[124] Ibrahim MM. Subcutaneous and visceral adipose tissue: structural and functional differences. Obes Rev 2010; 11(1): 11-8. [http://dx.doi.org/10.1111/j.1467-789X.2009.00623.x] [PMID: 19656312]

[125] Moreno-Navarrete JM, Fernández-Real JM. Adipocyte Differentiation. New York, NY: Springer New York 2011; pp. 17-38.

[126] Cuaranta-Monroy I, Simandi Z, Kolostyak Z, *et al.* Highly efficient differentiation of embryonic stem cells into adipocytes by ascorbic acid. Stem Cell Research. 2014; 13(1): 88-97.

[127] Han J, Lee JE, Jin J, *et al.* The spatiotemporal development of adipose tissue. Development 2011; 138(22): 5027-37. [http://dx.doi.org/10.1242/dev.067686] [PMID: 22028034]

[128] Lee Y-K, Cowan CA. Differentiation of white and brown adipocytes from human pluripotent stem cells. Methods in Enzymology. 2014; 533: pp. 35-47. [http://dx.doi.org/10.1016/B978-0-12-800280-3.00003-7]

[129] Dani C, Smith AG, Dessolin S, *et al.* Differentiation of embryonic stem cells into adipocytes *in vitro*. J Cell Sci 1997; 110(Pt 11): 1279-85. [PMID: 9202388]

[130] Mohsen-Kanson T, Hafner A-L, Wdziekonski B, *et al.* Differentiation of human induced pluripotent stem cells into brown and white adipocytes: role of Pax3. Stem Cells 2014; 32(6): 1459-67. [http://dx.doi.org/10.1002/stem.1607] [PMID: 24302443]

[131] Nishio M, Saeki K. Differentiation of Human Pluripotent Stem Cells into Highly Functional Classical Brown Adipocytes. Methods in Enzymology. 2014; 537: pp. 177-97. [http://dx.doi.org/10.1016/B978-0-12-411619-1.00010-0]

[132] Nishio M, Yoneshiro T, Nakahara M, *et al.* Production of functional classical brown adipocytes from human pluripotent stem cells using specific hemopoietin cocktail without gene transfer. Cell Metabolism 2012; 16(3): 394-406.

[133] Tseng Y-H, Kokkotou E, Schulz TJ, *et al.* New role of bone morphogenetic protein 7 in brown adipogenesis and energy expenditure. Nature 2008; 454(7207): 1000-4. [http://dx.doi.org/10.1038/nature07221] [PMID: 18719589]

[134] Hull CW. The birth of 3D printing: IRI achievement award address. Res Technol Manage 2015; 58(6): 25-9.

[135] Gross BC, Erkal JL, Lockwood SY, Chen C, Spence DM. Evaluation of 3D printing and its potential impact on biotechnology and the chemical sciences. Anal Chem 2014; 86(7): 3240-53. [http://dx.doi.org/10.1021/ac403397r] [PMID: 24432804]

[136] Willerth SM, Sakiyama-Elbert SE. Combining stem cells and biomaterial scaffolds for constructing tissues and cell delivery. Cambridge, MA: Harvard Stem Cell Institute 2008.

[137] Murphy SV, Atala A. 3D bioprinting of tissues and organs. Nat Biotechnol 2014; 32(8): 773-85. [http://dx.doi.org/10.1038/nbt.2958] [PMID: 25093879]

[138] Chia HN, Wu BM. Recent advances in 3D printing of biomaterials 2015; 9(1): 4-18.

[139] Richards DJ, Tan Y, Jia J, Yao H, Mei Y. 3D printing for tissue engineering. Isr J Chem 2013; 53(9-10): 805-14. [PMID: 26869728]

[140] Li J, He L, Zhou C, Zhou Y, *et al.* 3D printing for regenerative medicine: From bench to bedside. MRS Bull 2015; 40(2): 145-54. [http://dx.doi.org/10.1557/mrs.2015.5]

[141] Derby B. Inkjet printing of functional and structural materials: fluid property requirements, feature stability, and resolution. Annu Rev Mater Res 2010; 40(1): 395-414. [http://dx.doi.org/10.1146/annurev-matsci-070909-104502]

[142] Zein I, Hutmacher DW, Tan KC, Teoh SH. Fused deposition modeling of novel scaffold architectures for tissue engineering applications. Biomaterials 2002; 23(4): 1169-85. [http://dx.doi.org/10.1016/S0142-9612(01)00232-0] [PMID: 11791921]

[143] Stanton MM, Samitier J, Sánchez S. Bioprinting of 3D hydrogels. Lab Chip 2015; 15(15): 3111-5. [http://dx.doi.org/10.1039/C5LC90069G] [PMID: 26066320]

[144] Rutz AL, Hyland KE, Jakus AE, Burghardt WR, Shah RN. A multimaterial bioink method for 3D printing tunable, cell-compatible hydrogels. Adv Mater 2015; 27(9): 1607-14. [http://dx.doi.org/10.1002/adma.201405076] [PMID: 25641220]

[145] Hwang S, Reyes EI, Moon K-S, Rumpf RC, Kim NS. Thermo-mechanical characterization of metal/polymer composite filaments and printing parameter study for fused deposition modeling in the 3D printing process. Journal of Elec Materi 2014; 44(3): 771-7. [http://dx.doi.org/10.1007/s11664-014-3425-6]

[146] Annabi N, Tamayol A, Uquillas JA, *et al.* 25th anniversary article: Rational design and applications of hydrogels in regenerative medicine. Adv Mater 2014; 26(1): 85-123. [http://dx.doi.org/10.1002/adma.201303233] [PMID: 24741694]

[147] West JL, Hubbell JA. Polymeric biomaterials with degradation sites for proteases involved in cell migration. Macromolecules 1999; 32(1): 241-4. [http://dx.doi.org/10.1021/ma981296k]

[148] Ma PX. Biomimetic materials for tissue engineering. Adv Drug Deliv Rev 2008; 60(2): 184-98. [http://dx.doi.org/10.1016/j.addr.2007.08.041] [PMID: 18045729]

[149] Williams DF. On the mechanisms of biocompatibility. Biomaterials 2008; 29(20): 2941-53. [http://dx.doi.org/10.1016/j.biomaterials.2008.04.023] [PMID: 18440630]

[150] Fournier E, Passirani C, Montero-Menei CN, Benoit JP. Biocompatibility of implantable synthetic polymeric drug carriers: focus on brain biocompatibility. Biomaterials 2003; 24(19): 3311-31. [http://dx.doi.org/10.1016/S0142-9612(03)00161-3] [PMID: 12763459]

[151] Lutolf MP, Gilbert PM, Blau HM. Designing materials to direct stem-cell fate. Nature 2009; 462(7272): 433-41. [http://dx.doi.org/10.1038/nature08602] [PMID: 19940913]

[152] Talbot EL, Berson A, Brown PS, Bain CD. Evaporation of picoliter droplets on surfaces with a range of wettabilities and thermal conductivities. Phys Rev E Stat Nonlin Soft Matter Phys 2012; 85(6 Pt 1): 061604. [http://dx.doi.org/10.1103/PhysRevE.85.061604] [PMID: 23005106]

[153] Yao P, Fox PL. Aminoacyl-tRNA synthetases in medicine and disease. EMBO Mol Med 2013; 5(3): 332-43. [http://dx.doi.org/10.1002/emmm.201100626] [PMID: 23427196]

[154] Skorb EV, Andreeva DV. Surface Nanoarchitecture for bio-applications: self-regulating intelligent interfaces. Adv Funct Mater 2013; 23(36): 4483-506. [http://dx.doi.org/10.1002/adfm.201203884]

[155] Fedorovich NE, Alblas J, Hennink WE, Oner FC, Dhert WJ. Organ printing: the future of bone regeneration? Trends Biotechnol 2011; 29(12): 601-6. [http://dx.doi.org/10.1016/j.tibtech.2011.07.001] [PMID: 21831463]

[156] Bose S, Vahabzadeh S, Bandyopadhyay A. Bone tissue engineering using 3D printing. Mater Today 2013; 16(12): 496-504. [http://dx.doi.org/10.1016/j.mattod.2013.11.017]

[157] Li X, Akasaka T, Dunne N. Polymeric scaffolds for tissue engineering. Int J Polym Sci 2014; (Aug): 2014.

[158] Crapo PM, Gilbert TW, Badylak SF. An overview of tissue and whole organ decellularization processes. Biomaterials 2011; 32(12): 3233-43.

[http://dx.doi.org/10.1016/j.biomaterials.2011.01.057] [PMID: 21296410]

[159] Lee M, Wu BM. Recent advances in 3D printing of tissue engineering scaffolds. Computer-Aided Tissue Engineering. Humana Press 2012.
[http://dx.doi.org/10.1007/978-1-61779-764-4_15]

[160] Leukers B, Gulkan H, Irsen SH, Milz S, *et al.* Biocompatibility of ceramic scaffolds for bone replacement made by 3D printing. Materialwiss Werkstofftech 2005; 36(12): 781-7.
[http://dx.doi.org/10.1002/mawe.200500968]

[161] Gu Q, Hao J, Lu Y, Wang L, Wallace GG, Zhou Q. Three-dimensional bio-printing. Sci China Life Sci 2015; 58(5): 411-9.
[http://dx.doi.org/10.1007/s11427-015-4850-3] [PMID: 25921944]

[162] Tekin H, Sanchez JG, Landeros C, Dubbin K, Langer R, Khademhosseini A. Controlling spatial organization of multiple cell types in defined 3D geometries. Adv Mater Weinheim 2012; 24(41): 5543-7.

[163] Howard D, Buttery LD, Shakesheff KM, Roberts SJ. Tissue engineering: strategies, stem cells and scaffolds. J Anat 2008; 213(1): 66-72.
[http://dx.doi.org/10.1111/j.1469-7580.2008.00878.x] [PMID: 18422523]

[164] Fändrich F, Schulze M, Zehle G, Lange H, Ungefroren H. Stem cell-mediated tolerance inducing strategies in organ transplantation. Kidney Int 2004; 65(5): 1548-50.
[http://dx.doi.org/10.1111/j.1523-1755.2004.05408.x] [PMID: 15086889]

[165] Olabisi RM. Cell microencapsulation with synthetic polymers. J Biomed Mater Res A 2015; 103(2): 846-59.
[http://dx.doi.org/10.1002/jbm.a.35205] [PMID: 24771675]

[166] Turksen K. Bioprinting in Regenerative Medicine. Springer 2015.
[http://dx.doi.org/10.1007/978-3-319-21386-6]

[167] Cao H, Kuboyama N. A biodegradable porous composite scaffold of PGA/beta-TCP for bone tissue engineering. Bone 2010; 46(2): 386-95.
[http://dx.doi.org/10.1016/j.bone.2009.09.031] [PMID: 19800045]

[168] Ogawa T, Akazawa T, Tabata Y. *In vitro* proliferation and chondrogenic differentiation of rat bone marrow stem cells cultured with gelatin hydrogel microspheres for TGF-beta1 release. J Biomater Sci Polym Ed 2010; 21(5): 609-21.
[http://dx.doi.org/10.1163/156856209X434638] [PMID: 20338095]

[169] Nguyen TB, Min YK, Lee B-T. Nanoparticle biphasic calcium phosphate loading on gelatin-pectin scaffold for improved bone regeneration. Tissue Eng Part A 2015; 21(7-8): 1376-87.
[http://dx.doi.org/10.1089/ten.tea.2014.0313] [PMID: 25602709]

[170] Kanematsu A, Yamamoto S, Ozeki M, *et al.* Collagenous matrices as release carriers of exogenous growth factors. Biomaterials 2004; 25(18): 4513-20.
[http://dx.doi.org/10.1016/j.biomaterials.2003.11.035] [PMID: 15046942]

[171] Rosenzweig DH, Carelli E, Steffen T, Jarzem P, Haglund L. 3D-Printed ABS and PLA scaffolds for cartilage and nucleus pulposus tissue regeneration. Int J Mol Sci 2015; 16(7): 15118-35.
[http://dx.doi.org/10.3390/ijms160715118] [PMID: 26151846]

[172] Chang M-Y, Shiau A-L, Chen Y-H, Chang C-J, Chen HH, Wu C-L. Increased apoptotic potential and dose-enhancing effect of gold nanoparticles in combination with single-dose clinical electron beams on tumor-bearing mice. Cancer Sci 2008; 99(7): 1479-84.
[http://dx.doi.org/10.1111/j.1349-7006.2008.00827.x] [PMID: 18410403]

[173] Nair K, Gandhi M, Khalil S, *et al.* Characterization of cell viability during bioprinting processes. Biotechnol J 2009; 4(8): 1168-77.
[http://dx.doi.org/10.1002/biot.200900004] [PMID: 19507149]

[174] Xu T, Jin J, Gregory C, Hickman JJ, Boland T. Inkjet printing of viable mammalian cells. Biomaterials 2005; 26(1): 93-9.
[http://dx.doi.org/10.1016/j.biomaterials.2004.04.011] [PMID: 15193884]

[175] Chang R, Nam J, Sun W. Direct cell writing of 3D microorgan for *in vitro* pharmacokinetic model. Tissue Eng Part C Methods 2008; 14(2): 157-66.
[http://dx.doi.org/10.1089/ten.tec.2007.0392] [PMID: 18544030]

[176] Yu Y, Zhang Y, Martin JA, Ozbolat IT. Evaluation of cell viability and functionality in vessel-like bioprintable cell-laden tubular channels. J Biomech Eng 2013; 135(9): 91011.
[http://dx.doi.org/10.1115/1.4024575] [PMID: 23719889]

[177] Stern-Straeter J, Bonaterra GA, Kassner SS, *et al.* Characterization of human myoblast differentiation for tissue-engineering purposes by quantitative gene expression analysis. J Tissue Eng Regen Med 2011; 5(8): e197-206.
[http://dx.doi.org/10.1002/term.417] [PMID: 21370490]

[178] Li JP, de Wijn JR, van Blitterswijk CA, de Groot K. The effect of scaffold architecture on properties of direct 3D fiber deposition of porous Ti6Al4V for orthopedic implants. J Biomed Mater Res A 2010; 92(1): 33-42.
[http://dx.doi.org/10.1002/jbm.a.32330] [PMID: 19165798]

[179] Jeong CG, Hollister SJ. Mechanical, permeability, and degradation properties of 3D designed poly(1,8 octanediol-co-citrate) scaffolds for soft tissue engineering. J Biomed Mater Res B Appl Biomater 2010; 93(1): 141-9.
[PMID: 20091910]

[180] Do AV, Khorsand B, Geary SM, Salem AK. 3D Printing of Scaffolds for Tissue Regeneration Applications. Adv Healthc Mater 2015; 4(12): 1742-62.
[http://dx.doi.org/10.1002/adhm.201500168] [PMID: 26097108]

[181] Chang JW, Park SA, Park JK, *et al.* Tissue-engineered tracheal reconstruction using three-dimensionally printed artificial tracheal graft: preliminary report. Artif Organs 2014; 38(6): E95-E105.
[http://dx.doi.org/10.1111/aor.12310] [PMID: 24750044]

[182] Zopf DA, Hollister SJ, Nelson ME, Ohye RG, Green GE. Bioresorbable airway splint created with a three-dimensional printer. N Engl J Med 2013; 368(21): 2043-5.
[http://dx.doi.org/10.1056/NEJMc1206319] [PMID: 23697530]

[183] Park JH, Hong JM, Ju YM, *et al.* A novel tissue-engineered trachea with a mechanical behavior similar to native trachea. Biomaterials 2015; 62: 106-15.
[http://dx.doi.org/10.1016/j.biomaterials.2015.05.008] [PMID: 26041482]

[184] Zopf DA, Flanagan CL, Wheeler M, Hollister SJ, Green GE. Treatment of severe porcine tracheomalacia with a 3-dimensionally printed, bioresorbable, external airway splint. JAMA Otolaryngol Head Neck Surg 2014; 140(1): 66-71. [http://dx.doi.org/10.1001/jamaoto.2013.5644] [PMID: 24232078]

[185] Mozaffarian D, Benjamin EJ, Go AS, *et al.* Heart disease and stroke statistics--2015 update: a report from the American Heart Association. Circulation 2015; 131(4): e29-e322. [http://dx.doi.org/10.1161/CIR.0000000000000152] [PMID: 25520374]

[186] Koch MB, Davidsen M, Andersen LV, Juel K, Jensen GB, Galinier A, *et al.* Increasing prevalence despite decreasing incidence of ischaemic heart disease and myocardial infarction. A national register based perspective in Denmark, 1980-2009. Eur J Prev Cardiol 2015; 22(2): 189-95. [http://dx.doi.org/10.1177/2047487313509495] [PMID: 24165474]

[187] Lee VK, Lanzi AM, Haygan N, Yoo SS, Vincent PA, Dai G. Generation of multi-scale vascular network system within 3D hydrogel using 3D bio-printing technology. Cell Mol Bioeng 2014; 7(3): 460-72. [http://dx.doi.org/10.1007/s12195-014-0340-0] [PMID: 25484989]

[188] Gaetani R, Doevendans PA, Metz CH, *et al.* Cardiac tissue engineering using tissue printing technology and human cardiac progenitor cells. Biomaterials 2012; 33(6): 1782-90. [http://dx.doi.org/10.1016/j.biomaterials.2011.11.003] [PMID: 22136718]

[189] Pati F, Jang J, Ha DH, *et al.* Printing three-dimensional tissue analogues with decellularized extracellular matrix bioink. Nat Commun 2014; 5: 3935. [http://dx.doi.org/10.1038/ncomms4935] [PMID: 24887553]

[190] Jana S, Tranquillo RT, Lerman A. Cells for tissue engineering of cardiac valves. J Tissue Eng Regen Med 2015; 10(10): 804-24. [http://dx.doi.org/10.1002/term.2010] [PMID: 25712485]

[191] Duan B, Hockaday LA, Kang KH, Butcher JT. 3D bioprinting of heterogeneous aortic valve conduits with alginate/gelatin hydrogels. J Biomed Mater Res A 2013; 101(5): 1255-64. [http://dx.doi.org/10.1002/jbm.a.34420] [PMID: 23015540]

[192] Lund O, Chandrasekaran V, Grocott-Mason R, *et al.* Primary aortic valve replacement with allografts over twenty-five years: valve-related and procedure-related determinants of outcome. J Thorac Cardiovasc Surg 1999; 117(1): 77-90. [http://dx.doi.org/10.1016/S0022-5223(99)70471-X] [PMID: 9869760]

[193] Norotte C, Marga FS, Niklason LE, Forgacs G. Scaffold-free vascular tissue engineering using bioprinting. Biomaterials 2009; 30(30): 5910-7. [http://dx.doi.org/10.1016/j.biomaterials.2009.06.034] [PMID: 19664819]

[194] Hasan A, Khattab A, Islam MA, *et al.* Injectable Hydrogels for Cardiac Tissue Repair after Myocardial Infarction. Adv Sci 2015; 2(11) [http://dx.doi.org/10.1002/advs.201500122]

[195] Chen Y-C, Lin R-Z, Qi H, *et al.* Functional human vascular network generated in photocrosslinkable gelatin methacrylate hydrogels. Adv Funct Mater 2012; 22(10): 2027-39. [http://dx.doi.org/10.1002/adfm.201101662] [PMID: 22907987]

[196] Leo WJ, McLoughlin AJ, Malone DM. Effects of sterilization treatments on some properties of alginate solutions and gels. Biotechnol Prog 1990; 6(1): 51-3.
[http://dx.doi.org/10.1021/bp00001a008] [PMID: 1366434]

[197] Ventola CL. Medical applications for 3D printing: current and projected uses. P&T 2014; 39(10): 704-11.
[PMID: 25336867]

[198] Komae H, Sekine H, Dobashi I, *et al.* Three-dimensional functional human myocardial tissues fabricated from induced pluripotent stem cells. J Tissue Eng Regen Med 2015.
[http://dx.doi.org/10.1002/term.1995] [PMID: 25628251]

[199] Ozbolat IT, Yu Y. Bioprinting toward organ fabrication: challenges and future trends. IEEE Trans Biomed Eng 2013; 60(3): 691-9.
[http://dx.doi.org/10.1109/TBME.2013.2243912] [PMID: 23372076]

[200] Kusuma S, Shen Y-I, Hanjaya-Putra D, Mali P, Cheng L, Gerecht S. Self-organized vascular networks from human pluripotent stem cells in a synthetic matrix. Proc Natl Acad Sci USA 2013; 110(31): 12601-6.
[http://dx.doi.org/10.1073/pnas.1306562110] [PMID: 23858432]

[201] Shanks N, Greek R, Greek J. Are animal models predictive for humans? Philos Ethics Humanit Med 2009; 4(2): 2.
[http://dx.doi.org/10.1186/1747-5341-4-2] [PMID: 19146696]

[202] Mak IW, Evaniew N, Ghert M. Lost in translation: animal models and clinical trials in cancer treatment. Am J Transl Res 2014; 6(2): 114-8.
[PMID: 24489990]

[203] McGreevy JW, Hakim CH, McIntosh MA, Duan D. Animal models of Duchenne muscular dystrophy: from basic mechanisms to gene therapy. Dis Model Mech 2015; 8(3): 195-213.
[http://dx.doi.org/10.1242/dmm.018424] [PMID: 25740330]

CHAPTER 3

Hematopoiesis and Regenerative Medicine

Chanukya K. Colonne[1], **Veronica I. Antas**[1], **Jia Hao Yeo**[2], **Henry Williams**[1], **Chelsea Pilgrim**[1] and **Stuart T. Fraser**[1,2,*]

[1] *Discipline of Physiology, School of Medical Science, University of Sydney, Camperdown, New South Wales, Australia*

[2] *Discipline of Anatomy & Histology, School of Medical Science, University of Sydney, Camperdown, New South Wales, Australia*

Abstract: The hematopoietic, or blood-producing, system resides in the bone marrow of adult mammals. This system regulates the production of billions of new blood cells per day in healthy adult humans. Even slight perturbations of this production can lead to severe pathological conditions. One of the first applications of cellular regenerative medicine in clinical practice was the transplantation of bone marrow cells to generate a new, healthy blood production system in compromised patients. The success of bone marrow transplantation is dependent upon the potency of stem cell and progenitor populations within the adult mammalian bone marrow. The utility of hematopoietic stem cell (HSC) transplant has been extended to the treatment of a broad range of hematological diseases and disorders, as well as in the regeneration of the blood-producing tissue following radiation or chemotherapy. There is a strong push towards the development of vast numbers of mature blood cells *in vitro*. An *in vitro* system resulting in the consistent, large-scale production of patient-specific mature erythrocytes from HSCs or erythroid progenitors could alleviate the pressure felt by blood donation agencies. The cells that support blood cell production in the bone marrow and other organs, known collectively as the hematopoietic niche, are critical in blood cell lineage regeneration. The development of novel regenerative therapies to treat myelodysplastic syndromes, anemia, leukemia and other blood diseases deserves attention. Stem, progenitor and supportive cells within the hematopoietic tissues are essential elements of regenerative medicine. The utility, limitation and promise of these populations in regenerative medicine are described here.

* **Corresponding author Stuart T. Fraser:** Discipline of Anatomy & Histology, School of Medical Science, University of Sydney, Camperdown, New South Wales, Australia; Tel: +61 2 9036 3313; Fax: +61 2 9036 3316; E-mail: stuart.fraser@sydney.edu.au

Atta-ur-Rahman & Shazia Anjum (Eds.)

Keywords: Bioengineering, Blood, Bone marrow, Embryo, Hematopoiesis, Niches, Regenerative Medicine, Stem cells, Transplantation.

INTRODUCTION

A healthy adult human produces over 2 million new red blood cells per second or nearly 2 billion per day. This extraordinary rate of cellular production is dependent upon a hierarchy of stem, progenitor and maturing cell types forming in the hematopoietic organs. Hematopoietic stem cells (HSCs) are very rare cells capable of giving rise to all of the major blood cell types when transplanted. These characteristics have led to HSCs being one of the cell types most frequently transplanted for clinical treatment. Once HSCs have been transplanted from donor to recipient, these cells home to the bone marrow where they reside and give rise to the massive number of blood cells required to maintain human health. HSCs also give rise to a broad variety of blood cell lineages. These range from gas transporting erythrocytes (the most common form of blood cell in mammals) to megakaryocytes producing platelets; granulocytes (neutrophils, eosinophils and basophils) offering immune protection against bacteria, fungi and parasites; and the lymphocytes capable of forming an immunological memory of pathogen exposure. Specialized tissue macrophages and dendritic cells are also derived from HSCs. HSCs therefore have a profound capacity for regenerating the blood compartment, which comprises approximately one fifth of all cells in the healthy adult human. The remarkable capacity of HSCs to give rise to the entire hematopoietic system in recipient animals and patients has led to utilization of these cells in regenerative medicine. HSC transplantation can be performed from transplanting a patient's own stem cells (autologous transplants) or from transplanting a donor's cells (allogeneic transplants). Both these techniques are used commonly in medicine today and are often the only life saving or prolonging treatment available to these patients.

Significant challenges lie ahead in hematopoietic regenerative medicine, and we will discuss some of these in this chapter. In contrast to other stem cell types, such as embryonic or induced pluripotent stem cells, we cannot currently maintain HSCs *in vitro* in the stem cell state for an indefinite period. There are also significant challenges in finding appropriately matched HSCs for transplantation, and mitigating the many complications associated with HSC transplantation. Sourcing matched volunteers for blood donations for red blood cell and platelet transfusions is a challenge globally. In some clinical cases, the problem is due to defects in the hematopoietic niche, the supporting cells which maintain HSCs as stem cells. Developing novel systems to obtain patient-specific HSCs and progenitors as well as novel methods for modifying and enhancing the HSC niche, will dramatically improve the health outcomes of a vast number of patients suffering from hematopoietic and related diseases.

Here, we will discuss the link between blood cell generation and regenerative medicine. In contrast to other systems, the hematopoietic system is more diffusely scattered throughout the body. Regenerative therapeutic treatment will vary according to patient needs and may include regeneration of failing bone marrow, replacement of a deficient or defective thymus to produce T cells; re-construction of a spleen or other hematopoietic tissue; or the generation of mature circulating cells from an *ex vivo* or *in vitro* expanded progenitor population such as functional red blood cells, mature platelets or granulocytes to supplement immunodeficient individuals.

This discussion will proceed in four sections. Firstly, the hematopoietic system will be introduced. Secondly, the niches which regulate blood cell generation and constant production will be described. The application of HSC transplantation to the treatment of a range of human diseases is one of the major forms of regenerative medicine and will be discussed at length in the third section. Finally, focusing all of these studies onto future directions, we will discuss the application of novel technologies such as induced pluripotent stem cells and tissue engineering of blood producing tissues to the treatment of hematological disorders.

AN INTRODUCTION TO HEMATOPOIESIS

Hematopoiesis is the production of blood cells initiated in the early embryo and maintained throughout life. Maintenance of a homeostatic state requires the genesis and consistent life-long production of a broad palette of blood cell types.

Table **1** summarizes the blood cell types that are required for embryonic and adult mammalian homeostasis. A specific genetic program is required to encode the unique properties of each blood cell type. Each lineage requires distinct, though at times overlapping, external signals such as growth factors, cytokines, chemokines, binding to extracellular matrix components or adhesion molecules on neighboring cells, all leading to engagement of surface receptors and activation of downstream signaling processes. Mimicking these signals is crucial to the *in vitro* modeling of each lineage.

Table 1. Hematopoietic cell lineages and cell types generated in the developing mammalian embryo and adult.

Blood cell lineage	Cell type
Erythroid	primitive erythroblast; primitive erythroid, yolk sac definitive erythroid, fetal liver definitive erythroid, adult bone marrow definitive erythroid, Colony Forming Unit (CFU)-erythroid, blast forming unit-erythroid, reticulocyte, stress erythroblast.
Thrombocytic	primitive megakaryocyte, adult megakaryocyte, platelet, activated platelet, blood clot.
Granulocytic	common myeloid progenitor, CFU-granulocyte, myeloblast, neutrophil, eosinophil, basophil, activated granulocyte, myeloid-derived suppressor cell.
Macrophage	CFU-macrophage, monocyte, microglia, foetal liver macrophage, Kupffer cell, mesangial cell, histiocyte, marginal zone macrophage, osteoclasts, lymph node sinus macrophage, alveolar macrophage, erythroblastic island central macrophage, red pulp macrophage, peritoneal macrophage, thymic macrophage, tumour associated macrophage, M1/M2 activated macrophage, IL-10$^+$ regulatory macrophage.
Dendritic cell (DC)	myeloid DC, plasmacytoid DC, Langerhans cell, lymph node follicular dendritic cell, splenic DC, Peyer's patch DC.
B lymphoid	common lymphoid progenitor, Pro-B cell, CD5$^+$ B lymphocyte, naive B lymphocyte, activated B lymphocyte, plasma B cell, follicular B cell, marginal zone B cell, B-1 cell, regulatory B cell, IgA-producing B cells, memory B cell.
T lymphoid	pro-T cell, thymocyte, CD4$^+$ helper 1 and helper 2 T cells, CD8$^+$ cytotoxic T lymphocyte, FoxP3$^+$ regulatory T cell, Th17 T lymphocyte, Tf helper cells, gamma delta T lymphocyte, ROR gamma T cells, Natural Killer (NK) T cells, memory T cells.
Innate immune cell	NK, innate NKT, Nuocytes, ROR gamma cells, NKp48$^+$ innate immune cells.
Other lineages	mast cells, Peyer's patch inducer cells.

In adult mammals, the hematopoietic system is responsible for the production of the gas-transporting red blood cells (erythrocytes); the infection-fighting white blood cells (leukocytes); and the blood clotting platelets (thrombocytes). The three major arms of the hematopoietic system are the erythroid lineage (erythrocytes and platelets); the myeloid lineage (granulocytes and macrophages); and the lymphoid system (B, T, natural killer and innate lymphoid cells). As represented in Table **1**, each of these lineages then gives rise to a range of distinct mature cell types. For example, T lymphocyte precursors can give rise to helper T cells, cytotoxic T cells, regulatory T cells, ROR-gamma T cells, Interleukin-17 producing T cells and a memory T cell population. Generation of each blood cell type requires cues and activation of transcription factors. Loss of a specific population can often lead to pathological conditions as demonstrated by the lack of common lymphoid progenitors in severe combined immunodeficiency patients or the loss of $CD4^+$ helper T cells in HIV/AIDS patients. The number and function of mature blood cell types must also be appropriately regulated. An inability to do so can also result in human disease as seen in neutropenic individuals not producing required numbers of mature neutrophils, or the dyserythropoiesis observed in individuals with erythroid diseases such as sickle cell anemia and b-thalassemia.

Hematopoiesis is Dependent Upon the Hierarchical Production of Mature Cells from Rare Stem Cells

How can the mammalian adult produce this complex variety of cell types in such vast numbers on a constant basis? The adult hematopoietic system is hierarchically structured, progressing from uncommitted cells with self-renewal potential down to terminally differentiated, mature cells lacking self-renewal potential. In the classical bifurcation hierarchical model of hematopoiesis, self-renewing HSCs are at the top of the hierarchy, capable of producing all of the mature blood cells in the mammal [1, 2]. HSCs are the archetype of a self-renewing biological system, capable of both self-renewal and differentiation into all the types of blood cells, producing billions of blood cells that are required for our survival, each day, every day [3]. Self-renewal is defined by the replication of one undifferentiated stem cell and the concomitant generation of a more differentiated cell. If the daughter cells of stem cells were both stem cells themselves, the stem cell pool would expand at the expense of production of mature cell types. Conversely, if stem cell division led to the generation of two differentiating progenitor cells then the stem cell pool would rapidly be depleted, and mature cell numbers would gradually reduce until organ failure occurs. This takes place in bone marrow failure associated with conditions such as aplastic anemia. The potency of HSCs was best exemplified by a tour-de-force study in which it was demonstrated that a single HSC was able to reconstitute the entire long-term hematopoietic system of irradiated recipient mice [4]. We have attempted to summarize the current thinking of the generation of mature blood cell types from HSCs in Fig. (**1**).

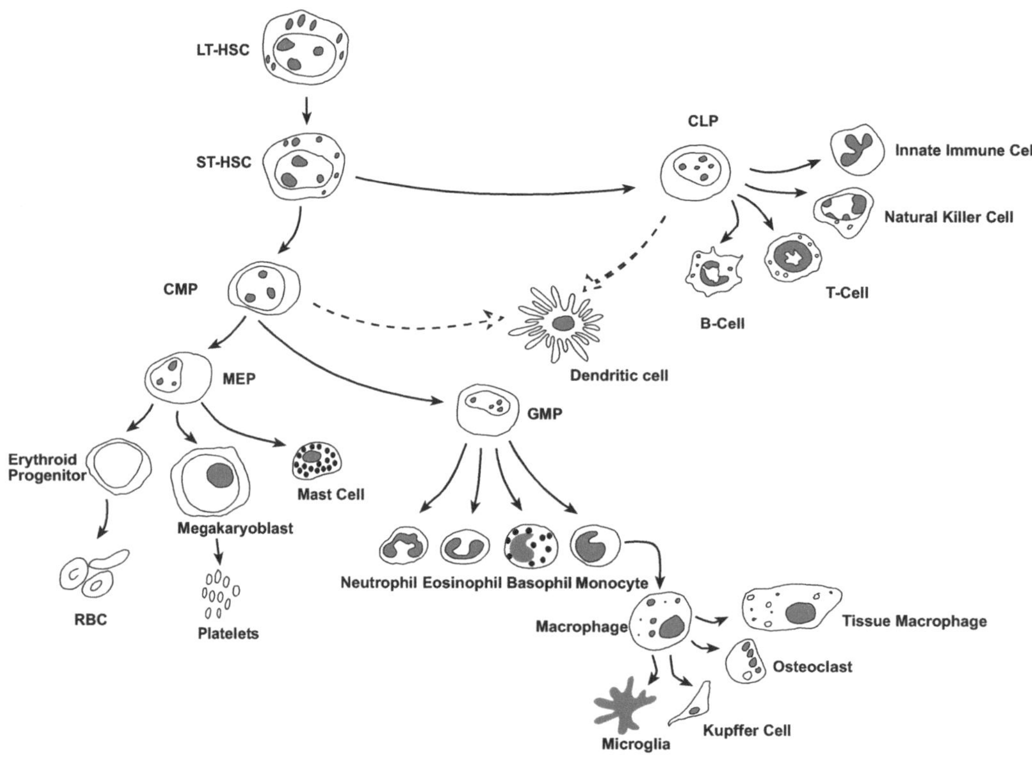

Fig. (1). The hematopoietic hierarchy found in the healthy adult mouse bone marrow and peripheral tissues.

HSCs produce all forms of mature terminally differentiated blood cells [3, 5]. These HSCs were first identified in 1961 by Till and McCulloch, as cells with the capability to rescue an individual from radiation damage [6]. Proliferative cell colonies were observed in the spleens of lethally irradiated mice, following treatment with bone marrow cells obtained from healthy donor mice. These primary colonies retained their ability to form further colonies once injected into secondary hosts, and gave rise to differentiated blood cells of multiple blood lineages. This was shortly followed by the demonstration that these colonies were derived from single cells (namely HSCs) which were capable of self-renewal, proliferation and differentiation into myeloerythroid and lymphoid cells [3, 7].

HSCs give rise to progenitors committed to different arms of the hematopoietic system such as common myeloid (CMPs) and common lymphoid progenitors (CLPs) (Fig. **1**) [1]. CMPs give rise to megakaryocyte-erythrocyte progenitors (MEPs) and granulocyte-macrophage progenitors (GMPs). These lineage-restricted progenitors then give rise to more committed progenitor types such as erythroblasts, megakaryoblasts, myeloblasts and lymphoblasts. According to this model the commitment of multipotent progenitor cells to either myeloid or lymphoid lineages is an early step in lineage commitment [8]. Some studies suggest that the loss of lymphoid differentiation potential could occur as one of the first lineage commitment steps [9, 10].

Identifying and Characterising Hematopoietic Stem Cells

Flow cytometry and fluorescence-activated cell sorting (FACS) are useful means for enriching rare populations from complex mixtures. A broad range of monoclonal antibodies are available commercially which bind to specific hematopoietic cell types. However, currently, we have no single antibody to identify hematopoietic stem and progenitor cells. The typical strategy for identifying HSCs is to exclude terminally mature cells according to their expression of mature lineage surface markers such as: CD19 for B lymphocytes; CD3 for T lymphocytes; Gr-1 for granulocytes and Mac-1 for macrophages. These markers are termed the lineage markers (Lin). Committed or mature hematopoietic cells make up the vast majority of BM mononuclear cells. Once these are excluded by surface antigen expression, the remaining cells can be further subdivided according to the expression of markers that enrich for stem cell activity. Surface antigen such as Stem cell antigen-1 (Sca1) and the c-Kit receptor for stem cell factor have been used to enrich for HSC activity. This has led to the coining of the terms LSK ($Lin^-Sca\text{-}1^+c\text{-}Kit^+$) or KSL ($c\text{-}Kit^+Sca\text{-}1^+Lin^-$), a surface phenotype which identifies a heterogeneous population enriched for stem cell activity [11 - 13].

Reconstitution assays are utilized to characterize HSC activity. Functional *in vivo* assays are considered the gold standard for the detection of HSC activity. Long term reconstituting HSCs are defined as having the capability of reconstituting the entire lineage of hematopoietic cells in a lethally irradiated mouse for at least 4 months after transplantation [4, 14, 15]. HSCs must retain their self-renewal and differentiation potential when injected into secondary and tertiary hosts [16]. However LSKs are a heterogeneous population of cells [8] with varying degrees of self-renewal [8, 17], potency [18, 19] and lifespan [4, 20]. Reconstitution assays have demonstrated that HSCs can be characterized based on their reconstitution time periods, as short term (ST), intermediate term (IT), and long term (LT) HSCs [21]. HSCs with the capability of reconstituting the hematopoietic system for more than 12 months after transplantation are termed LT-HSCs, less than 6 months ST-HSCs, more than 6 months but less than 12 months IT-HSCs [8]. These HSCs were found to have differing antigen expression profiles, with LT-HSCs found in the $CD150^{+}$ $CD48^{-}$ $CD244^{-}$ $Thy1.1^{low}$ $CD34^{-/low}$ $Flt3^{-}$ LSK subset, ST-HSCs enriched in the $CD34^{+}$ $Flt3^{-}$ LSK subset, and IT-HSCs found in the $CD49b^{high}$ Rhi^{low} $CD34^{-}$ $Flt3^{-}$ $CD34^{-}$ LSK subset of HSCs. ST-HSCs were shown to be unable to reconstitute the hematopoietic system of irradiated mice following secondary transplantation, while reconstitution activity of LT-HSCs remained unchanged post secondary transplantation [8, 21]. Meanwhile IT-HSCs were characterized by the loss of one lineage following injection into a secondary host [21].

Embryonic Hematopoiesis

Blood production is initiated in the mammalian embryo shortly after gastrulation. Indeed, blood cells are the first mature cell type to arise in the developing mammalian conceptus (the term used to describe the combined embryo proper, extra-embryonic yolk sac (YS) and the placenta, a fusion of maternal and embryonic tissues). Hematopoiesis occurs in multiple waves and in numerous different anatomical sites or niches, which has been extensively investigated in the mouse (Fig. **2**) [22, 23]. The initial wave of blood production is termed 'primitive' as it does not directly contribute to the blood producing tissues of the adult. This occurs in the YS, an extra-embryonic membrane surrounding and encapsulating the mouse conceptus at mouse embryonic day 7.5 (E7.5) [24, 25]. The primary role of this primitive wave is to produce primitive erythrocytes for tissue oxygenation as the embryo rapidly develops, though it also gives rise to macrophages and megakaryocytes [22, 26]. These transient primitive erythrocytes (EryP) are rapidly replaced by the second wave of hematopoiesis in the YS (E8.5), which is termed 'definitive' (adult-type), which results in erythroid, megakaryocyte, and several myeloid lineages [27, 28]. The third wave results from HSCs produced within the ventral aspect of the dorsal aorta of the aorta-gonad-mesonephros (AGM) region of the embryo body (E10.5), YS, and placenta [26, 29]. This is followed by the definitive hematopoietic progenitors migrating to and colonising the fetal liver, thymus, spleen, and finally bone marrow [30, 31]. The HSCs colonizing the bone marrow then go on to produce blood cells throughout the postnatal life of the mammal [26].

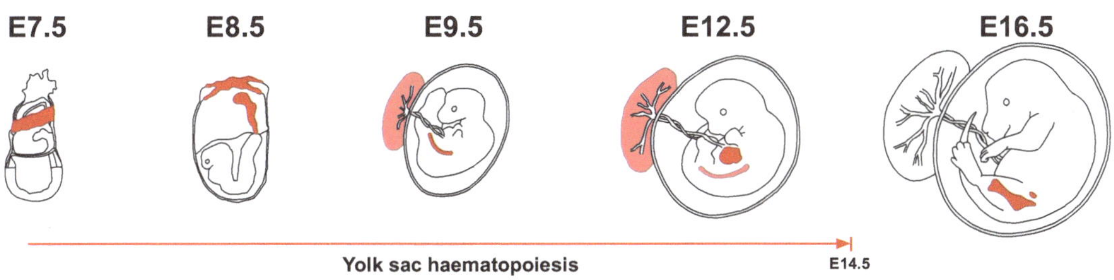

Fig. (2). The hematopoietic developmental timeline of embryogenesis. Schematic representations of mouse embryos (not to scale) are shown at different embryonic days (E) of development. The hematopoietic tissues active at each stage are shaded in red.

Although adult HSC transplants are widely used, limitations of this therapy (especially the low availability of HLA-matched donors and related complications of therapy) have led researchers to seek for other potential sources of hematopoietic stem and progenitor cells. Here, we will explore the main hematopoietic embryonic tissues, together with the potential use of embryonic/fetal tissues as alternative sources of transplantable hematopoietic cells. Some characteristics of embryonic HSCs, such as their active cycling properties (as opposed to the highly quiescent adult HSCs), make these cells potentially valuable in clinical settings [32].

The Yolk Sac is the Initial Site of Blood Production

The YS is a bi-layered structure formed from mesodermal and endodermal germ layers, and surround the developing rodent embryo [33]. The mesodermal layer migrates and interacts with the endodermal layer between embryonic day E7 and E7.5 when the mesoderm then forms mesodermal cell masses [34]. Mesodermal cell masses are epithelial cells that are believed to support the transition of endothelial cells to hematopoietic cells, forming blood islands [35]. This process is aided by the endodermal layer which is comprised of the parietal and visceral endoderm. These layers allow transport of nutrients between the mother and the fetus during early stages of development. The nutrients that stay contained in the endoderm promote a positive environment for blood islands to form in the underlying mesodermal layer (Fig. **3**) [36].

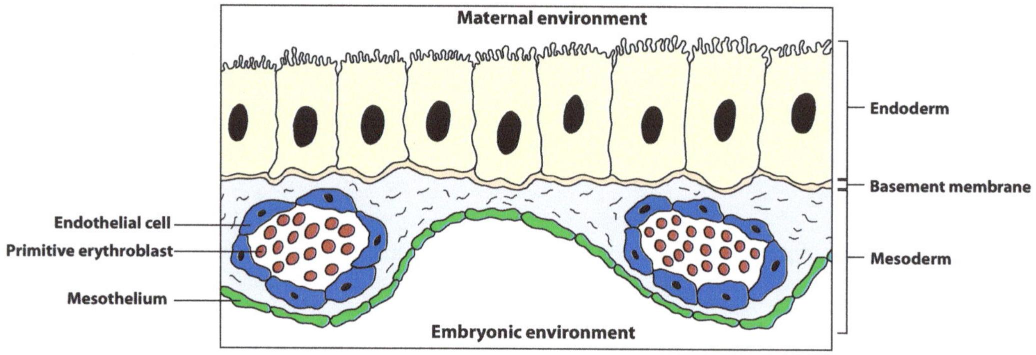

Fig. (3). Schematic representation of mouse yolk sac shows the close association between the visceral endoderm and developing mesoderm-derived populations forming the first hematopoietic niche.

As previously mentioned the first two waves of haematopoiesis occur in the YS. The primitive erythrocytes along with primitive macrophages and megakaryocytes are formed in the blood islands of the mesoderm at E7.5, and after the first heart beat at E8.25 they enter the embryonic circulation [25, 28, 35, 37]. In the circulation, the primitive erythroid cells enucleate, continue to expand in number and accumulate hemoglobin while in the circulation [38, 39]. The second wave results in the production of definitive erythroid progenitors at around E8.25, and their numbers expands until E10.5 when they enter the circulation to populate the fetal liver [40]. The fetal liver acts as a site for further multiplication for these cells and as a site of storage before they move to the bone marrow [27, 41].

While AGM hematopoietic cells have long-term reconstitution properties, YS hematopoietic cells only have limited reconstitution potential [42]. This leads to the suggestion that although the YS is the first site of hematopoiesis during development it does not produce cells that are capable of long-term reconstitution [29, 42]. However, YS hematopoietic cells have long term repopulation potential if injected into newborn mice suggesting that there is another step that needs to occur *in vivo* for the HSCs to mature [43] and transgenic labelling systems have supported the idea that the YS is a source of definitive HSCs [44].

The Dorsal Aorta as a Critical Hematopoietic Niche

The para-aortic splanchnopleura (P-Sp)/aorta-gonad-mesonephros (AGM) region is located in the caudal half of mouse and human embryos, and comprises the dorsal aorta and its surrounding tissues [42, 45 - 47]. In mouse embryos, dorsal aortic hematopoietic activity is characterized by the production of 'immature HSCs' at around E9.0, followed by generation of fully competent, mature HSCs *in situ* at around E10.5 [46, 48, 49]. Similar to the mouse, human hematopoietic stem and progenitor cells are also produced in the AGM region starting at around the 27th day of gestation [47]. Human HSCs emerge in the AGM region well before they can be detected in other sites, such as yolk sac, liver and placenta [50].

Hematopoietic cells produced in the dorsal aorta are derived from a specialized subset of endothelial cells named hemogenic endothelial cells. In both mouse and humans, this endothelial-to-hematopoietic transition is evidenced by the presence of intraluminal groups (or clusters) of hematopoietic cells in the dorsal aorta [47, 51 - 53]. Generation of HSCs is restricted to the ventral aspect of the dorsal aorta, possibly due to an inductive effect of ventral tissues surrounding this vessel [54]. Although the AGM has not been directly used for human clinical transplantation purposes yet, its stroma is supportive of HSC activity and can be used for hematopoietic culture [55]. HSC repopulating- and colony-forming potentials are preserved when liver or adult bone marrow HSCs are co-cultured with AGM stromal cells [56]. Different types of AGM cells have distinct supportive abilities.

For example, AGM endothelial cells support HSC maintenance and differentiation [57], while cells from the urogenital region maintain HSC reconstitution potential without inducing HSCs [58]. Hence, AGM cells may be used to culture and/or expand hematopoietic cells prior to transplantation to patients.

The Placenta is a Blood-production Site

In vitro culture studies showed that the embryonic component of mouse placenta could generate hematopoietic progenitors and HSCs [59 - 61]. Mouse allantois and chorion, two precursor tissues of the placenta, can also generate myeloid and definitive erythroid cells in culture [62]. The human placenta is also an important source of hematopoietic progenitors and HSCs, and from week 6 of gestation onwards, multi-lineage, high-level repopulating HSCs are present in close association with the placental vasculature [63]. In contrast to the mouse, the human placenta only shows HSC activity long after the emergence of HSCs in the AGM [50, 60, 61]. The mouse placenta lacks HSCs at birth while the human placenta remains a rich source of HSCs at the time of delivery [60, 63]. Therefore, investigators have been studying the potential use of the human placenta as a source of hematopoietic cells for regenerative medicine. This extra-embryonic tissue can be a very valuable source of cells for clinical applications, because placental cells have great phenotypic plasticity and multi-lineage differentiation capacity as well as good engraftment ability and immunomodulatory properties [64, 65]. The human placenta at birth is an easily obtained organ and there is little ethical concern about its use, making it an attractive option for clinical applications [66]. The human term placenta also contains other stem cells that could also be used for clinical purposes, such as trophoblastic, epithelial and mesenchymal stem cells [65, 66].

The Fetal Liver: A Niche for Expansion of Blood Stem and Progenitor Cells

The mouse fetal liver is not a site of HSC generation but rather is colonized by hematopoietic cells produced in other locations such as the yolk sac and AGM [67]. The mouse hepatic environment supports HSCs and hematopoietic cell expansion and terminal maturation [68 - 70]. The human fetal liver also supports hematopoietic maturation and proliferation and becomes a major hematopoietic organ at around the 5th week of gestation [71 - 74]. Given the high proliferative capacity of hepatic hematopoietic cells, the human fetal liver has been studied as an alternative source of hematopoietic progenitors and HSCs [75 - 77]. Initial studies suggest that the fetal liver cells can be successfully used for transplantation [78 - 80]. When fetal liver cells, umbilical cord blood and adult bone marrow were ranked according to their engraftment abilities, hepatic cells would be at the top of the list, followed by cord blood and adult bone marrow cells [81]. One disadvantage of using the fetal liver as a source of hematopoietic cells is the limited number of cells obtained in a typical harvest [76], which often restricts the clinical use of fetal liver to in utero transplantation [80, 82]. Hence, recent research efforts have been directed towards the achievement of fetal liver blood cell expansion in culture while maintaining multi-lineage differentiation potential [76].

The Fetal Spleen: A Newly Recognized Fetal Blood Producing Niche

Similar to the fetal liver, mouse fetal spleen (FS) is also seeded by erythromyeloid, lymphoid progenitors and HSCs generated in other tissues [83 - 85]. Distinct erythroblast populations and their progeny can be found in this organ [83, 85, 86]. Committed hematopoietic progenitors can differentiate into B, natural killer (NK) and myeloid cells in the FS [83]. However, the mouse FS does not support HSC expansion [83, 85]. In contrast, the human fetal spleen is not a major hematopoietic organ [87]. Very little (if any) B cell generation occurs in the second-semester human spleen. B cells produced in the bone marrow migrate to the fetal spleen to complete their maturation [88, 89]. The human fetal spleen contains a small number of erythroid cells [90]. In contrast to purified endothelial cells from other blood-producing organs, endothelial cells from the human fetal spleen are unable generate hematopoietic cells in culture [91].

THE HEMATOPOIETIC NICHES

While the HSC remains the best characterized somatic stem cell [92], the methodology of attaining their *in vitro* expansion continues to elude us, thereby limiting its utility in regenerative medicine. The maintenance of HSCs and the regulation of their self-renewal and differentiation *in vivo* is thought to be mediated by the microenvironment they reside in - known as the HSC niche [93]. This term was first coined by Schofield in 1978, for the specific HSC bone marrow microenvironment, where he proposed that HSCs are in intimate contact with bone and that the cellular interactions between these groups of cells were responsible for the unlimited self-renewal capacity and inhibition of the maturation of HSCs [93]. The study of these HSC niches may hold the key to unlocking the puzzle of *in vitro* expansion of HSCs, thereby paving the path to utilizing HSCs in therapeutic regenerative medicine.

Embryonic Hematopoietic Niches

Hematopoietic cells do not develop independently of their surroundings. Signalling pathways, soluble factors, cytokines and cell interactions contribute to create a specialized microenvironment, or niche, that favors hematopoietic development [93]. During hematopoietic development, niche components such as shear stress, biomechanical forces, temperature and oxygen levels act as critical mediators of hematopoietic activity [23, 94, 95]. Endothelial cells are also important components of embryonic hematopoietic niches, since some of these cells (hemogenic endothelial cells) are able to give rise to hematopoietic stem and progenitor cells [43, 48, 60, 96 - 98]. Additionally, tissues located in the proximity of hematopoietic niches can regulate blood cell generation. This is exemplified by the AGM region, which receives hematopoietic-inductive stimuli from adjacent ventral tissues and suppressive signals from dorsal regions [54]. Hematopoietic activity in the yolk sac is under a similar pattern of regulation, as visceral endoderm signalling is critical for activation of primitive hematopoiesis [99].

The Adult Bone Marrow Hematopoietic Niche

In adult mammals, the bone marrow serves as the main HSC niche, therefore elucidating its mechanics and architecture may provide us with a way of reconstructing this niche *in vivo*, allowing expansion of HSCs outside the mammalian body. There seems to be at least two anatomically distinct bone marrow niches, namely the endosteal and perivascular niches of the bone marrow, where HSCs receive and integrate regulatory cues from neighboring cells, extracellular matrix components and soluble factors [23, 30, 95, 100].

Morphological evidence for the endosteal niche was demonstrated when cells with HSC or hematopoietic progenitor phenotype were shown to closely localize to the endosteal lining of bone marrow cavities in trabecular regions of long bones, whereas the more differentiated hematopoietic progenitors were demonstrated to be localized mainly to the central perivascular bone marrow region [101 - 105]. Fig. (**4**) summarizes the findings we describe in this section.

The unique mesenchymal bone-forming cells, osteoblasts, are a critical component of the hypoxic and free calcium ion rich environment that is the endosteal bone marrow niche [100, 106, 107]. When osteoblasts are conditionally ablated, a reversible progressive bone loss is accompanied by a decrease in both bone marrow cellularity and number of HSCs, along with the induction of extramedullary hematopoiesis [108, 109]. Transgenic mice expressing an osteoblast specific constitutively active form of parathyroid hormone, an important player in calcium homeostasis, bone formation and resorption, exhibited a simultaneous increase in the number of both osteoblasts and HSC in the bone marrow. Furthermore *in vitro* maintenance of HSCs was greater when supported by stromal cells isolated from these transgenic mice [110]. Mice lacking BMP1A (BMP receptor 1A, normally expressed on osteoblasts lining the endosteum) in the bone marrow stroma, showed a simultaneous increase in both the number of N-cadherin positive osteoblasts, and the number of HSCs that were in direct contact with these osteoblasts *via* N-cadherin [102].

Osteoblasts in the endosteal niche are involved in many cooperative interactions with the surrounding cells and stroma. Osteoblasts utilize cytokines, chemokines, adhesion molecules and transcription factors to regulate many interactions enabling the HSC to self-renewal and differentiate. Osteoblasts mediate interactions such as the Jagged interactions with Notch receptors [110, 111]; the interaction of angiopoietin-1 at the osteoblast surface with Tie-2 on stem cells [112]; the interplay between stem cell factor (SCF) and its receptor c-Kit [113]; as well as mediating the functions of CXCL12 [114], adrenalin [115] and osteopontin [116, 117].

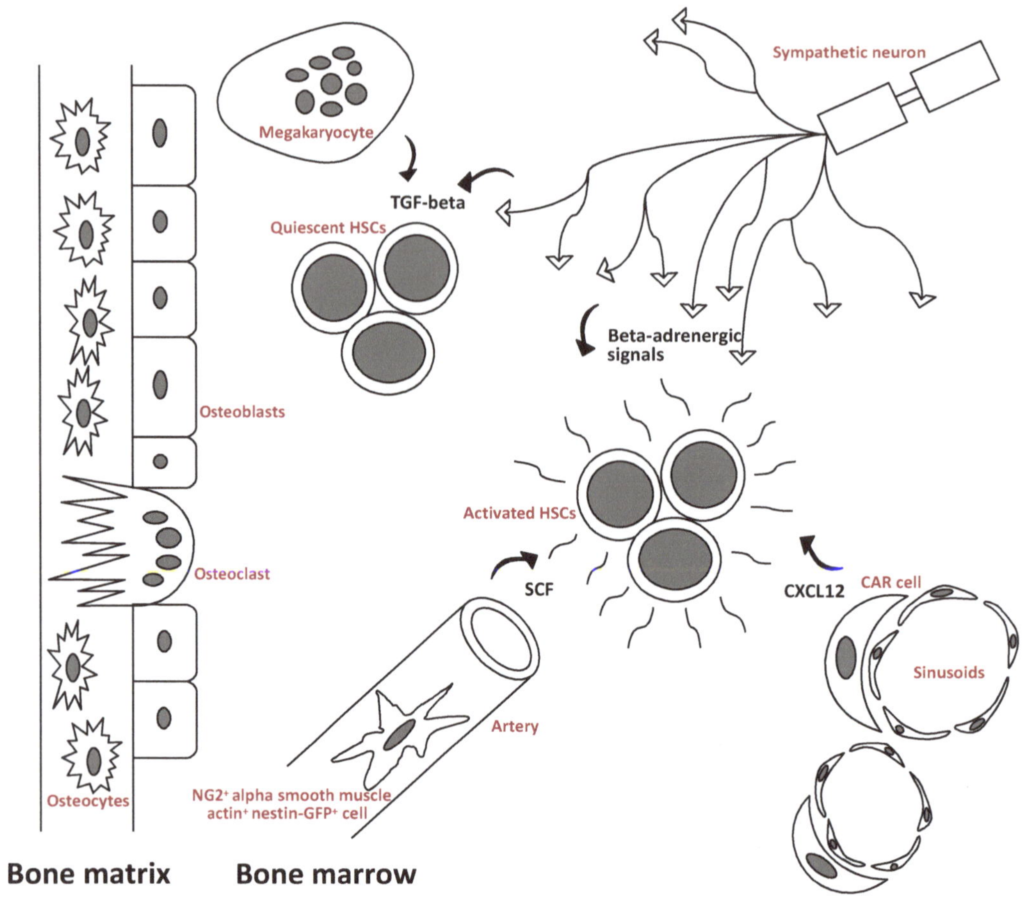

Fig. (4). A simplified view of the mouse adult bone marrow hematopoietic niche.

The perivascular niche was first discovered in 2006, where HSCs were shown to localize near blood vessels in normal bone marrow [101]. This niche includes both small arterioles and sinusoids, and both these groups of vessels have mesenchymal stromal cells localizing near them. These stromal cells contain essentially all of the mesenchymal stem cell (MSC) activity of the bone marrow, capable of differentiating into osteolineage cells, adipocytes, and chondrocytes [118]. These cells secrete soluble factors critical for the regulation of HSC maintenance such as CXCL12 [119, 120] and are involved indirectly in the sympathetic nerve regulation of HSC maintenance by expression of β3-adrenergic receptors [118]. The HSCs adjacent to the sinusoids interact with CAR cells

(CXCL12 abundant reticular cells), which surround sinusoidal endothelial cells adjacent to the endosteum [119]. These cells are involved in SCF and CXCL12 production, and the loss of the interaction between CXCL12 and CXCR4 (receptor for the ligand) results in a severe reduction of HSC numbers [119]. Imaging studies have also revealed that the quiescent HSCs associate specifically with small arterioles that are preferentially found in endosteal bone marrow [121]. These arterioles are unique in that they are surrounded exclusively by mesenchymal stromal cells, rare cells expressing NG2 (CSPG4), alpha-smooth muscle actin, and nestin-GFP, and are indispensable for maintaining HSC quiescence [121]. Sympathetic nerves ensheathed by autonomous, non-myelinating Schwann cells are also part of this niche, making direct contact with HSCs and lying parallel to blood vessels [122]. Schwann cells maintain HSC dormancy by activation of TGF-β/Smad signalling [122].

Megakaryocytes have also been proposed to play a role in the regulation of multiple niches in the bone marrow [123 - 125]. Megakaryocytes maintain HSC quiescence through TGF-beta signalling and support the plasma cell niche *via* IL-6 and APRIL secretion [123]. FoxP3-expressing T-regulatory cells (T-regs) have been found to co-localize with HSCs on the endosteal surface of the bone marrow although the biological relevance to HSC maintenance is still being investigated [126]. Given that the molecular interactions between HSCs and the cellular components of these niches seem to dictate the balance between HSC self-renewal and differentiation, the future application of HSCs for therapeutic use will likely require a three-dimensional reconstruction of a stem cell niche.

The bone marrow also acts as a niche for lineage-specific hematopoietic cell development. Erythroblastic islands (EBI) are erythroid niches consisting of a central macrophage surrounded by developing erythroid progenitors. These macrophages are proposed to act as "nurse cells" by providing iron and other growth factors essential to erythropoiesis [127 - 129]. The central macrophage of the EBI engulfs erythroid nuclei and degrades them, aiding terminal differentiation of erythroid cells [70]. The formation of EBIs is required for retention of erythroblasts in the bone marrow, and promotes the survival and proliferation of erythroblasts [130, 131]. Recently, we have shown that heme oxygenase-1 deletion, which leads to anemic mice, also disrupts the formation of these EBIs [132].

The bone marrow also contains stromal cells to support B lymphopoiesis. B-cell progenitors attach to CXCL12-expressing cells in the bone marrow and then re-locate and attach to IL-7-expressing cells as they mature into more mature pro-B-cells progenitors [133]. After naive B-cells are sensitized by antigens in the peripheral lymphoid organs, the resultant plasma cells migrate to the bone marrow which serves as a unique niche for these mature cells [133]. This niche offers prerequisite soluble factors from other hematopoietic lineage cells such as basophils, eosinophils and megakaryocytes [134 - 136].

Adult Peripheral Hematopoietic Niches: The Thymus Provides a Niche for T Cell Maturation

The thymus is a specialized peripheral lymphoid organ that features multiple microenvironments playing individual roles in supporting T-cell development. The outer thymic cortex possesses microenvironments that promote the differentiation of lymphoid progenitor cells into T-cells lineage, and is where positive selection of immature thymocytes based on their distinct T-cell receptor (TCR) rearrangements occur [137, 138]. The thymic cortex consists of basket-like globular thymic epithelial stromal cells expressing β5t, which are classified as thymic nurse cells. These thymoproteasome-containing thymic nurse cells interact with developing $CD4^+$ $CD8^+$ thymocytes that undergo the secondary TCRα rearrangement [139]. Within the thymic cortex, there is also a blood-thymus barrier that protects thymocytes from exposure to other antigens [140]. The inner thymic medulla offers microenvironments that generate regulatory T cells and verify self-tolerance by deleting self-reactive T cells known as negative selection [137, 138]. Interleukin (IL)-15^+ MHC class-II^+ thymic epithelial cells present in the thymic medulla (but not the thymic cortex) are thought to support stromal cells for CD8 T-cell and NK-T cell maturation [141]. Hassall's corpuscles exclusive to the thymic medulla, express thymic stromal lymphopoietin critical for thymopoiesis and is suggested to have a critical role in dendritic cell-mediated generation of $CD4^+$ $CD25^+$ regulatory T cells within the thymus [142].

In the adult thymus, IL-7^{high} cells are found in the cortico-medullary junction and display traits of both cortical and medullary thymic epithelial cells. These IL-7^{high} stromal cells co-express chemokines (CCL19, CCL25, CXCL12) and cytokines (IL-15) in the fetal thymus that are critical for thymopoiesis [143]. It is unclear if the cortico-medullary junction represents a separate thymic niche of T-cell development. Thymic nurse cells can also be found located within the cortico-medullary junction. They express cytokeratins five and eight (K5 and K8), and the transcription factor Trp-63 found in the thymic epithelial progenitor subset [144]. This further suggests a probable role for the cortico-medullary junction in regulating the microenvironment and the stromal thymic epithelial cells.

The thymus is at its largest and most active during the neonatal and pre-adolescent periods. Loss of the thymus at an early age through genetic mutation (as in DiGeorge Syndrome) results in severe immunodeficiency and a high susceptibility to infection. After puberty, the thymus begins to atrophy and involute. The thymic stroma is gradually replaced by adipose tissue, and the microarchitecture and integrity of the thymic microenvironment is severely disrupted [145]. The thymus is largely degenerated in the elderly and consists of mostly adipose tissue. While this degeneration may not greatly affect healthy individuals, recovery from treatments such as chemotherapy and radiotherapy is severely compromised [146].

Adult Peripheral Hematopoietic Niches: The Splenic Niche During Stress

The spleen is the largest lymphoid organ in the body and performs both immune and hematopoietic functions. In addition to large numbers of lymphocytes and erythrocytes, the spleen contains vascular and lymphatic endothelial cells, marginal zone macrophages, follicular dendritic cells, fibroblastic reticular cells, marginal reticular cells and red pulp fibroblasts. The white pulp consists of a thick accumulation of lymphocytes surrounding an artery, which acts as a site for B cell proliferation. The red pulp contains large numbers of erythrocytes that it filters and degrades. The blood flows through a network of channels called sinusoids, where macrophages and dendritic cells screen antigens in the blood and mount an immune response. The spleen is also involved in erythropoiesis particularly during the response to anemic stress in a process termed stress erythropoiesis

[147]. Stress erythropoiesis is achieved *via* the migration and subsequent proliferation of erythroid progenitors in extra-medullary sites such as the spleen, which act as secondary sites of erythrocyte production [147]. This process is a critical phase in the recovery from haemorrhage in mice and is supported *via* BMP4 and Hedgehog signalling acting together with Epo, SCF, and hypoxia [148 - 150]. The depletion of CD169^{+} splenic macrophages after bone marrow transplant in mice delays reticulocytosis and hematocrit recovery and abrogated the BMP4-dependent stress erythropoiesis induced by phenylhydrazine further demonstrating the spleen as an erythroid niche in stress erythropoiesis [151]. Stress erythropoiesis manifests as a dramatic increase in spleen size in mice (splenomegaly) and may also play a role in the splenomegaly noted in chronic haemolytic anaemic conditions such as thalassemia and other haemoglobinopathies. Understanding the mechanisms regulating stress erythropoiesis could provide a means of recapitulating this process under controlled conditions, by isolating erythroblasts from patients, and inducing their proliferation in a splenic niche environment *ex vivo* therefore enabling large scale production of mature erythrocytes for blood transfusion.

Table 2. The hematopoietic organs and tissues of the mammalian embryo, fetus and adult laboratory mouse.

Organ/Tissue	Resident hematopoietic cells.
Yolk sac	primitive erythroid, early macrophages, early megakaryocytes, definitive erythroid, lymphoid progenitors, HSC, microglial progenitors.
Allantois	erythroid progenitors, HSC.
Placenta	hematopoietic progenitors and stem cells.
Dorsal aorta	hematopoietic progenitors and stem cells.
Vitelline artery	hematopoietic progenitors and stem cells.
Omphalomesenteric artery	hematopoietic progenitors and stem cells.
Fetal liver	multilineage progenitors, erythroid progenitors, erythroblastic island macrophages, HSC.
Fetal spleen	multilineage progenitors, erythroid progenitors, HSC.
Bone marrow	HSC, multilineage progenitors, erythroblastic island macrophages, memory lymphocytes, megakaryocytes, mast cells.
Circulating blood	reticulocytes, erythrocytes, lymphocytes, NK cells, neutrophils, basophils, eosinophils, monocytes, platelets.

(Table 2) contd.....

Organ/Tissue	Resident hematopoietic cells.
Thymus	thymocytes (developing T lymphocytes), thymic B cells, thymic macrophages, thymic dendritic cells.
Spleen	lymphocytes, aged and damaged erythrocytes, dendritic cells, macrophages.
Liver (stress)	erythroid progenitors, Kupffer cells.
Lymph nodes	B, T lymphocytes, macrophages, dendritic cells.
Peyer's patches	Peyer's patch inducer cells, B, T lymphocytes, macrophages, dendritic cells.
Interfollicular zones	B, T lymphocytes, macrophages, dendritic cells.
NALT/BALT/GALT	B, T lymphocytes, macrophages, dendritic cells.

HEMATOPOIETIC STEM CELL TRANSPLANTATION

A Classical Case Study in Regenerative Medicine

Hematopoietic stem cell transplantation (HSCT) is an umbrella term used to cover all types of transplants that utilize HSCs including; peripheral blood stem cell transplant (PBSCT); umbilical cord stem cell transplant; and bone marrow transplant (BMT). Transplantation from allogeneic donors (immunologically matched donors) have taken place for over 50 years, and is an essential part of the treatment of life threatening hematological malignancies such as acute leukemias, lymphomas and multiple myelomas. However, with developing technological advances HSCT is being utilized more and more for the treatment of non-malignant hematological conditions, such as aplastic anemia and thalassemia. Autologous HSCT involves the intravenous infusion of a patient's own HSCs and is widely used to rescue patients from bone marrow failure usually due to iatrogenic causes such as high dose chemotherapy and radiotherapy used to treat cancer. This allows replacement of the patient bone marrow stem cells, and the production of new healthy blood cells and platelets. HSCT is the most widely used form of stem cell therapy today, and was one of the first clinically meaningful examples of regenerative medicine.

A History of Bone Marrow Transplantation

The first reported use of human bone marrow for therapeutic purposes was in a 19-year old female with gold-induced aplastic anemia who received intravenous transfusions of ABO matched marrow from her brother in 1938 [152]. In 1940 a 42-year old man with idiopathic aplastic anemia received an intra-sternal transfusion of bone marrow from his brother with group O blood [153]. However, as both these attempts employed infusions of only a few millilitres of bone marrow, they yielded no productive results [154]. This was followed by further failed attempts by Rekers and colleagues in the classified laboratories of the Atomic Energy Commission in Rochester, NY, during the World War II period [155]. This group carried out trials of bone marrow infusion from healthy dogs into irradiated dogs. Surprisingly, no significant effect to pancytopenia or survival of these animals was observed as the exposure level needed to provide sufficient immunosuppression to allow engraftment was at least double than the amount used. Following World War II and the nuclear bombing of Hiroshima and Nagasaki, a number of studies demonstrated that the myelosuppressive effects of radiation could be overcome. Irradiated mice whose spleens were shielded could be protected from the lethal effects of radiation: intra-peritoneal injection of cells from the spleen could achieve the same effect [156]. The lethal effects of irradiation could be overcome by grafting or injecting hematopoietic tissue from normal mice into irradiated animals [157, 158]. The development of "radiation chimaeras" was clear evidence indicating that the 'irradiation protective effect' was achieved *via* the infusion of the cellular component of the bone marrow [159]. The marrow of lethally irradiated mice given infusions of syngeneic marrow marked by the T6 chromosome was made up of cells of the cytogenetic type of the donor [159].

In 1956, Barnes and colleagues reported of the treatment of leukemia in mice by irradiation and marrow transplantation [160]. These studies in experimental animals were soon applied to clinical medicine. In 1957, a group in New York, led by E. Donnall Thomas, the father of bone marrow transplantation, reported the first clinical case study series of human bone marrow recipients [161]. The bone marrow for infusion for these patients was obtained from either fetal or adult cadavers and only one patient showed a transient marrow graft. Syngeneic transplants on two sets of identical twins who suffered from terminal leukemia were soon attempted [162]. The patient was given whole body lethal irradiation followed by intravenous infusion of bone marrow from the respective healthy twin. While this resulted in prompt recovery of bone marrow function, the remission from leukemia was only achieved for 7 weeks in one twin, and 12 weeks in the other. This study however, did demonstrate that compatible bone marrow transfusion can restore marrow function in humans after lethal irradiation, and that this can be used to treat even advanced level leukemia.

Attempts at clinically meaningful bone marrow transplantation were thwarted by the lack of understanding of transplantation immunology, and resulting issues with rejection. This was made evident by the compilation of over 200 attempts at allogeneic marrow transplantation by 1970, all of which were unsuccessful [163]. This let to the re-examination of animal models to try resolving these issues of rejection. The syndrome of 'Runt disease' was described in newborn mice given allogeneic cells in 1953 [164]. The principles established by this syndrome would later on be discovered to be the same as for 'secondary disease' in irradiated mice, and finally to be the same as for the now well characterized graft-*versus*-host-disease (GVHD) in humans. Early studies in dogs provided evidence of the importance of matching leukocyte antigens [165]. A growing understanding of HLA typing, paved the way for further work on the clinical application of bone marrow grafting. This led to the first successful allogeneic graft in 1968 in patients with severe combined immunodeficiency (SCID) and Wiskott-Aldrich syndrome [166 - 168]. All three patients were alive and well 25 years later marking the success of this powerful therapy [169].

Sources of Hematopoietic Stem Cells

The discovery in the 1980s that peripheral blood stem cells (PBSCs) were capable of reconstituting the hematopoietic system was a medical breakthrough [170 - 172]. Before this discovery, HSCs for transplantation came from the bone marrow, aspirated repeatedly from the posterior iliac crests from the donor under general or local anesthesia, a procedure far more difficult and risky in comparison [173]. Bone marrow HSCs emigrate into the peripheral circulation and home back to the bone marrow. For PBSC transplantation to be successful, adequate numbers of HSCs must be harvested from the donor. Initially, chemotherapy was used to increase the yield of PBSC harvested, due to the observation that there are higher numbers of circulating HSCs in patients following chemotherapy treatment [174,

175]. However, hematopoietic growth factors were found to improve HSC mobilization, with granulocyte colony-stimulating factor (G-CSF), granulocyte-macrophage colony-stimulating factor (GM-CSF), and plerixafor (a reversible antagonist of CXCR4), being approved for the use of stem cell mobilization. Currently, these agents are used for stem cell mobilization, either by themselves or alongside chemotherapy depending on expected yield and disease factors [176].

The Human Umbilical Cord: A Source of Therapeutic Stem Cell Populations

Umbilical cord blood can be used as a source of stem cells in situations where transplantation is urgent and suitable donors are not found [177]. The human umbilical cord starts to develop around the fifth week of gestation. Umbilical cord blood contains hematopoietic stem and multi-potent cells (colony - forming unit - granulocyte, erythroid, macrophage, megakaryocyte (CFU-GEMM)), as well as erythroid (burst - forming unit - erythroid (BFU-E)) and granulocyte–macrophage (colony forming unit - granulocyte, macrophage (CFU-GM)) progenitor cells [178, 179]. Different components of the umbilical cord are enriched with mesenchymal stem cells and cord-blood embryonic-like stem cells [66, 180, 181]. With the ease of obtaining umbilical cords and their rich stem cell activity, the cord blood has been proposed as a major source of stem cells for regenerative medicine [182].

The first report of a successful human umbilical cord transplant dates back to 1988, when a child with Fanconi's anemia received umbilical cord blood from an HLA-identical sibling [183]. Umbilical cords are immediately available and can be used in patients who lack a clinically suitable bone marrow donor [184]. Umbilical cord is superior to bone marrow transplant in several aspects. Firstly, there is a higher frequency of hematopoietic progenitor cells in cord blood than in the bone marrow [185]. Umbilical cord transplantation requires less stringent human leukocyte antigen (HLA) matching than bone marrow transplantation [184]. As a result, cord blood transplants have a lower incidence and severity of the immunological reaction known as graft-*versus*-host disease (GVHD) compared to bone marrow transplants [186]. Additionally, cord blood stem and progenitor cells have a higher proliferation potential than their bone marrow counterparts [187]. However there are disadvantages in the use of umbilical cords for transplants. The umbilical cord contains a limited number of hematopoietic stem and progenitor cells, which can lead to engraftment failure. This limitation could be addressed by using *ex vivo* culture to expand the numbers of hematopoietic stem and progenitor cells prior to transplantation [188]. Furthermore, compared to bone marrow transplants, cord blood transplant has a longer hematopoietic recovery time, with a prolonged neutrophil engraftment time. This results in a longer duration of neutropenic complications to the patient, such as severe mucositis and an increased infection risk and resulting mortality [186].

Complications of Hematopoietic Stem Cell Transplantation

While stem cell transplantation may seem the cure for a myriad of incurable diseases, the current protocols are not without significant morbidity and mortality. Complications are mainly due to the transplanted tissue (graft) response to the patient (host) tissue, and the toxic effects of the chemoradiotherapy and immunosuppressants [189]. Early complications include mucositis, sinusoidal obstruction syndrome, transplant related lung injury, and infections. Of these, mucositis is the most common, developing in 60-85% of patients undergoing stem cell transplantation [190]. It is caused by the myeloablative preparative regimens, and affects the oral and intestinal mucosa, leading to painful oropharyngeal ulceration, dysphagia, and pain requiring treatment with intravenous opioids [173]. Sinusoidal obstruction syndrome, resulting from the obstruction of the hepatic circulation due to the sloughing of the sinusoidal endothelium, is the next most common acute problem, and is potentially fatal [173]. Transplantation lung injury has a mortality rate exceeding 60 percent, and risk factors include total body irradiation, allogeneic transplantation, and acute GVHD [173]. Infections are a major issue, and result from the combination of the damage to the epithelium of the gut, mouth, and skin, and immunosuppressive effect of the preparative regimens [173, 191]. GVHD is the most important acute complication post allogeneic transplant however, resulting in devastating damage to the skin, gut, and liver. Given that a large proportion of post-transplant complications are directly related to the transplanted graft response to host tissue and the post transplant immunosuppression, it would seem logical that if there were a way to develop HSCs and long term progenitors from a patient's own tissues, correct any genetic abnormalities, and re-transplant them in the patient, the majority of these adverse complications could be evaded.

FUTURE DIRECTIONS AND THE POTENTIAL OF REGENERATIVE MEDICINE

Generating Hematopoietic Stem and Progenitor Cells from Pluripotent Stem Cells *in vitro*

Pluripotent Stem Cells

Tissue engineering and the *in vitro* production of clinically useful cell types was greatly enhanced by the development of stem cells which could be maintained, in their stem cell state, indefinitely in culture. Currently, we are unable to maintain HSCs indefinitely; however there are some cell types which do have this activity. These cells are pluripotent, meaning they are capable of giving rise to all of the cell lineages in the adult mammalian body *via* embryonic and fetal intermediate stages. These cell types include embryonic stem (ES) and induced pluripotent stem (iPS) cells. Both can form any cell type in the adult mammalian system though do not contribute to the placenta. Here, we will discuss the utility of these remarkable cells in the formation of the adult hematopoietic system.

Embryonic Stem Cells

Embryonic stem (ES) cells are derived from very early stage embryos. ES cells are cell lines which are obtained by the *ex vivo* culture of blastocysts on supportive stromal layers. The inner cell mass grows out from the blastocyst and develops into cells that can be maintained as undifferentiated stem cells indefinitely. Mouse ES cell lines have been found to be particularly useful as they can be modified genetically and re-introduced to blastocysts which can then be implanted into mouse pseudo-pregnant host mothers to give rise to genetically altered offspring. ES cells are a useful *in vitro* model of development as they can be induced to differentiate spontaneously *in vitro* depending upon the signals they receive from the environment.

ES cells were first differentiated *in vitro* to hematopoietic cells by Gordon Keller and colleagues in the early 1990s [192 - 194]. The relative ease and utility of this system meant that with low initial numbers of ES cells (~tens of thousands), cultures of hematopoietic progenitors of the erythroid and myeloid systems could be generated *in vitro* within approximately 7-10 days. This became a useful model of embryonic hematopoiesis as the need for animal experimentation could be avoided. Cultures of ES cells differentiating into hematopoietic cells could be scaled up to generate millions of cells at intermediate stages, allowing for a fine dissection of the genetic and signalling pathways needed for the embryo to develop its own hematopoietic system. Additionally, ES cells can be readily genetically manipulated to lose gene function or express lineage-specific transgenes or markers. Again, this makes the *in vitro* differentiation of ES cells a very useful system for identifying critical points and processes in embryonic blood production.

Keller and colleagues identified a progenitor cell type capable of giving rise to both primitive erythroid and definitive hematopoietic lineages [195]. Soon research groups around the world were reporting the generation of a range of different hematopoietic lineages from mouse (and later human) ES cells. Primitive erythroid cells were generated by a number of groups from mouse ES and human ES cells [195 - 199]. Addition of the hormone thrombopoietin significantly enhanced the production of megakaryocytes from ES cultures [200]. Myeloid progenitors were readily produced. The use of stromal cell layers such as OP9 was shown to be beneficial in the expansion of ES-derived hematopoietic progenitors. Mature neutrophils were formed from mouse and human ES cells [201, 202]. ES cell cultures were used to demonstrate the role of the transcription factor Pu-1 in macrophage formation [203]. Functional macrophages were generated from human ES cells [204]. Allostimulatory dendritic cells have been generated *in vitro* from ES cells [205, 206]. The addition of interleukin-7 to OP9 cultures led to the production of B lymphocytes from mouse ES cells [207]. Expression of the notch ligand delta-like on the surface of OP9 stromal cells induced the differentiation of T lymphocytes from ES cells in culture, mimicking the thymic microenvironment [208]. Remarkably, these ES-derived T lymphocytes could reconstitute immune function in immunodeficient mice, offering a conceptual pathway to ES-derived therapies for human immunodeficiencies such as severe combined immunodeficiency [208].

Initial studies with transplantation of ES-derived hematopoietic progenitors into recipient mice suggested that no long-term reconstitution of the bone marrow could be obtained from ES cultures [209]. However ectopic expression of the transcription factor HoxB4 in ES cells led to the induction of cells with long-term hematopoietic reconstituting activity [210]. HoxB4-transduced ES-derived hematopoietic progenitors can give rise to functional components of the immune system following transplantation [211]. However, hematopoietic reconstitution was not completely normal when HoxB4 was over-expressed leading to concerns for clinical utility as HoxB4 is linked to leukemia development [212].

Induced Pluripotent Stem Cells and Hematopoietic Development

The remarkable ability of ES cells to maintain the ability to give rise to all adult cell types (pluripotency) led Shinya Yamanaka and colleagues to ask whether the ectopic expression of ES cell-specific genes in terminally differentiated mature cells could lead to re-programming of those mature cells to a pluripotent state. In 2006, Takahashi and Yamanaka reported that they were indeed able to induce pluripotency in mature cells by over-expression of 4 genes found in ES cells [213]. This induction of pluripotency led to the now routine production of pluripotent stem cells from mice, rats and humans as well as many other species [214]. Induced pluripotent stem (iPS) cells have a massive potential for improving human health through regenerative medicine. iPS cells generated from patient samples can be induced to differentiate into a range of mature cell types which could be used for transplantation [215 - 217]. The first clinical report of the use of iPS came from Japan, where iPS cells were induced from a female patient. The patient-specific iPS were then differentiated into a sheet of retinal pigment epithelial cells and transplanted back into the patient's retina. Whether this has improved the macular degeneration in the patient has yet to be reported.

As iPS cells are similar to ES cells in pluripotency, the ability of iPS cells to generate hematopoietic cells *in vitro* has been examined. A range of hematopoietic cell types have been generated from human iPS cells [218]. These include; hemoglobin-expressing erythroid cells [219]; functionally mature neutrophils and macrophage [220]; platelet-generating megakaryocytes [221, 222]; T lymphocytes [218] as well as Natural Killer (NK) cells derived from SCID patient gene corrected iPS cells [223]. Amabile and colleagues reported generating a long-term reconstituting hematopoietic progenitor population isolated from teratomas generated from iPS cells [224]. While in this case, the teratomas may have given rise to a useful cell type, it highlights the ability of pluripotent stem cells to form aggressive multilineage tumours termed teratomas. This cancer-forming activity of ES and iPS cells is a major health concern for the application of pluripotent cells to clinical settings and must be considered before use of iPS-derived cells for therapeutic purposes.

Regenerative Medicine for Red Blood Cell Transfusion: Generation of Mature Erythroid Cells from Patient-derived Hematopoietic Progenitors

Blood transfusion is commonly practiced in modern medicine throughout the world, and can be lifesaving in surgeries, trauma, cancer treatment and haematological pathologies. Currently, the only source of blood for transfusion purposes is through voluntary human donations, and there are massive disparities in the donor to recipient ratios throughout the world. The worst affected are countries of lower socio-economical background, with approximately 48% of the world's blood donations occurring in high income countries, where only 15% of the world's population reside in. In fact, 82 countries of low and middle income report receiving an insufficient supply of less than 10 donations per 1000 people. Furthermore an ageing population with increasing transfusion requirements, alongside decreasing fertility rates reducing the donor-eligible population, is predicted to result in a global insufficiency of blood supply as early as 2050, even in developed countries [225]. Another major barrier to blood transfusion is blood group compatibility. Currently there are at least 30 blood groups that include most of the 308 recognized antigens, in addition to ABO and rhesus antigens, critical to transfusion compatibility [226]. Receiving transfusions, especially multiple transfusions, that are not perfectly compatible can lead to the formation of antibodies against immunogenic antigens not present in that patient, resulting in an increased difficulty to find matched donors for these patients [227, 228]. This is especially a concern for patients with chronic transfusion requirements, such as β-thalassemia or sickle cell patients [229].

Due to these problems attempts to go "cell-free" have been trialed aimed at producing synthetic cell free hemoglobin-based carriers and perfluorocarbon-based emulsions. Unfortunately, these products have been associated with increased risks of death and myocardial infarction without clinical benefit [230, 231]. Other technologies such as antigen masking or enzymatic cleavage to generate the 'O negative' universal donor cells have also been proposed recently, but are still under development [232]. Regular screening for blood born diseases is not routine in 39 countries, increasing risk of infection with HIV and hepatitis [229]. Given these issues, and the massive demand for blood products, there is significant effort put into generating blood cells *ex vivo*.

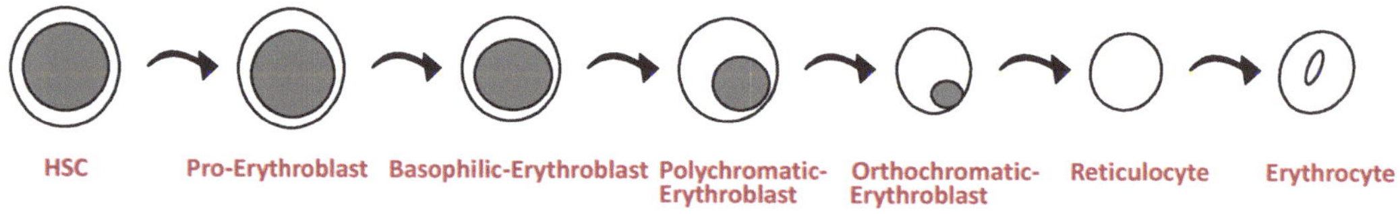

Fig. (5). Schematic representation of erythroid differentiation. HSC: hematopoietic stem cell.

Over two billion erythrocytes are produced each day in humans *via* erythropoiesis. Replicating this process *ex vivo* requires the understanding of how this process occurs *in vivo*. For human transfusion grade production of erythrocytes, the cultured cells need to have a phenotype and functionality similar to normal erythrocytes, including important aspects of erythropoiesis such as hemoglobin synthesis, enucleation, and deformable morphology. The earliest erythroid restricted progenitor is the BFU-e which gives rise to the CFU-e [229]. Both of these populations are able to give rise to erythroid colonies in methylcellulose colony forming cultures. Erythroblasts develop through a number of stages, gradually decreasing cell volume, increasing hemoglobin synthesis, and condensing their nucleus, such that by the reticulocyte stage, the nucleus is completely lost, in a process known as enucleation. Reticulocytes enter the bloodstream and become further condensed in size, *via* the loss of additional plasma membranes and cellular organelles, to attain the classical biconcave shape of an erythrocyte. Erythropoiesis as discussed previously, occurs normally in the bone marrow niche in adults, and is governed by a number of cellular and extracellular interactions (Fig. **5**).

Producing erythroid cells *ex vivo* can be approached by utilising differentiation of primary multipotent hematopoietic stem and progenitor cells (HSPCs) or pluripotent stem cells (ESCs or iPSCs), or by employing immortalised erythroid precursors. HSPCs can be differentiated into the erythroid lineage utilising the three steps of commitment, expansion, and maturation. Many successful attempts at generating reticulocytes from cord blood, peripheral blood HSPCs [233 - 235] and pluripotent stem cells have been published [199, 236, 237]. The challenges facing all of these approaches include; generation of adequate numbers of erythrocytes to meet transfusion requirements, problems with inducing the correct type of hemoglobin, and reliance on murine feeders and other animal products such as animal serum in order to induce appropriate maturation and enucleation [229].

Utilizing Primary HSPCs as a Source for *Ex vivo* Erythroid Cell Generation

Utilizing primary HSPCs to produce erythroid cells *ex vivo* can be done employing either peripheral blood HSPCs (PB-HSPCs) or cord blood HSPCs (CB-HSPCs). Evidence for generating erythrocytes from HSPCs *ex vivo* was reported in 1998 but it was the Douay group that demonstrated the first proof of principle for transfusion of *in vitro* generated erythrocytes in humans [234, 238]. This group pioneered the *ex vivo* generation of erythrocytes from cord blood HSCs, albeit with very low rates of enucleation. However, this was improved by co-culturing PB-HSPCs and CB-HSPCs with mouse MS-5 cells, achieving 90% terminal maturation of erythrocytes [233, 234]. Despite this, a protocol avoiding the use of adherent feeder layers is desirable for large scale production of erythroid cells, as it reduces the complexity and expense of using another cell type and is better suited for scale-up employing existing bioreactor technology [229]. Such a feeder free protocol was able to achieve >70% enucleation rates [235] and was later successfully adapted for use in an agitated bioreactor system leading to very high amplification levels in the order of 1.7 x10^6 fold and up to 90% enucleation efficiency [239].

Peripheral blood provides low numbers of HSPCs which can be increased by mobilization technologies utilizing G-CSF. Cord blood provides an enriched source of CD34$^+$ HSPCs although the total number of cells are limited [229]. PB-HSPCs give rise to predominantly adult hemoglobin-expressing erythrocytes which also have elevated fetal hemoglobin levels whereas CB-HSPCs give rise to predominantly fetal hemoglobin erythrocytes [233, 234, 240]. The amplification potential of PB-HSPCs are relatively low (120,000-fold) compared with that of CB-HSPCs (>2 million fold) [234]. The regenerative potential of CB-HSPCs are such that a single donated CB unit can be directed to yield 500 transfusable erythroid units [239]. The downside of CB-HSPCs is that these cells are not immortal and there are a limited number of CB banks worldwide. This means that CB-HSPCs would probably not benefit patients with rare blood groups, and those chronic transfusion dependent patients with alloimmunity [229].

Pluripotent Stem Cells as a Source for *Ex Vivo* Erythroid Cell Generation

Mouse and human ESCs have been successfully utilised to generate enucleated erythrocytes [199, 237, 241]. Erythropoietin (EPO) and other factors present in feeder lines or human plasma seem to play a significant role in the enucleation process, helping to dramatically increase the percentage of enucleated cells [199, 236, 242]. Human iPSCs have also been successfully utilized to produce enucleated erythrocytes, and similar to ESCs, EPO and other factors present in human plasma play an important role in this process. Feeder cells lines aided enucleation [236, 237, 243, 244]. The protocols were further enhanced by increasing culture time and adding heparin and insulin resulting in 26% enucleation rates in a feeder cell-free maturation system [244]. Both ESC- and iPSC-derived erythroblasts and erythrocytes express embryonic and fetal hemoglobins which presents a practical issue when considering transfusion [199, 237, 244]. Technological improvements such as increasing culture times and enforcing transgene expression of RUNX1a have led to higher levels of adult β-hemoglobin production [199, 245].

The clinical applicability of ESC-derived erythrocytes is limited due to the ethical considerations surrounding the use of human embryos. Currently, there is no identified NIH-approved-for-research human ESC line with an O RhD negative type and the limited number of human ESC lines may not be able to provide for all blood groups, especially the rare blood types [229]. In contrast, it has been proposed that only three carefully selected iPSC lines may suffice to provide for all transfusion-dependent patients in France [246]. Utilising iPSCs also opens the door to autotransfusion. The goal would be to use a patient's own somatic cells to generate iPSCs and erythroid progenitors, which would allow for the proliferation of potentially an unlimited number of ready-made, 100% immune-matched erythroid cells, available for transplantation into the patient whenever needed. This would result in the mitigation of all the current issues surrounding transfusion. This includes alloimmunization and incompatibility risks (*e.g.* immediate and delayed hemolytic transfusion reactions, non-hemolytic transfusion reactions, transfusion related acute lung injury, urticarial and anaphylactic reactions) as well as non-immunological complications (*e.g.* risk of contracting blood borne viruses, parasites, bacteria, and prions). It would also be possible to correct genetic defects in iPSCs to produce functional adult erythrocytes for patients with hemoglobinopathies, such as sickle cell and β-thalassemia, as discussed later on in this chapter. Additionally, a relatively novel idea is transdifferentiation, reprogramming human somatic cells (*e.g.* fibroblasts) directly to erythrocytes, thereby bypassing the HSPC stage. While this idea has been utilized to generate reticulocytes, the potential of this method to generate large numbers of erythrocytes has not yet been studied [247]. Recently, healthy mature erythroid cells have been generated from iPS cells derived from patients with pyruvate kinase deficiency. Following iPSC generation, gene editing was performed to correct the genetic defect. Healthy mature erythroid cells were then generated *in vitro* without the pyruvate kinase deficiency. Development of superior protocols improving erythrocyte yields from human iPSCs could lead to this technology being utilised in the clinical setting to treat patients with genetic disorders of the hematopoietic system.

New Directions to Improve Yield of *Ex Vivo* Erythrocyte Production

Mimicking stress erythropoiesis, as described previously, has potential to improve the production yield of erythrocytes *ex vivo* or *in vitro*. Utilizing SCF, EPO, insulin-like growth factor-1, and a glucocorticoid receptor activator such as dexamethasone, can lead to proliferation of erythroid precursors in the order of 100,000-fold [248]. However, the problem with this approach has been the failure of generating a sufficient yield of erythrocytes to match transfusion demands. Amplification yields of CB-derived erythroid precursors are restricted to below 10 billion-fold [248]. Peripheral blood-derived erythroid precursor's expansion is restricted to below 10-20 fold [249]. Other groups have attempted to establish immortal human erythroid precursor lines [250]. While defects in terminal differentiation and maturation were initially a technical hurdle [251, 252]. this was overcome and successful generation of immortalized mouse erythroblast lines able to produce enucleated erythrocytes, were produced utilizing mouse ESCs [241, 253]. Another idea tested was using the "Yamanaka" factors and other factors commonly used for reprogramming somatic cells to iPSCs and immortalize human cord blood-erythroblasts [254]. These cells were able to preserve their ability to differentiate into enucleated erythrocytes even following 12 months of culture, expanding at least 10^65 fold whilst maintaining their cell identity and karyotype. Optimizing protocols for increasing expansion fold, shifting technologies to feeder free and xeno-free protocols, switching hemoglobin from fetal to adult types, minimizing associated costs, and forming practical protocols that are compatible with large scale production lines, will be required before *ex vivo* erythrocyte generation can be applied for transfusion.

Application of Hematopoietic Regenerative Medicine to Human Disease

Regenerative Medicine Utilizing Cell-based Gene Therapies; The Future is Here

Gene therapy is the introduction of exogenous genetic material into cells for therapeutic purposes, resulting in a personalized form of regenerative medicine that can potentially treat a number of otherwise incurable genetic diseases. The hematopoietic system presents an ideal system that is amenable to gene therapies due to blood cells being derived from a common progenitor cell, the HSC, which is capable of reconstituting the entire circulatory system of an organism. The self-renewal activity and multipotency of HSCs can be harnessed to gene therapy to amplify the genetically modified cells into large numbers of cells of multiple cell lineages *in vivo*. Development of this technique could potentially replace the need for allogeneic HSC transplantation for non-malignant hematopoietic diseases, avoiding a large number of complications that stem from immune mismatch, as the patient's own cells could be genetically modified and re-transplanted into the patient. A number of human clinical trials have already been commenced to treat hematological conditions such as; severe combined immunodeficiency (SCID) [255 - 257]; X-linked chronic granulomatous disease [258 - 260]; Wiskott-Aldrich Syndrome [255, 261]; β-thalassemia as well as the storage disorders adrenoleukodystrophy and metachromatic leukodystrophy [262, 263].

β-thalassemia results from mutations in one or both of the β-globin chains. When both genes are affected, there is no production of β-globin, leading to severe anemia, jaundice, severe hepatosplenomegaly and radiological abnormalities due to extramedullary hematopoiesis. The severity of the anemia is such that long term blood transfusions are required, leading to problems with iron overload such as cardiac failure, liver failure and skin pigmentation and also transfusion transmitted infections. With the advent of iron chelation therapy, iron overload complications markedly reduced, and the survival of patients have improved [264]. A study conducted in the United Kingdom in 2000, demonstrated that about 50% of patients with β-thalassemia major die before the age of 35 years, and proposed that iron chelation therapy is too burdensome for full adherence [265]. HSC transplantation remains the only curative therapeutic direction for β-thalassemia major [266, 267]. Umbilical cord blood transplants have also been successfully utilized in the treatment of β-thalassemia major and have increased the safety and availability of HLA-identical related donor transplantation, while lowering the probabilities of acute and chronic GVHD [268, 269].

β-globin gene transfer and expression has been successful in thalassemia mouse models, despite the early problems with instability of vectors, low titers and variable expression [270 - 272]. Lentiviral vectors have been utilised to transfer the human β-globin gene sequence and its promoter regions into murine stem cells [273, 274] as well as into long term repopulating human hematopoietic stem cells [275, 276]. The first successful trial was performed in 2007, where three severe compound heterozygote patients with β-E thalassemia received a lentiviral vector with a corrected b-globin gene [277]. One out of the three patients became transfusion independent, marking the first successful β-thalassemia gene correction, while out of the remaining two patients, one failed to engraft requiring reinfusion of backup bone marrow, and the other had a successful engraftment, but remained transfusion dependent due to low vector copy number [277]. iPSCs can be generated from skin fibroblasts, amniotic fluid, or chorionic villus sampling of patients with β-thalassemia, and be differentiated into erythropoietic cells that synthesize hemoglobin [278, 279]. Homologous recombination has been successfully utilized to correct mutations in iPSCs derived from a β-thalassemia patient, sickle cell mouse model and sickle cell patient, with the patient derived iPSCs providing proof of principle that the β-globin gene can be repaired in human iPSCs [280 - 282].

Bio-engineering Hematopoietic Cells, Tissues and Organs for Regenerative Medicine

Recent advances in tissue engineering, additive manufacturing and stem cell reprogramming have led to the development of new fields in which all of these technologies are combined to generate new patient-tailored cells, tissues and organs. With great potential in the treatment of chronic disease or traumatic conditions with loss of entire tissues or organs, this new field of stem cell-driven regenerative medicine is garnering considerable attention. In this section, we aim to provide an overview of how these technologies are being applied to the treatment of hematological disorders largely from a cell biologist's perspective.

Bioengineering the Primary Hematopoietic Organ in Adults - Factors to Consider for an Ex Vivo Bone Marrow Model

Considering the remarkable productivity of the adult bone marrow in blood cell generation and the frequency of pathologies interfering with this productivity, an *ex vivo* model of the bone marrow microenvironment is a promising research and therapeutic direction. An *ex vivo* model of the bone marrow will enable studies in examining the role of accessory cells in HSC homing and engraftment, screening new drugs and dissecting the mechanisms in pathological/malignancy conditions. Having such a model would also facilitate in driving towards individualized medicine and therapy. It would potentially have translational medical applications such as efficiently expanding HSCs for HSCT or controlled production of mature blood cells for transfusion. It also promises an ectopic bone marrow which could provide a healthy environment when the native bone marrow niche is damaged.

However, there is still a limited availability of appropriate *ex vivo* or *in vitro* models that mimic complex niche organization completely. An ideal replicate of the bone marrow should closely resemble both the bone marrow structure and function *in vivo*.

The traditional approach was long-term bone marrow cultures (also known as Dexter cultures) [283]. These static cultures were carried out in plastic flasks, with hematopoietic cells attaching to a layer of stromal cells to imitate the bone marrow heterogeneity. However, these cultures are incapable of predicting *in vivo* tissue functions accurately, leading to the development of techniques to culture cells in bioreactors. These approaches include; fixed bed bioreactors, stirred suspension bioreactors, rotating wall vessels and perfusion chambers [284 - 289]. These bioreactors allow parameters (including nutrient delivery) to be tightly regulated and allowed higher cell densities than the static cultures.

A good *ex vivo* model relies heavily on the properties of the scaffold substitute. There are 5 main types of scaffolds: hydrogels, macroporous scaffolds, fibrous mesh, interpenetrating networks and microspheres. Each scaffold has its own advantages for *ex vivo* expansion and differentiation of hematopoietic progenitors and precursors. Modifications by changing attachment surfaces were reported to optimise these cultures. These biomaterials can be further manipulated by modifying surface chemistry (achieved by adding functional groups or oligopeptides, or coating with extracellular matrix) topographical surfaces, and mechanical properties (such as matrix elasticity and shear stress) [290]. In addition to physical factors such as space and mechano-related stress, there might be a synergistic effect of both the extracellular matrix and the degradation by-product. A low oxygen tension (hypoxic) microenvironment is required for HSC quiescence and cell renewal. Generating such a hypoxic gradient between the endosteal and the vascular niches would be typically challenging. Other exogenous conditions should be considered such as providing essential growth factors that are not primarily secreted in the bone marrow such as erythropoietin and balancing physicochemical conditions such as dissolved gases and ionic concentrations. The delivery of nutrients varies between numerous *ex vivo* models.

As the marrow is a 3-dimensional (3D) structure, a better representation of *in vivo* bone marrow would include both spatial and chemical complexity. A combination of bioreactors and scaffolds derived from various sources has been employed. These scaffold materials include macroporous collagen carriers, cellulose porous microspheres, silk and various synthetic polymers [291 - 293]. Engineering of stromal cells such as mesenchymal progenitors, osteoblasts and endothelial cells onto a 3D organizational structure also provides a more ideal microenvironment for supporting hematopoietic cells *in vitro* [294, 295]. 3D culture systems however have their disadvantages. Controlling or predicting cell seeding in the 3D scaffold, and hence desired cell-to-cell and cell-to-ECM interactions is hard to manipulate. As compared to 2D cultures, there is also much more difficulty in analyzing cells *in situ* (such as live-cell imaging or fluorescence imaging) and harvesting cells from these 3D-culture material for analyzing protein or gene expression. Furthermore, spatial gradients of important physicochemical parameters, such as the concentration of soluble factors, pO2 and pH, are harder to manipulate in these 3D cultures [284, 290]. A major issue with 3D culture systems was the absence of fluid flow to represent how cells interact with circulating blood and the hemodynamics involved. Hence, the microfluidic "organ-on-a-chip" system was engineered to attempt to overcome these limitations.

3D Bioprinting and "Organ-on-a-chip" Technologies

3D bioprinting was developed in 1986 by Charles W. Hull as "stereolithography" [296]. These printing technologies designed mainly for printing molten plastics and metals were improvised, and adapted to print fragile and fastidious living biological materials to form 3D scaffolds [297]. Despite the different bioprinting technologies available, the parameters that biologists are interested in are: surface resolution (distance between "droplet of ink"), cell densities, cell viability and the type of biological materials (especially the printability of the material) used for printing. The main challenge is to fabricate the microarchitecture of the extracellular matrix (ECM) components, integrated with multiple cell types, in sufficient resolution to recapitulate biological functions. One such approach is by building mini-tissue from multi-cellular blocks [297]. The aim is not to replicate an entire living organ but to synthesize small functional units that recapitulate their biological functions, mimicking the entire tissue-level or organ-level function. Examples of such mini-tissues are the "organ-on-a-chip" technologies, where 3D bioprinting accurately reproduce functional tissue units maintained by a microfluidic network. One representative example is the "bone marrow-on-a-chip" (eBM) containing artificial bone and living marrow [298].

Organs-on-chips are microfluidic devices used for culturing cells. The word 'chip' stems from a modified form of printing (photolithographic etching) used to manufacture computer microchips. They contain micrometer sized chambers continuously perfused with fluids by microfluidic systems usually made of a silicone rubber, poly-dimethyl-siloxane (or known as PDMS). These chips representing different organs can be linked *via* fluids, to mimic physiological interactions between different organs [297, 299]. It was envisioned that these biomimetic microsystems could be linked together by a microfluidic circulatory system as a "human"-on-a-chip concept [299, 300]. This microfabricating/printing technology was applied to fabricate an *ex vivo* bone marrow model - that is the bone marrow-on-a-chip (eBM). A carefully engineered scaffold device was first implanted into subcutaneous mesenteric regions of a mouse to generate bone marrow [301]. The device was engineered to minimize adipocytic marrow generation that inhibits hematopoiesis. The graft used contained identical key cellular (both stromal and hematopoietic progenitors) and molecular components to the *in vivo* mouse bone marrow: CXCL12 expression on endosteal cells and blood vessels, CXCR4 expression by lymphocytic clusters in both endosteal and perivascular niche, Nestin$^+$ and lectin receptor$^+$ expression on cells in the perivascular regions; and CD31 expression on the vascular endothelium [298]. This graft was surgically removed after 8 weeks and re-engineered to incorporate microfluidics for culturing the cells *in vitro*. The eBM was cultured with optimised serum-free medium supplemented with cytokines. The authors show that the eBM retains fully functional HSCs after 4 days of culture *in vitro*. The eBM also exhibited similar cellular response to the *in vivo* mouse bone marrow in response to clinical relevant doses of gamma-radiation and G-CSF administration [298]. Though this study was limited to a 7 day culture, it has demonstrated the successful generation of an *ex vivo* model that recapitulates marrow functions.

An approach aiming towards an ectopic bone marrow to produce protoplatelets has also been studied. Despite differences in engineering technicalities, this microfluidic platelet bioreactor-on-a-chip successfully recapitulates hemodynamic vascular shear stress and the marrow microenvironment, triggering protoplatelet production. This model demonstrated higher protoplatelet yields (approximately 90% protoplatelet-producing megakaryocytes) compared to static cultures (approximately 10% protoplatelet-producing megakaryocytes) from human-induced pluripotent stem-cell derived megakaryocytes [302].

Similar microfluidic bone marrow devices have been engineered to study disease development and mechanisms. One study attempted to culture multiple myeloma cells, a notoriously difficult task due to their survival dependence on the BM niche. However this was made possible with the use of a microfluidic platform to emulate the dynamic physiology of the BM niche, by using an eight chamber device to facilitate the interactions between the human fetal osteoblast cell line, hFOB 1.19, and mononuclear cells from BM aspirates of 3 multiple myeloma patients [303]. Such devices can facilitate testing personalized therapeutics, avoid use of animal models, and aid in the study of the biology of such diseases.

However, in all "organ-on-a-chip" attempts to recapitulate hematopoietic tissue function, the migratory behavior of hematopoietic cells was ignored. A main characteristic of the hematopoietic system, both in development and pathology, is the migration of hematopoietic cells. As mentioned in previous sections, hematopoiesis occurs from organ-to-organ during development. The recruitment of leukocytes during disease and inflammation, the migration of plasma cells into marrow as memory plasma cells, and the perpetual cycling nature of HSCs in and out of the bone marrow, further reflects the nomadic nature of hematopoietic cells. This migration involves many protein-protein interactions which may play important roles in development and disease and therefore should be taken into account when attempting to form an *ex vivo* organ-on-a-chip. Furthermore, the distribution and recruitment of immune cells into non-lymphoid organs was also not reflected in these attempts. Currently, these organs-on-a-chip devices are not able to replicate the immune system and the immune responses that occur physiologically. Other future directions should also include engineering innervation into these biomimetic microfluidic devices to mimic the regulation by the nervous system; using iPSCs or CRISRP/Cas9 technology to genetically modify specific hematopoietic progenitors to study normal and malignant hematopoiesis; and engineering biomimetic "organ-on-a-chip" devices that recapitulate embryonic and fetal hematopoiesis. Currently, these technologies have only been used in improvising cultures of hematopoietic cells *in vitro*.

Engineering Hematopoietic and Lymphoid Organs from Pluripotent Stem Cells

The loss of one or more of the many hematopoietic cell types shown in Table **1** can lead to human pathologies. In some of the cases, an entire organ may be missing, resulting in devastating consequences to the affected individual. Here we will discuss potential technologies to re-engineer these affected organs.

Bioengineering the Thymus

Athymia (absence of a thymus) can lead to T lymphocyte immunodeficiency. Thymus defects may arise following chemotherapy, during the natural involution of the thymus in aging, due to congenital defects, trauma or following the development of autoimmune syndromes targeting the thymus [304]. For example, athymia has been reported in DiGeorge Syndrome [305]. A lack of a supportive hematopoietic or lymphoid structure or niche results in defects in blood cell development, in this case specifically defects in T lymphocyte physiology. Repairing or regenerating the tissue architecture defective in these patients offers a novel method to allow hematopoietic and immune cell reconstitution [304]. However applying pluripotent stem cell differentiation culture systems to the regeneration of tissue architecture requires an understanding of the developmental biological processes leading to the initial formation of the defective organ [306].

The thymic microenvironment is a combination of distinct thymic epithelial populations, dendritic cells, macrophages and endothelium all leading to the homing of T lymphoid progenitors, expansion of progenitor numbers, and the development of immune tolerance. Purified thymic epithelial progenitor cells form functional thymic tissue *in vivo* when injected into athymic mice [307]. This thymic epithelial progenitor could therefore be applied to the regeneration of functional thymic tissue [308]. An "implantable" thymic microenvironment has been developed in which thymic epithelial cells and thymic mesenchymal cells are aggregated to form a functional thymic unit. Hematopoietic progenitors are then able to migrate to these implanted microenvironments and generate mature functional human T lymphocytes [309]. Human pluripotent stem cell lines can be induced to differentiate into the anterior foregut endoderm, the endodermal-population responsible for the formation of the thyroid, parathyroid, trachea, lung and the epithelial components of the thymus [304]. Functional thymic epithelial cells have been generated from hESC which are capable of reconstituting the thymic microenvironment when transplanted into T cell deficient mice [310].

Bioengineering the Spleen

The spleen is a large organ with two distinct functions. Firstly, the white pulp is the largest lymphoid tissue in the body and responsible for the immune responses to blood-borne pathogens. Secondly, the red pulp is the major site for clearance of aged or damaged erythrocytes from the circulation. Loss of a functional spleen is a relatively common phenomenon. This can be due to trauma leading to splenectomies, or autosplenectomy where hematological diseases such as sickle cell anemia or β-thalassemia result in splenic fibrosis to the level where the spleen is essentially rendered nonfunctional by the body. Following loss of a functional spleen, patients become susceptible to gram-negative bacteria and show abnormal hematological parameters due to loss of abnormal erythrocyte clearance. The development of a new spleen could therefore offer many patients improved health outcomes, quality of life and possibly reduction in health costs. However, the regeneration activity of splenic tissue is poor, and tissue-engineered spleen have had limited success [311]. Recent success has focused upon the development of external "spleens" able to remove pathogens from circulating blood. An extracorporeal fluidic device has been engineered to function as an artificial spleen (biospleen). This biospleen is capable of removing *Staphylococcus aureus* and *Escherichia coli* and lethal doses of endotoxins in living rats, and multiple microorganisms and their endotoxins from extracted human blood [312]. This was further microengineered with the organ-on-a-chip technology, capable of removing malaria-infected erythrocytes from extracted mouse blood [313].

Artificial Lymph Nodes

Though small in size, lymph nodes are highly potent areas of immune function. It has been proposed that "artificial lymph nodes" could be generated from stem cell and progenitor populations such as lymph node mesenchymal progenitors which would form the capsule and reticular network which in turn could be populated by migrating hematopoeitic progenitors [311, 314]. A vascular network with the appropriate architecture, and the endothelial cell types unique to the lymph node, such as high endothelial vessels, would also be required. Attempts at generating "artificial lymph nodes" have been considerably more successful than attempts at forming artificial spleens. Bioreactors have been utilized in attempts to generate human lymph node-like structures *in vitro* [315]. Matrices such as collagen/polyurethane and collagen sponges have been used to begin to model the lymph node architecture for eventual 3D bioprinting into a lymph node-type structure. Artificial lymphoid tissue has been developed by encapsulating a thymic stromal cell line in a biocompatible scaffold composed of collagen and polyurethane in a sponge-like matrix [316]. This allows for the migration of lymphoid populations into the matrix and formation of lymphoid architecture and immune responses in "lymph node-like organoids" [316]. Artificial lymph nodes have been generated in immunodeficient mice following implantation of biocompatible scaffolds containing stromal cells [317]. These artificial lymph nodes result in improved immune function in recipient mice [317]. However, with the large number of lymph nodes within the human body (over 180 are estimated); the urgency to generate an artificial structure mimicking their activity is less than for an organ such as the thymus or spleen. Tissue-engineered intestinal structures ("neointestines") have been posited as therapeutic material for patients with short bowel syndrome [318]. When implanted, these neointestinal structures attracted host immune cells including T, B and NK cells as well as tissue-resident macrophages [319].

CONCLUDING REMARKS

The hematopoietic system is responsible for producing over a quarter of all the cells in our bodies, prevents us from hemorrhaging to death and fights infections on a daily basis. It is the first functionally mature system to appear in the developing mammalian embryo and occupies a multitude of tissues and organs in the adult body. It is also largely based on a hierarchy with rare hematopoietic stem cells at its apex. Genetic modification or repair of these stem cells can therefore lead to profound whole-body changes. These properties place the hematopoietic system as a target for regenerative medical approaches. By combining stem cell biology, reprogramming, *in vitro* differentiation and tissue engineering, regenerative medicine can be applied to improve the outcomes of patients with currently untreatable hematological conditions.

CONFLICT OF INTEREST

The authors confirm that they have no conflict of interest to declare for this publication.

ACKNOWLEDGEMENTS

We firstly offer our apologies to any researchers whose work has not been described in this lengthy chapter. We also thank the members of the Laboratory of Blood Cell Development, University of Sydney. This laboratory has been supported by funding from the Sydney Medical School, the Bosch Institute Translational Grant-in-Aid, the Feline Health Research Fund, the NHMRC University of Sydney Equipment Grants and the National Breast Cancer Foundation. CKC was supported by a Sydney Medical School Postgraduate Scholarship. JYH is supported by the NWG Macintosh Scholarship (Discipline of Anatomy & Histology, University of Sydney). STF is the Sesquicentennial Senior Lecturer in Molecular Embryology and gratefully acknowledges the support of the Disciplines of Physiology, Anatomy & Histology, University of Sydney.

REFERENCES

[1] Akashi K, Traver D, Miyamoto T, Weissman IL. A clonogenic common myeloid progenitor that gives rise to all myeloid lineages. Nature 2000; 404(6774): 193-7.
[http://dx.doi.org/10.1038/35004599] [PMID: 10724173]

[2] Orkin SH, Zon LI. Hematopoiesis: an evolving paradigm for stem cell biology. Cell 2008; 132(4): 631-44.
[http://dx.doi.org/10.1016/j.cell.2008.01.025] [PMID: 18295580]

[3] Becker AJ, McCULLOCH EA, Till JE. Cytological demonstration of the clonal nature of spleen colonies derived from transplanted mouse marrow cells. Nature 1963; 197: 452-4.

[http://dx.doi.org/10.1038/197452a0] [PMID: 13970094]

[4] Osawa M, Hanada K, Hamada H, Nakauchi H. Long-term lymphohematopoietic reconstitution by a single CD34-low/negative hematopoietic stem cell. Science 1996; 273(5272): 242-5. [http://dx.doi.org/10.1126/science.273.5272.242] [PMID: 8662508]

[5] Smith LG, Weissman IL, Heimfeld S. Clonal analysis of hematopoietic stem-cell differentiation *in vivo*. Proc Natl Acad Sci USA 1991; 88(7): 2788-92. [http://dx.doi.org/10.1073/pnas.88.7.2788] [PMID: 1672767]

[6] Till JE, McCULLOCH EA. A direct measurement of the radiation sensitivity of normal mouse bone marrow cells. Radiat Res 1961; 14(2): 213-22. [http://dx.doi.org/10.2307/3570892] [PMID: 13776896]

[7] Till JE, McCulloch EA, Siminovitch L. A stochastic model of stem cell proliferation, based on the growth of spleen colony-forming cells. Proc Natl Acad Sci USA 1964; 51(1): 29-36. [http://dx.doi.org/10.1073/pnas.51.1.29] [PMID: 14104600]

[8] Ema H, Morita Y, Suda T. Heterogeneity and hierarchy of hematopoietic stem cells. Exp Hematol 2014; 42(2): 74-82.e2. [http://dx.doi.org/10.1016/j.exphem.2013.11.004] [PMID: 24269919]

[9] Kawamoto H, Ohmura K, Fujimoto S, Katsura Y. Emergence of T cell progenitors without B cell or myeloid differentiation potential at the earliest stage of hematopoiesis in the murine fetal liver. J Immunol 1999; 162(5): 2725-31. [PMID: 10072517]

[10] Adolfsson J, Månsson R, Buza-Vidas N, *et al.* Identification of Flt3+ lympho-myeloid stem cells lacking erythro-megakaryocytic potential a revised road map for adult blood lineage commitment. Cell 2005; 121(2): 295-306. [http://dx.doi.org/10.1016/j.cell.2005.02.013] [PMID: 15851035]

[11] Ikuta K, Weissman IL. Evidence that hematopoietic stem cells express mouse c-kit but do not depend on steel factor for their generation. Proc Natl Acad Sci USA 1992; 89(4): 1502-6. [http://dx.doi.org/10.1073/pnas.89.4.1502] [PMID: 1371359]

[12] Spangrude GJ, Heimfeld S, Weissman IL. Purification and characterization of mouse hematopoietic stem cells. Science 1988; 241(4861): 58-62. [http://dx.doi.org/10.1126/science.2898810] [PMID: 2898810]

[13] Ogawa M, Matsuzaki Y, Nishikawa S, *et al.* Expression and function of c-kit in hemopoietic progenitor cells. J Exp Med 1991; 174(1): 63-71. [http://dx.doi.org/10.1084/jem.174.1.63] [PMID: 1711568]

[14] Avagyan S, Amrani YM, Snoeck H-W. Identification and *in vivo* analysis of murine hematopoietic stem cells. Methods Enzymol 2010; 476: 429-47. [http://dx.doi.org/10.1016/S0076-6879(10)76023-7] [PMID: 20691879]

[15] Harrison DE, Jordan CT, Zhong RK, Astle CM. Primitive hemopoietic stem cells: direct assay of most productive populations by competitive repopulation with simple binomial, correlation and covariance calculations. Exp Hematol 1993; 21(2): 206-19. [PMID: 8425559]

[16] Domen J, Weissman IL. Self-renewal, differentiation or death: regulation and manipulation of hematopoietic stem cell fate. Mol Med Today 1999; 5(5): 201-8. [http://dx.doi.org/10.1016/S1357-4310(99)01464-1] [PMID: 10322312]

[17] Guenechea G, Gan OI, Dorrell C, Dick JE. Distinct classes of human stem cells that differ in proliferative and self-renewal potential. Nat Immunol 2001; 2(1): 75-82. [http://dx.doi.org/10.1038/83199] [PMID: 11135582]

[18] Muller-Sieburg CE, Sieburg HB, Bernitz JM, Cattarossi G. Stem cell heterogeneity: implications for aging and regenerative medicine. Blood 2012; 119(17): 3900-7. [http://dx.doi.org/10.1182/blood-2011-12-376749] [PMID: 22408258]

[19] Copley MR, Beer PA, Eaves CJ. Hematopoietic stem cell heterogeneity takes center stage. Cell Stem Cell 2012; 10(6): 690-7. [http://dx.doi.org/10.1016/j.stem.2012.05.006] [PMID: 22704509]

[20] Morrison SJ, Wandycz AM, Akashi K, Globerson A, Weissman IL. The aging of hematopoietic stem cells. Nat Med 1996; 2(9): 1011-6. [http://dx.doi.org/10.1038/nm0996-1011] [PMID: 8782459]

[21] Yamamoto R, Morita Y, Ooehara J, *et al.* Clonal analysis unveils self-renewing lineage-restricted progenitors generated directly from hematopoietic stem cells. Cell 2013; 154(5): 1112-26. [http://dx.doi.org/10.1016/j.cell.2013.08.007] [PMID: 23993099]

[22] Baron MH, Isern J, Fraser ST. The embryonic origins of erythropoiesis in mammals. Blood 2012; 119(21): 4828-37. [http://dx.doi.org/10.1182/blood-2012-01-153486] [PMID: 22337720]

[23] Al-Drees MA, Yeo JH, Boumelhem BB, *et al.* Stem Cells Int 2015; 2015: 1-14. [http://dx.doi.org/10.1155/2015/571893]

[24] Isern J, He Z, Fraser ST, *et al.* Single-lineage transcriptome analysis reveals key regulatory pathways in primitive erythroid progenitors in the mouse embryo. Blood 2011; 117(18): 4924-34. [http://dx.doi.org/10.1182/blood-2010-10-313676] [PMID: 21263157]

[25] Fraser ST. The modern primitives: applying new technological approaches to explore the biology of the earliest red blood cells. ISRN Hematol 2013; 2013(8): 568928. [PMID: 24222861]

[26] Orkin SH, Zon LI. SnapShot: hematopoiesis. Cell 2008; 132(4): 712. [http://dx.doi.org/10.1016/j.cell.2008.02.013] [PMID: 18295585]

[27] Palis J, Robertson S, Kennedy M, Wall C, Keller G. Development of erythroid and myeloid progenitors in the yolk sac and embryo proper of the mouse. Development 1999; 126(22): 5073-84. [PMID: 10529424]

[28] Fraser ST, Midwinter RG, Berger BS, *et al.* Heme Oxygenase-1: A critical link between iron metabolism, erythropoiesis, and development. Advances in hematology. 2011; 2011: 1-6.

[29] Medvinsky A, Rybtsov S, Taoudi S. Embryonic origin of the adult hematopoietic system: advances and questions. Development 2011; 138(6): 1017-31. [http://dx.doi.org/10.1242/dev.040998] [PMID: 21343360]

[30] Kumaravelu P, Hook L, Morrison AM, *et al.* Quantitative developmental anatomy of definitive haematopoietic stem cells/long-term repopulating units (HSC/RUs): role of the aorta-gonad-mesonephros (AGM) region and the yolk sac in colonisation of the mouse embryonic liver. Development 2002; 129(21): 4891-9.
[PMID: 12397098]

[31] Cumano A, Godin I. Ontogeny of the hematopoietic system. Annu Rev Immunol 2007; 25(1): 745-85.
[http://dx.doi.org/10.1146/annurev.immunol.25.022106.141538] [PMID: 17201678]

[32] Babovic S, Eaves CJ. Hierarchical organization of fetal and adult hematopoietic stem cells. Exp Cell Res 2014; 329(2): 185-91.
[http://dx.doi.org/10.1016/j.yexcr.2014.08.005] [PMID: 25128815]

[33] Jollie WP. Development, morphology, and function of the yolk-sac placenta of laboratory rodents. Teratology 1990; 41(4): 361-81.
[http://dx.doi.org/10.1002/tera.1420410403] [PMID: 2187257]

[34] Bielinska M, Wilson DB. Induction of yolk sac endoderm in GATA-4-deficient embryoid bodies by retinoic acid. Mech Dev 1997; 65(1-2): 43-54.
[http://dx.doi.org/10.1016/S0925-4773(97)00053-1] [PMID: 9256344]

[35] Palis J, Yoder MC. Yolk-sac hematopoiesis: the first blood cells of mouse and man. Exp Hematol 2001; 29(8): 927-36.
[http://dx.doi.org/10.1016/S0301-472X(01)00669-5] [PMID: 11495698]

[36] Sugiyama D, Inoue-Yokoo T, Fraser ST, Kulkeaw K, Mizuochi C, Horio Y. Embryonic regulation of the mouse hematopoietic niche. ScientificWorldJournal 2011; 11: 1770-80.
[http://dx.doi.org/10.1100/2011/598097] [PMID: 22125435]

[37] McGrath KE, Koniski AD, Malik J, Palis J. Circulation is established in a stepwise pattern in the mammalian embryo. Blood 2003; 101(5): 1669-76.
[http://dx.doi.org/10.1182/blood-2002-08-2531] [PMID: 12406884]

[38] Kingsley PD, Malik J, Fantauzzo KA, Palis J. Yolk sac-derived primitive erythroblasts enucleate during mammalian embryogenesis. Blood 2004; 104(1): 19-25.
[http://dx.doi.org/10.1182/blood-2003-12-4162] [PMID: 15031208]

[39] Fraser ST, Isern J, Baron MH. Maturation and enucleation of primitive erythroblasts during mouse embryogenesis is accompanied by changes in cell-surface antigen expression. Blood 2007; 109(1): 343-52.
[http://dx.doi.org/10.1182/blood-2006-03-006569] [PMID: 16940424]

[40] McGrath KE, Frame JM, Fromm GJ, *et al.* A transient definitive erythroid lineage with unique regulation of the β-globin locus in the mammalian embryo. Blood 2011; 117(17): 4600-8.
[http://dx.doi.org/10.1182/blood-2010-12-325357] [PMID: 21378272]

[41] Lux CT, Yoshimoto M, McGrath K, Conway SJ, Palis J, Yoder MC. All primitive and definitive hematopoietic progenitor cells emerging before E10 in the mouse embryo are products of the yolk sac. Blood 2008; 111(7): 3435-8.
[http://dx.doi.org/10.1182/blood-2007-08-107086] [PMID: 17932251]

[42] Cumano A, Ferraz JC, Klaine M, Di Santo JP, Godin I. Intraembryonic, but not yolk sac hematopoietic

precursors, isolated before circulation, provide long-term multilineage reconstitution. Immunity 2001; 15(3): 477-85.
[http://dx.doi.org/10.1016/S1074-7613(01)00190-X] [PMID: 11567637]

[43] Fraser ST, Ogawa M, Yu RT, Nishikawa S, Yoder MC, Nishikawa S. Definitive hematopoietic commitment within the embryonic vascular endothelial-cadherin(+) population. Exp Hematol 2002; 30(9): 1070-8.
[http://dx.doi.org/10.1016/S0301-472X(02)00887-1] [PMID: 12225799]

[44] Samokhvalov IM, Samokhvalova NI, Nishikawa S. Cell tracing shows the contribution of the yolk sac to adult haematopoiesis. Nature 2007; 446(7139): 1056-61.
[http://dx.doi.org/10.1038/nature05725] [PMID: 17377529]

[45] Godin I, Dieterlen-Lièvre F, Cumano A. Emergence of multipotent hemopoietic cells in the yolk sac and paraaortic splanchnopleura in mouse embryos, beginning at 8.5 days postcoitus. Proc Natl Acad Sci USA 1995; 92(3): 773-7.
[http://dx.doi.org/10.1073/pnas.92.3.773] [PMID: 7846049]

[46] Medvinsky A, Dzierzak E. Definitive hematopoiesis is autonomously initiated by the AGM region. Cell 1996; 86(6): 897-906.
[http://dx.doi.org/10.1016/S0092-8674(00)80165-8] [PMID: 8808625]

[47] Tavian M, Coulombel L, Luton D, Clemente HS, Dieterlen-Lièvre F, Péault B. Aorta-associated CD34+ hematopoietic cells in the early human embryo. Blood 1996; 87(1): 67-72.
[PMID: 8547678]

[48] de Bruijn MF, Speck NA, Peeters MC, Dzierzak E. Definitive hematopoietic stem cells first develop within the major arterial regions of the mouse embryo. EMBO J 2000; 19(11): 2465-74.
[http://dx.doi.org/10.1093/emboj/19.11.2465] [PMID: 10835345]

[49] Taoudi S, Medvinsky A. Functional identification of the hematopoietic stem cell niche in the ventral domain of the embryonic dorsal aorta. Proc Natl Acad Sci USA 2007; 104(22): 9399-403.
[http://dx.doi.org/10.1073/pnas.0700984104] [PMID: 17517650]

[50] Ivanovs A, Rybtsov S, Welch L, Anderson RA, Turner ML, Medvinsky A. Highly potent human hematopoietic stem cells first emerge in the intraembryonic aorta-gonad-mesonephros region. J Exp Med 2011; 208(12): 2417-27.
[http://dx.doi.org/10.1084/jem.20111688] [PMID: 22042975]

[51] Mizuochi C, Fraser ST, Biasch K, *et al.* Intra-aortic clusters undergo endothelial to hematopoietic phenotypic transition during early embryogenesis. 2012.
[http://dx.doi.org/10.1371/journal.pone.0035763]

[52] North T, Gu TL, Stacy T, *et al.* Cbfa2 is required for the formation of intra-aortic hematopoietic clusters. Development 1999; 126(11): 2563-75.
[PMID: 10226014]

[53] Yokomizo T, Ng CE, Osato M, Dzierzak E. Three-dimensional imaging of whole midgestation murine embryos shows an intravascular localization for all hematopoietic clusters. Blood 2011; 117(23): 6132-4.
[http://dx.doi.org/10.1182/blood-2011-02-334037] [PMID: 21505195]

[54] Peeters M, Ottersbach K, Bollerot K, *et al.* Ventral embryonic tissues and Hedgehog proteins induce early AGM hematopoietic stem cell development. Development 2009; 136(15): 2613-21. [http://dx.doi.org/10.1242/dev.034728] [PMID: 19570846]

[55] Matsuoka S, Tsuji K, Hisakawa H, *et al.* Generation of definitive hematopoietic stem cells from murine early yolk sac and paraaortic splanchnopleures by aorta-gonad-mesonephros region-derived stromal cells. Blood 2001; 98(1): 6-12. [http://dx.doi.org/10.1182/blood.V98.1.6] [PMID: 11418454]

[56] Oostendorp RA, Harvey KN, Kusadasi N, *et al.* Stromal cell lines from mouse aorta-gonad--mesonephros subregions are potent supporters of hematopoietic stem cell activity. Blood 2002; 99(4): 1183-9. [http://dx.doi.org/10.1182/blood.V99.4.1183] [PMID: 11830464]

[57] Ohneda O, Fennie C, Zheng Z, *et al.* Hematopoietic stem cell maintenance and differentiation are supported by embryonic aorta-gonad-mesonephros region-derived endothelium. Blood 1998; 92(3): 908-19. [PMID: 9680359]

[58] Oostendorp RA, Robin C, Steinhoff C, *et al.* Long-term maintenance of hematopoietic stem cells does not require contact with embryo-derived stromal cells in cocultures. Stem Cells 2005; 23(6): 842-51. [http://dx.doi.org/10.1634/stemcells.2004-0120] [PMID: 15917480]

[59] Gekas C, Dieterlen-Lièvre F, Orkin SH, Mikkola HK. The placenta is a niche for hematopoietic stem cells. Dev Cell 2005; 8(3): 365-75. [http://dx.doi.org/10.1016/j.devcel.2004.12.016] [PMID: 15737932]

[60] Gekas C, Dieterlen-Lièvre F, Orkin SH, Mikkola HK. The placenta is a niche for hematopoietic stem cells. Dev Cell 2005; 8(3): 365-75. [http://dx.doi.org/10.1016/j.devcel.2004.12.016] [PMID: 15737932]

[61] Ottersbach K, Dzierzak E. The murine placenta contains hematopoietic stem cells within the vascular labyrinth region. Dev Cell 2005; 8(3): 377-87. [http://dx.doi.org/10.1016/j.devcel.2005.02.001] [PMID: 15737933]

[62] Zeigler BM, Sugiyama D, Chen M, Guo Y, Downs KM, Speck NA. The allantois and chorion, when isolated before circulation or chorio-allantoic fusion, have hematopoietic potential. Development 2006; 133(21): 4183-92. [http://dx.doi.org/10.1242/dev.02596] [PMID: 17038514]

[63] Robin C, Bollerot K, Mendes S, *et al.* Human placenta is a potent hematopoietic niche containing hematopoietic stem and progenitor cells throughout development. Cell Stem Cell 2009; 5(4): 385-95. [http://dx.doi.org/10.1016/j.stem.2009.08.020] [PMID: 19796619]

[64] Bailo M, Soncini M, Vertua E, *et al.* Engraftment potential of human amnion and chorion cells derived from term placenta. Transplantation 2004; 78(10): 1439-48. [http://dx.doi.org/10.1097/01.TP.0000144606.84234.49] [PMID: 15599307]

[65] Parolini O, Alviano F, Bagnara GP, *et al.* Concise review: isolation and characterization of cells from human term placenta: outcome of the first international Workshop on Placenta Derived Stem Cells. Stem Cells 2008; 26(2): 300-11.

[http://dx.doi.org/10.1634/stemcells.2007-0594] [PMID: 17975221]

[66] Tsagias N, Koliakos I, Lappa M, Karagiannis V, Koliakos GG. Placenta perfusion has hematopoietic and mesenchymal progenitor stem cell potential. Transfusion 2011; 51(5): 976-85. [http://dx.doi.org/10.1111/j.1537-2995.2011.03077.x] [PMID: 21382046]

[67] Kieusseian A, Brunet de la Grange P, Burlen-Defranoux O, Godin I, Cumano A. Immature hematopoietic stem cells undergo maturation in the fetal liver. Development 2012; 139(19): 3521-30. [http://dx.doi.org/10.1242/dev.079210] [PMID: 22899849]

[68] Christensen JL, Wright DE, Wagers AJ, Weissman IL. Circulation and chemotaxis of fetal hematopoietic stem cells. PLoS Biol 2004; 2(3): E75. [http://dx.doi.org/10.1371/journal.pbio.0020075] [PMID: 15024423]

[69] Ema H, Nakauchi H. Expansion of hematopoietic stem cells in the developing liver of a mouse embryo. Blood 2000; 95(7): 2284-8. [PMID: 10733497]

[70] Isern J, Fraser ST, He Z, Baron MH. The fetal liver is a niche for maturation of primitive erythroid cells. Proc Natl Acad Sci USA 2008; 105(18): 6662-7. [http://dx.doi.org/10.1073/pnas.0802032105] [PMID: 18445648]

[71] Peschle C, Migliaccio AR, Migliaccio G, *et al.* Embryonic----Fetal Hb switch in humans: studies on erythroid bursts generated by embryonic progenitors from yolk sac and liver. Proc Nat Acad Sci USA 1984; 81(8): 2416-0.

[72] Durand C, Dzierzak E. Embryonic beginnings of adult hematopoietic stem cells. Haematologica 2005; 90(1): 100-8. [PMID: 15642676]

[73] Tavassoli M, Minguell JJ. Homing of hemopoietic progenitor cells to the marrow. Proc Soc Exp Biol Med 1991; 196(4): 367-73. [http://dx.doi.org/10.3181/00379727-196-43201] [PMID: 2008437]

[74] Tavian M, Hallais MF, Péault B. Emergence of intraembryonic hematopoietic precursors in the pre-liver human embryo. Development 1999; 126(4): 793-803. [PMID: 9895326]

[75] Golfier F, Bárcena A, Cruz J, Harrison M, Muench M. Mid-trimester fetal livers are a rich source of CD34+/++ cells for transplantation. Bone Marrow Transplant 1999; 24(5): 451-61. [http://dx.doi.org/10.1038/sj.bmt.1701940] [PMID: 10482927]

[76] Rollini P, Faes-Van't Hull E, Kaiser S, Kapp U, Leyvraz S. Phenotypic and functional analysis of human fetal liver hematopoietic stem cells in culture. Stem Cells Dev 2007; 16(2): 281-96. [http://dx.doi.org/10.1089/scd.2006.0096] [PMID: 17521239]

[77] Weekx SF, Van Bockstaele DR, Plum J, *et al.* CD34++ CD38- and CD34+ CD38+ human hematopoietic progenitors from fetal liver, cord blood, and adult bone marrow respond differently to hematopoietic cytokines depending on the ontogenic source. Exp Hematol 1998; 26(11): 1034-42. [PMID: 9766443]

[78] Gale RP. Fetal liver transplantation in aplastic anemia and leukemia Fetal liver transplantation. Dordrecht: Springer Netherlands 1987; pp. 89-94.

[http://dx.doi.org/10.1007/978-94-009-3365-1_9]

[79] Lucarelli G, Izzi T, Porcellini A, *et al.* Fetal liver transplantation in 2 patients with acute leukaemia after total body irradiation. Scand J Haematol 1982; 28(1): 65-71. [http://dx.doi.org/10.1111/j.1600-0609.1982.tb02122.x] [PMID: 6122263]

[80] Touraine JL. Transplantation of fetal liver stem cells into patients and into human fetuses, with induction of immunologic tolerance. Transplant Proc 1993; 25(1 Pt 2): 1012-3. [PMID: 8442024]

[81] Holyoake TL, Nicolini FE, Eaves CJ. Functional differences between transplantable human hematopoietic stem cells from fetal liver, cord blood, and adult marrow. Exp Hematol 1999; 27(9): 1418-27. [http://dx.doi.org/10.1016/S0301-472X(99)00078-8] [PMID: 10480433]

[82] Flake AW, Zanjani ED. In utero hematopoietic stem cell transplantation: ontogenic opportunities and biologic barriers. Blood 1999; 94(7): 2179-91. [PMID: 10498587]

[83] Desanti GE, Cumano A, Golub R. Identification of CD4int progenitors in mouse fetal spleen, a source of resident lymphoid cells. J Leukoc Biol 2008; 83(5): 1145-54. [http://dx.doi.org/10.1189/jlb.1107755] [PMID: 18270249]

[84] Godin I, Garcia-Porrero JA, Dieterlen-Lièvre F, Cumano A. Stem cell emergence and hemopoietic activity are incompatible in mouse intraembryonic sites. J Exp Med 1999; 190(1): 43-52. [http://dx.doi.org/10.1084/jem.190.1.43] [PMID: 10429669]

[85] Bertrand JY, Desanti GE, Lo-Man R, Leclerc C, Cumano A, Golub R. Fetal spleen stroma drives macrophage commitment. Development 2006; 133(18): 3619-28. [http://dx.doi.org/10.1242/dev.02510] [PMID: 16914502]

[86] Yanai N, Matsuya Y, Obinata M. Spleen stromal cell lines selectively support erythroid colony formation. Blood 1989; 74(7): 2391-7. [PMID: 2804371]

[87] Wolf BC, Luevano E, Neiman RS. Evidence to suggest that the human fetal spleen is not a hematopoietic organ. AJCP 1998; 80(2): 140-4. [http://dx.doi.org/10.1093/ajcp/80.2.140]

[88] Asma GE, Langlois van den Bergh R, Vossen JM. Development of pre-B and B lymphocytes in the human fetus. Clin Exp Immunol 1984; 56(2): 407-14. [PMID: 6610515]

[89] Lee J, Monson NL, Lipsky PE. The V lambda J lambda repertoire in human fetal spleen: evidence for positive selection and extensive receptor editing. J Immunol 2000; 165(11): 6322-33. [http://dx.doi.org/10.4049/jimmunol.165.11.6322] [PMID: 11086069]

[90] Sharma S, Pati HP, Ahuja RK, Takkar D, Kochupillai V. Haemopoietic cell composition of human fetal liver, spleen and thymus. Med Oncol 1997; 14(2): 99-101. [http://dx.doi.org/10.1007/BF02990954] [PMID: 9330269]

[91] Oberlin E, Tavian M, Blazsek I, Péault B. Blood-forming potential of vascular endothelium in the human embryo. Development 2002; 129(17): 4147-57.

[PMID: 12163416]

[92] Weissman IL. Stem cells: units of development, units of regeneration, and units in evolution. Cell 2000; 100(1): 157-68.
[http://dx.doi.org/10.1016/S0092-8674(00)81692-X] [PMID: 10647940]

[93] Schofield R. The relationship between the spleen colony-forming cell and the haemopoietic stem cell. Blood Cells 1978; 4(1-2): 7-25.
[PMID: 747780]

[94] Adamo L, Naveiras O, Wenzel PL, *et al.* Biomechanical forces promote embryonic haematopoiesis. Nature 2009; 459(7250): 1131-5.
[http://dx.doi.org/10.1038/nature08073] [PMID: 19440194]

[95] Wang LD, Wagers AJ. Dynamic niches in the origination and differentiation of haematopoietic stem cells. Nat Rev Mol Cell Biol 2011; 12(10): 643-55.
[http://dx.doi.org/10.1038/nrm3184] [PMID: 21886187]

[96] Nishikawa SI, Nishikawa S, Kawamoto H, *et al. in vitro* generation of lymphohematopoietic cells from endothelial cells purified from murine embryos. Immunity 1998; 8(6): 761-9.
[http://dx.doi.org/10.1016/S1074-7613(00)80581-6] [PMID: 9655490]

[97] Nishikawa SI, Nishikawa S, Hirashima M, Matsuyoshi N, Kodama H. Progressive lineage analysis by cell sorting and culture identifies FLK1+VE-cadherin+ cells at a diverging point of endothelial and hemopoietic lineages. Development 1998; 125(9): 1747-57.
[PMID: 9521912]

[98] Zovein AC, Hofmann JJ, Lynch M, *et al.* Fate tracing reveals the endothelial origin of hematopoietic stem cells. Cell Stem Cell 2008; 3(6): 625-36.
[http://dx.doi.org/10.1016/j.stem.2008.09.018] [PMID: 19041779]

[99] Belaoussoff M, Farrington SM, Baron MH. Hematopoietic induction and respecification of A-P identity by visceral endoderm signaling in the mouse embryo. Development 1998; 125(24): 5009-18.
[PMID: 9811585]

[100] Adams GB, Chabner KT, Alley IR, *et al.* Stem cell engraftment at the endosteal niche is specified by the calcium-sensing receptor. Nature 2006; 439(7076): 599-603.
[http://dx.doi.org/10.1038/nature04247] [PMID: 16382241]

[101] Kiel MJ, Morrison SJ. Maintaining hematopoietic stem cells in the vascular niche. Immunity 2006; 25(6): 862-4.
[http://dx.doi.org/10.1016/j.immuni.2006.11.005] [PMID: 17174928]

[102] Zhang J, Niu C, Ye L, *et al.* Identification of the haematopoietic stem cell niche and control of the niche size. Nature 2003; 425(6960): 836-41.
[http://dx.doi.org/10.1038/nature02041] [PMID: 14574412]

[103] Lord BI, Testa NG, Hendry JH. The relative spatial distributions of CFUs and CFUc in the normal mouse femur. Blood 1975; 46(1): 65-72.
[PMID: 1131427]

[104] Askenasy N, Farkas DL. *In vivo* imaging studies of the effect of recipient conditioning, donor cell phenotype and antigen disparity on homing of haematopoietic cells to the bone marrow. Br J Haematol

2003; 120(3): 505-15.
[http://dx.doi.org/10.1046/j.1365-2141.2003.04114.x] [PMID: 12580970]

[105] Nilsson SK, Johnston HM, Coverdale JA. Spatial localization of transplanted hemopoietic stem cells: inferences for the localization of stem cell niches. Blood 2001; 97(8): 2293-9.
[http://dx.doi.org/10.1182/blood.V97.8.2293] [PMID: 11290590]

[106] Adams GB, Scadden DT. The hematopoietic stem cell in its place. Nat Immunol 2006; 7(4): 333-7.
[http://dx.doi.org/10.1038/ni1331] [PMID: 16550195]

[107] Nombela-Arrieta C, Pivarnik G, Winkel B, *et al.* Quantitative imaging of haematopoietic stem and progenitor cell localization and hypoxic status in the bone marrow microenvironment. Nat Cell Biol 2013; 15(5): 533-43.
[http://dx.doi.org/10.1038/ncb2730] [PMID: 23624405]

[108] Visnjic D, Kalajzic I, Gronowicz G, *et al.* Conditional ablation of the osteoblast lineage in Col2.3deltatk transgenic mice. J Bone Miner Res 2001; 16(12): 2222-31.
[http://dx.doi.org/10.1359/jbmr.2001.16.12.2222] [PMID: 11760835]

[109] Visnjic D, Kalajzic Z, Rowe DW, Katavic V, Lorenzo J, Aguila HL. Hematopoiesis is severely altered in mice with an induced osteoblast deficiency. Blood 2004; 103(9): 3258-64.
[http://dx.doi.org/10.1182/blood-2003-11-4011] [PMID: 14726388]

[110] Calvi LM, Adams GB, Weibrecht KW, *et al.* Osteoblastic cells regulate the haematopoietic stem cell niche. Nature 2003; 425(6960): 841-6.
[http://dx.doi.org/10.1038/nature02040] [PMID: 14574413]

[111] Mancini SJ, Mantei N, Dumortier A, Suter U, MacDonald HR, Radtke F. Jagged1-dependent Notch signaling is dispensable for hematopoietic stem cell self-renewal and differentiation. Blood 2005; 105(6): 2340-2.
[http://dx.doi.org/10.1182/blood-2004-08-3207] [PMID: 15550486]

[112] Arai F, Hirao A, Ohmura M, *et al.* Tie2/angiopoietin-1 signaling regulates hematopoietic stem cell quiescence in the bone marrow niche. Cell 2004; 118(2): 149-61.
[http://dx.doi.org/10.1016/j.cell.2004.07.004] [PMID: 15260986]

[113] Thorén LA, Liuba K, Bryder D, *et al.* Kit regulates maintenance of quiescent hematopoietic stem cells. J Immunol 2008; 180(4): 2045-53.
[http://dx.doi.org/10.4049/jimmunol.180.4.2045] [PMID: 18250409]

[114] Sugiyama D, Tsuji K. Definitive hematopoiesis from endothelial cells in the mouse embryo; a simple guide. Trends Cardiovasc Med 2006; 16(2): 45-9.
[http://dx.doi.org/10.1016/j.tcm.2005.11.006] [PMID: 16473761]

[115] Katayama Y, Battista M, Kao W-M, *et al.* Signals from the sympathetic nervous system regulate hematopoietic stem cell egress from bone marrow. Cell 2006; 124(2): 407-21.
[http://dx.doi.org/10.1016/j.cell.2005.10.041] [PMID: 16439213]

[116] Stier S, Ko Y, Forkert R, *et al.* Osteopontin is a hematopoietic stem cell niche component that negatively regulates stem cell pool size. J Exp Med 2005; 201(11): 1781-91.
[http://dx.doi.org/10.1084/jem.20041992] [PMID: 15928197]

[117] Nilsson SK, Johnston HM, Whitty GA, *et al.* Osteopontin, a key component of the hematopoietic stem

cell niche and regulator of primitive hematopoietic progenitor cells. Blood 2005; 106(4): 1232-9. [http://dx.doi.org/10.1182/blood-2004-11-4422] [PMID: 15845900]

[118] Méndez-Ferrer S, Michurina TV, Ferraro F, *et al.* Mesenchymal and haematopoietic stem cells form a unique bone marrow niche. Nature 2010; 466(7308): 829-34. [http://dx.doi.org/10.1038/nature09262] [PMID: 20703299]

[119] Sugiyama T, Kohara H, Noda M, Nagasawa T. Maintenance of the hematopoietic stem cell pool by CXCL12-CXCR4 chemokine signaling in bone marrow stromal cell niches. Immunity 2006; 25(6): 977-88. [http://dx.doi.org/10.1016/j.immuni.2006.10.016] [PMID: 17174120]

[120] Sacchetti B, Funari A, Michienzi S, *et al.* Self-renewing osteoprogenitors in bone marrow sinusoids can organize a hematopoietic microenvironment. Cell 2007; 131(2): 324-36. [http://dx.doi.org/10.1016/j.cell.2007.08.025] [PMID: 17956733]

[121] Kunisaki Y, Bruns I, Scheiermann C, *et al.* Arteriolar niches maintain haematopoietic stem cell quiescence. Nature 2013; 502(7473): 637-43. [http://dx.doi.org/10.1038/nature12612] [PMID: 24107994]

[122] Yamazaki S, Ema H, Karlsson G, *et al.* Nonmyelinating Schwann cells maintain hematopoietic stem cell hibernation in the bone marrow niche. Cell 2011; 147(5): 1146-58. [http://dx.doi.org/10.1016/j.cell.2011.09.053] [PMID: 22118468]

[123] Nakamura-Ishizu A, Suda T. Aging of the hematopoietic stem cells niche. Int J Hematol 2014; 100(4): 317-25. [http://dx.doi.org/10.1007/s12185-014-1641-8] [PMID: 25096220]

[124] Olson TS, Caselli A, Otsuru S, *et al.* Megakaryocytes promote murine osteoblastic HSC niche expansion and stem cell engraftment after radioablative conditioning. Blood 2013; 121(26): 5238-49. [http://dx.doi.org/10.1182/blood-2012-10-463414] [PMID: 23667055]

[125] Zhao M, Perry JM, Marshall H, *et al.* Megakaryocytes maintain homeostatic quiescence and promote post-injury regeneration of hematopoietic stem cells. Nat Med 2014; 20(11): 1321-6. [http://dx.doi.org/10.1038/nm.3706] [PMID: 25326798]

[126] Fujisaki J, Wu J, Carlson AL, *et al. In vivo* imaging of Treg cells providing immune privilege to the haematopoietic stem-cell niche. Nature 2011; 474(7350): 216-9. [http://dx.doi.org/10.1038/nature10160] [PMID: 21654805]

[127] Bessis MC. L'ilot erythroblastique. Unite functionelle de la moelle osseuse. Rev Hematol (Paris) 1958; 1958(13): 8-11.

[128] Bessis MC, Breton-Gorius J. Iron metabolism in the bone marrow as seen by electron microscopy: a critical review. Blood 1962; 19(6): 635-63. [PMID: 13868561]

[129] Chasis JA, Mohandas N. Erythroblastic islands: niches for erythropoiesis. Blood 2008; 112(3): 470-8. [http://dx.doi.org/10.1182/blood-2008-03-077883] [PMID: 18650462]

[130] Chow A, Lucas D, Hidalgo A, *et al.* Bone marrow CD169+ macrophages promote the retention of hematopoietic stem and progenitor cells in the mesenchymal stem cell niche. J Exp Med 2011; 208(2): 261-71.

[http://dx.doi.org/10.1084/jem.20101688] [PMID: 21282381]

[131] Ramos P, Casu C, Gardenghi S, *et al.* Macrophages support pathological erythropoiesis in polycythemia vera and beta-thalassemia. Nat Med ; (): 1-11.

[132] Fraser ST, Midwinter RG, Coupland LA, *et al.* Heme oxygenase-1 deficiency alters erythroblastic island formation, steady-state erythropoiesis and red blood cell lifespan in mice. Haematologica 2015; 100(5): 601-10.
[http://dx.doi.org/10.3324/haematol.2014.116368] [PMID: 25682599]

[133] Tokoyoda K, Egawa T, Sugiyama T, Choi BI, Nagasawa T. Cellular niches controlling B lymphocyte behavior within bone marrow during development. Immunity 2004; 20(6): 707-18.
[http://dx.doi.org/10.1016/j.immuni.2004.05.001] [PMID: 15189736]

[134] Rodriguez Gomez M, Talke Y, Goebel N, Hermann F, Reich B, Mack M. Basophils support the survival of plasma cells in mice. J Immunol 2010; 185(12): 7180-5.
[http://dx.doi.org/10.4049/jimmunol.1002319] [PMID: 21068399]

[135] Chu VT, Fröhlich A, Steinhauser G, *et al.* Eosinophils are required for the maintenance of plasma cells in the bone marrow. Nat Immunol 2011; 12(2): 151-9.
[http://dx.doi.org/10.1038/ni.1981] [PMID: 21217761]

[136] Winter O, Moser K, Mohr E, *et al.* Megakaryocytes constitute a functional component of a plasma cell niche in the bone marrow. Blood 2010; 116(11): 1867-75.
[http://dx.doi.org/10.1182/blood-2009-12-259457] [PMID: 20538807]

[137] Anderson G, Takahama Y. Thymic epithelial cells: working class heroes for T cell development and repertoire selection. Trends Immunol 2012; 33(6): 256-63.
[http://dx.doi.org/10.1016/j.it.2012.03.005] [PMID: 22591984]

[138] Klein L, Hinterberger M, Wirnsberger G, Kyewski B. Antigen presentation in the thymus for positive selection and central tolerance induction. Nat Rev Immunol 2009; 9(12): 833-44.
[http://dx.doi.org/10.1038/nri2669] [PMID: 19935803]

[139] Nakagawa Y, Ohigashi I, Nitta T, *et al.* Thymic nurse cells provide microenvironment for secondary T cell receptor α rearrangement in cortical thymocytes. Proc Natl Acad Sci USA 2012; 109(50): 20572-7.
[http://dx.doi.org/10.1073/pnas.1213069109] [PMID: 23188800]

[140] Bubanovic IV. Failure of blood-thymus barrier as a mechanism of tumor and trophoblast escape. Med Hypotheses 2003; 60(3): 315-20.
[http://dx.doi.org/10.1016/S0306-9877(02)00450-4] [PMID: 12581602]

[141] Cui G, Hara T, Simmons S, *et al.* Characterization of the IL-15 niche in primary and secondary lymphoid organs *in vivo*. Proc Natl Acad Sci USA 2014; 111(5): 1915-20.
[http://dx.doi.org/10.1073/pnas.1318281111] [PMID: 24449915]

[142] Watanabe N, Wang Y-H, Lee HK, *et al.* Hassall's corpuscles instruct dendritic cells to induce CD4+CD25+ regulatory T cells in human thymus. Nature 2005; 436(7054): 1181-5.
[http://dx.doi.org/10.1038/nature03886] [PMID: 16121185]

[143] Alves NL, Richard-Le Goff O, Huntington ND, *et al.* Characterization of the thymic IL-7 niche *in vivo*. Proc Natl Acad Sci USA 2009; 106(5): 1512-7.

[http://dx.doi.org/10.1073/pnas.0809559106] [PMID: 19164539]

[144] Hendrix TM, Chilukuri RV, Martinez M, *et al.* Cellular Immunology. Cell Immunol 2010; 261(2): 81-92.
[http://dx.doi.org/10.1016/j.cellimm.2009.11.004] [PMID: 20035931]

[145] Sutherland JS, Goldberg GL, Hammett MV, *et al.* Activation of thymic regeneration in mice and humans following androgen blockade. J Immunol 2005; 175(4): 2741-53.
[http://dx.doi.org/10.4049/jimmunol.175.4.2741] [PMID: 16081852]

[146] Chidgey A, Dudakov J, Seach N, Boyd R. Impact of niche aging on thymic regeneration and immune reconstitution. Semin Immunol 2007; 19(5): 331-40.
[http://dx.doi.org/10.1016/j.smim.2007.10.006] [PMID: 18024073]

[147] Socolovsky M. Molecular insights into stress erythropoiesis. Curr Opin Hematol 2007; 14(3): 215-24.
[http://dx.doi.org/10.1097/MOH.0b013e3280de2bf1] [PMID: 17414210]

[148] Perry JM, Harandi OF, Paulson RF. BMP4, SCF, and hypoxia cooperatively regulate the expansion of murine stress erythroid progenitors. Blood 2007; 109(10): 4494-502.
[http://dx.doi.org/10.1182/blood-2006-04-016154] [PMID: 17284534]

[149] Lenox LE, Perry JM, Paulson RF. BMP4 and Madh5 regulate the erythroid response to acute anemia. Blood 2005; 105(7): 2741-8.
[http://dx.doi.org/10.1182/blood-2004-02-0703] [PMID: 15591122]

[150] Wu D-C, Paulson RF. Hypoxia regulates BMP4 expression in the murine spleen during the recovery from acute anemia. PLoS One 2010; 5(6): e11303.
[http://dx.doi.org/10.1371/journal.pone.0011303] [PMID: 20585586]

[151] Chow A, Huggins M, Ahmed J, *et al.* CD169$^+$ macrophages provide a niche promoting erythropoiesis under homeostasis and stress. Nat Med 2013; 19(4): 429-36.
[http://dx.doi.org/10.1038/nm.3057] [PMID: 23502962]

[152] Osgood EE, Riddle MC. Aplastic anemia treated with daily transfusions and intravenous marrow; case report. Ann Intern Med 1939; 13(2): 357-67.

[153] Morrison M, Samwick AA. Intramedullary (sternal) transfusion of human bone marrow: Preliminary report. J Am Med Assoc 1940; 115(20): 1708-11.
[http://dx.doi.org/10.1001/jama.1940.02810460040010]

[154] Thomas ED, Blume KG, Forman SJ, Appelbaum FR. Thomas' hematopoietic cell transplantation. John Wiley & Sons 2004; p. 457.

[155] Rekers PE, Coulter MP, Warren SL. Effect of transplantation of bone marrow into irradiated animals. Arch Surg 1950; 60(4): 635-67.
[http://dx.doi.org/10.1001/archsurg.1950.01250010656001]

[156] Jacobson LO, Simmons EL, Marks EK, Robson MJ, Bethard WF, Gaston EO. The role of the spleen in radiation injury and recovery. J Lab Clin Med 1950; 35(5): 746-70.
[PMID: 15422209]

[157] Lorenz E, Uphoff D, Reid TR, Shelton E. Modification of irradiation injury in mice and guinea pigs by bone marrow injections. J Natl Cancer Inst 1951; 12(1): 197-201.
[PMID: 14874130]

[158] Lorenz E, Congdon C, Uphoff D. Modification of acute irradiation injury in mice and guinea-pigs by bone marrow injections. Radiology 1952; 58(6): 863-77. [http://dx.doi.org/10.1148/58.6.863] [PMID: 14941986]

[159] Ford CE, Hamerton JL, Barnes DW, Loutit JF. Cytological identification of radiation-chimaeras. Nature 1956; 177(4506): 452-4. [http://dx.doi.org/10.1038/177452a0] [PMID: 13309336]

[160] Barnes DW, Corp MJ, Loutit JF, Neal FE. Treatment of murine leukaemia with X rays and homologous bone marrow; preliminary communication. BMJ 1956; 2(4993): 626-7. [http://dx.doi.org/10.1136/bmj.2.4993.626] [PMID: 13356034]

[161] Thomas ED, Lochte HL Jr, Lu WC, Ferrebee JW. Intravenous infusion of bone marrow in patients receiving radiation and chemotherapy. N Engl J Med 1957; 257(11): 491-6. [http://dx.doi.org/10.1056/NEJM195709122571102] [PMID: 13464965]

[162] Thomas ED, Lochte HL Jr, Cannon JH, Sahler OD, Ferrebee JW. Supralethal whole body irradiation and isologous marrow transplantation in man. J Clin Invest 1959; 38: 1709-16. [http://dx.doi.org/10.1172/JCI103949] [PMID: 13837954]

[163] Bortin MM. A compendium of reported human bone marrow transplants. Transplantation. LWW 1970; 9(6): 571-87.

[164] Billingham RE, Brent L, Medawar PB. Actively acquired tolerance of foreign cells. Nature 1953; 172(4379): 603-6. [http://dx.doi.org/10.1038/172603a0] [PMID: 13099277]

[165] Storb R, Rudolph RH, Thomas ED. Marrow grafts between canine siblings matched by serotyping and mixed leukocyte culture. J Clin Invest 1971; 50(6): 1272-5. [http://dx.doi.org/10.1172/JCI106605] [PMID: 4931083]

[166] Gatti RA, Meuwissen HJ, Allen HD, Hong R, Good RA. Immunological reconstitution of sex-linked lymphopenic immunological deficiency. Lancet 1968; 2(7583): 1366-9. [http://dx.doi.org/10.1016/S0140-6736(68)92673-1] [PMID: 4177932]

[167] Bach FH, Albertini RJ, Joo P, Anderson JL, Bortin MM. Bone-marrow transplantation in a patient with the Wiskott-Aldrich syndrome. Lancet 1968; 2(7583): 1364-6. [http://dx.doi.org/10.1016/S0140-6736(68)92672-X] [PMID: 4177931]

[168] De Koning J, Van Bekkum DW, Dicke KA, Dooren LJ, Rádl J, Van Rood JJ. Transplantation of bone-marrow cells and fetal thymus in an infant with lymphopenic immunological deficiency. Lancet 1969; 1(7608): 1223-7. [http://dx.doi.org/10.1016/S0140-6736(69)92112-6] [PMID: 4182410]

[169] Bortin MM, Bach FH, van Bekkum DW, Good RA, van Rood JJ. 25th anniversary of the first successful allogeneic bone marrow transplants. Bone Marrow Transplant 1994; 14(2): 211-2. [PMID: 7994234]

[170] Abrams RA, Glaubiger D, Appelbaum FR, Deisseroth AB. Result of attempted hematopoietic reconstitution using isologous, peripheral blood mononuclear cells: a case report. Blood 1980; 56(3): 516-20. [PMID: 6105897]

[171] Juttner CA, To LB, Haylock DN, Branford A, Kimber RJ. Circulating autologous stem cells collected in very early remission from acute non-lymphoblastic leukaemia produce prompt but incomplete haemopoietic reconstitution after high dose melphalan or supralethal chemoradiotherapy. Br J Haematol 1985; 61(4): 739-45.
[http://dx.doi.org/10.1111/j.1365-2141.1985.tb02888.x] [PMID: 2867773]

[172] Kessinger A, Armitage JO, Landmark JD, Weisenburger DD. Reconstitution of human hematopoietic function with autologous cryopreserved circulating stem cells. Exp Hematol 1986; 14(3): 192-6.
[PMID: 2868909]

[173] Copelan EA. Hematopoietic stem-cell transplantation. N Engl J Med 2006; 354(17): 1813-26.
[http://dx.doi.org/10.1056/NEJMra052638] [PMID: 16641398]

[174] Lohrmann H-P, Schreml W, Fliedner TM, Heimpel H. Reaction of human granulopoiesis to high-dose cyclophosphamide therapy. Blut 1979; 38(1): 9-16.
[http://dx.doi.org/10.1007/BF01082923] [PMID: 758963]

[175] Richman CM, Weiner RS, Yankee RA. Increase in circulating stem cells following chemotherapy in man. Blood 1976; 47(6): 1031-9.
[PMID: 1276467]

[176] Motabi IH, DiPersio JF. Blood Reviews. YBLRE 2012; 26(6): 267-78.

[177] Almici C, Carlo-Stella C, Wagner JE, Rizzoli V. Umbilical cord blood as a source of hematopoietic stem cells: from research to clinical application. Haematologica 1995; 80(5): 473-9.
[PMID: 8566892]

[178] Broxmeyer HE, Douglas GW, Hangoc G, *et al.* Human umbilical cord blood as a potential source of transplantable hematopoietic stem/progenitor cells. Proc Natl Acad Sci USA 1989; 86(10): 3828-32.
[http://dx.doi.org/10.1073/pnas.86.10.3828] [PMID: 2566997]

[179] Nakahata T, Ogawa M. Hemopoietic colony-forming cells in umbilical cord blood with extensive capability to generate mono- and multipotential hemopoietic progenitors. J Clin Invest 1982; 70(6): 1324-8.
[http://dx.doi.org/10.1172/JCI110734] [PMID: 7174797]

[180] Conconi MT, Di Liddo R, Tommasini M. Phenotype and differentiation potential of stromal populations obtained from various zones of human umbilical cord: An overview. Open Tissue Eng Regen Med J 2011; 4: 6-20.
[http://dx.doi.org/10.2174/1875043501104010006]

[181] McGuckin CP, Forraz N, Baradez MO, *et al.* Production of stem cells with embryonic characteristics from human umbilical cord blood. Cell Prolif 2005; 38(4): 245-55.
[http://dx.doi.org/10.1111/j.1365-2184.2005.00346.x] [PMID: 16098183]

[182] Forraz N, McGuckin CP. The umbilical cord: a rich and ethical stem cell source to advance regenerative medicine. Cell Prolif 2011; 44 (Suppl. 1): 60-9.
[http://dx.doi.org/10.1111/j.1365-2184.2010.00729.x] [PMID: 21481046]

[183] Gluckman E, Broxmeyer HA, Auerbach AD, *et al.* Hematopoietic reconstitution in a patient with Fanconi's anemia by means of umbilical-cord blood from an HLA-identical sibling. N Engl J Med 1989; 321(17): 1174-8.

[http://dx.doi.org/10.1056/NEJM198910263211707] [PMID: 2571931]

[184] Smith AR, Wagner JE. Alternative haematopoietic stem cell sources for transplantation: place of umbilical cord blood. Br J Haematol 2009; 147(2): 246-61.
[http://dx.doi.org/10.1111/j.1365-2141.2009.07828.x] [PMID: 19796274]

[185] Broxmeyer HE. Questions to be answered regarding umbilical cord blood hematopoietic stem and progenitor cells and their use in-transplantation. Transfusion 1995; 35(8): 694-702.
[http://dx.doi.org/10.1046/j.1537-2995.1995.35895357903.x] [PMID: 7631412]

[186] Laughlin MJ, Eapen M, Rubinstein P, *et al.* Outcomes after transplantation of cord blood or bone marrow from unrelated donors in adults with leukemia. N Engl J Med 2004; 351(22): 2265-75.
[http://dx.doi.org/10.1056/NEJMoa041276] [PMID: 15564543]

[187] Piacibello W, Sanavio F, Garetto L, *et al.* Extensive amplification and self-renewal of human primitive hematopoietic stem cells from cord blood. Blood 1997; 89(8): 2644-53.
[PMID: 9108381]

[188] Shpall EJ, Quinones R, Giller R, *et al.* Transplantation of *ex vivo* expanded cord blood. Biol Blood Marrow Transplant 2002; 8(7): 368-76.
[http://dx.doi.org/10.1053/bbmt.2002.v8.pm12171483] [PMID: 12171483]

[189] Mauch P, Constine L, Greenberger J, *et al.* Hematopoietic stem cell compartment: acute and late effects of radiation therapy and chemotherapy. Int J Radiat Oncol Biol Phys 1995; 31(5): 1319-39.
[http://dx.doi.org/10.1016/0360-3016(94)00430-S] [PMID: 7713791]

[190] Villa A, Sonis ST. Mucositis: pathobiology and management. Curr Opin Oncol 2015; 27(3): 159-64.
[http://dx.doi.org/10.1097/CCO.0000000000000180] [PMID: 25774860]

[191] Nucci M, Anaissie E. Infections after high-dose chemotherapy and autologous hematopoietic stem cell transplantation. 2015; 49-61.
[http://dx.doi.org/10.1007/978-3-662-44000-1_4]

[192] Wiles MV, Keller G. Multiple hematopoietic lineages develop from embryonic stem (ES) cells in culture. Development 1991; 111(2): 259-67.
[PMID: 1893864]

[193] Schmitt RM, Bruyns E, Snodgrass HR. Hematopoietic development of embryonic stem cells *in vitro*: cytokine and receptor gene expression. Genes Dev 1991; 5(5): 728-40.
[http://dx.doi.org/10.1101/gad.5.5.728] [PMID: 1709130]

[194] Burkert U, von Rüden T, Wagner EF. Early fetal hematopoietic development from *in vitro* differentiated embryonic stem cells. New Biol 1991; 3(7): 698-708.
[PMID: 1721533]

[195] Kennedy M, Firpo M, Choi K, *et al.* A common precursor for primitive erythropoiesis and definitive haematopoiesis. Nature 1997; 386(6624): 488-93.
[http://dx.doi.org/10.1038/386488a0] [PMID: 9087406]

[196] Choi K, Kennedy M, Kazarov A, Papadimitriou JC, Keller G. A common precursor for hematopoietic and endothelial cells. Development 1998; 125(4): 725-32.
[PMID: 9435292]

[197] Fujimoto T, Ogawa M, Minegishi N, *et al.* Step-wise divergence of primitive and definitive haematopoietic and endothelial cell lineages during embryonic stem cell differentiation. Genes Cells 2001; 6(12): 1113-27.
[http://dx.doi.org/10.1046/j.1365-2443.2001.00490.x] [PMID: 11737271]

[198] Chao R, Gong X, Wang L, Wang P, Wang Y. CD71(high) population represents primitive erythroblasts derived from mouse embryonic stem cells. Stem Cell Res (Amst) 2015; 14(1): 30-8.
[http://dx.doi.org/10.1016/j.scr.2014.11.002] [PMID: 25485690]

[199] Olivier EN, Qiu C, Velho M, Hirsch RE, Bouhassira EE. Large-scale production of embryonic red blood cells from human embryonic stem cells. Exp Hematol 2006; 34(12): 1635-42.
[http://dx.doi.org/10.1016/j.exphem.2006.07.003] [PMID: 17157159]

[200] Uzan G, Prandini MH, Rosa JP, Berthier R. Hematopoietic differentiation of embryonic stem cells: an *in vitro* model to study gene regulation during megakaryocytopoiesis. Stem Cells 1996; 14(S1) (Suppl. 1): 194-9.
[http://dx.doi.org/10.1002/stem.5530140725] [PMID: 11012221]

[201] Yokoyama Y, Suzuki T, Sakata-Yanagimoto M, *et al.* Derivation of functional mature neutrophils from human embryonic stem cells. Blood 2009; 113(26): 6584-92.
[http://dx.doi.org/10.1182/blood-2008-06-160838] [PMID: 19321863]

[202] Lieber JG, Webb S, Suratt BT, *et al.* The *in vitro* production and characterization of neutrophils from embryonic stem cells. Blood 2004; 103(3): 852-9.
[http://dx.doi.org/10.1182/blood-2003-04-1030] [PMID: 14525782]

[203] Henkel GW, McKercher SR, Yamamoto H, Anderson KL, Oshima RG, Maki RA. PU.1 but not ets-2 is essential for macrophage development from embryonic stem cells. Blood 1996; 88(8): 2917-26.
[PMID: 8874188]

[204] Subramanian A, Guo B, Marsden MD, *et al.* Macrophage differentiation from embryoid bodies derived from human embryonic stem cells. J Stem Cells 2009; 4(1): 29-45.
[PMID: 20498689]

[205] Vodyanik MA, Slukvin II. Directed differentiation of human embryonic stem cells to dendritic cells Methods in molecular biology. Totowa, NJ: Humana Press 2007; pp. 275-93.

[206] Fairchild PJ, Brook FA, Gardner RL, *et al.* Directed differentiation of dendritic cells from mouse embryonic stem cells. Curr Biol 2000; 10(23): 1515-8.
[http://dx.doi.org/10.1016/S0960-9822(00)00824-1] [PMID: 11114519]

[207] Schroeder T, Fraser ST, Ogawa M, *et al.* Recombination signal sequence-binding protein Jkappa alters mesodermal cell fate decisions by suppressing cardiomyogenesis. Proc Natl Acad Sci USA 2003; 100(7): 4018-23.
[http://dx.doi.org/10.1073/pnas.0438008100] [PMID: 12655061]

[208] Schmitt TM, de Pooter RF, Gronski MA, Cho SK, Ohashi PS, Zúñiga-Pflücker JC. Induction of T cell development and establishment of T cell competence from embryonic stem cells differentiated *in vitro*. Nat Immunol 2004; 5(4): 410-7.
[http://dx.doi.org/10.1038/ni1055] [PMID: 15034575]

[209] Müller AM, Dzierzak EA. ES cells have only a limited lymphopoietic potential after adoptive transfer

into mouse recipients. Development 1993; 118(4): 1343-51.
[PMID: 8269860]

[210] Kyba M, Perlingeiro RC, Daley GQ. HoxB4 confers definitive lymphoid-myeloid engraftment potential on embryonic stem cell and yolk sac hematopoietic progenitors. Cell 2002; 109(1): 29-37. [http://dx.doi.org/10.1016/S0092-8674(02)00680-3] [PMID: 11955444]

[211] Chan K-M, Bonde S, Klump H, Zavazava N. Hematopoiesis and immunity of HOXB4-transduced embryonic stem cell-derived hematopoietic progenitor cells. Blood 2008; 111(6): 2953-61. [http://dx.doi.org/10.1182/blood-2007-10-117366] [PMID: 18056836]

[212] Pilat S, Carotta S, Schiedlmeier B, *et al.* HOXB4 enforces equivalent fates of ES-cell-derived and adult hematopoietic cells. Proc Natl Acad Sci USA 2005; 102(34): 12101-6. [http://dx.doi.org/10.1073/pnas.0505624102] [PMID: 16093308]

[213] Takahashi K, Yamanaka S. Induction of pluripotent stem cells from mouse embryonic and adult fibroblast cultures by defined factors. Cell 2006; 126(4): 663-76. [http://dx.doi.org/10.1016/j.cell.2006.07.024] [PMID: 16904174]

[214] Deng Y, Liu Q, Luo C, *et al.* Generation of induced pluripotent stem cells from buffalo (*Bubalus bubalis*) fetal fibroblasts with buffalo defined factors. Stem Cells Dev 2012; 21(13): 2485-94. [http://dx.doi.org/10.1089/scd.2012.0018] [PMID: 22420535]

[215] Hirami Y, Osakada F, Takahashi K, *et al.* Generation of retinal cells from mouse and human induced pluripotent stem cells. Neurosci Lett 2009; 458(3): 126-31. [http://dx.doi.org/10.1016/j.neulet.2009.04.035] [PMID: 19379795]

[216] Burridge PW, Keller G, Gold JD, Wu JC. Production *of de novo* cardiomyocytes: human pluripotent stem cell differentiation and direct reprogramming. Cell Stem Cell 2012; 10(1): 16-28. [http://dx.doi.org/10.1016/j.stem.2011.12.013] [PMID: 22226352]

[217] Karumbayaram S, Novitch BG, Patterson M, *et al.* Directed differentiation of human-induced pluripotent stem cells generates active motor neurons. Stem Cells 2009; 27(4): 806-11. [http://dx.doi.org/10.1002/stem.31] [PMID: 19350680]

[218] Focosi D, Amabile G, Di Ruscio A, Quaranta P, Tenen DG, Pistello M. Induced pluripotent stem cells in hematology: current and future applications. Blood Cancer J 2014; 4(5): e211. [http://dx.doi.org/10.1038/bcj.2014.30] [PMID: 24813079]

[219] Chang K-H, Bonig H, Papayannopoulou T. Generation and characterization of erythroid cells from human embryonic stem cells and induced pluripotent stem cells: An overview. 2011; (5391): 1-10.

[220] Lachmann N, Ackermann M, Frenzel E, *et al.* Large-scale hematopoietic differentiation of human induced pluripotent stem cells provides granulocytes or macrophages for cell replacement therapies. Stem Cell Rep 2015; 4(2): 282-96. [http://dx.doi.org/10.1016/j.stemcr.2015.01.005] [PMID: 25680479]

[221] Feng Q, Shabrani N, Thon JN, *et al.* Scalable generation of universal platelets from human induced pluripotent stem cells. Stem Cell Rep 2014; 3(5): 817-31. [http://dx.doi.org/10.1016/j.stemcr.2014.09.010] [PMID: 25418726]

[222] Nakamura S, Takayama N, Hirata S, *et al.* Expandable megakaryocyte cell lines enable clinically applicable generation of platelets from human induced pluripotent stem cells. Cell Stem Cell 2014;

14(4): 535-48.
[http://dx.doi.org/10.1016/j.stem.2014.01.011] [PMID: 24529595]

[223] Menon T, Firth AL, Scripture-Adams DD, *et al.* Lymphoid regeneration from gene-corrected SCID-X1 subject-derived iPSCs. Cell Stem Cell 2015; 16(4): 367-72.
[http://dx.doi.org/10.1016/j.stem.2015.02.005] [PMID: 25772073]

[224] Amabile G, Welner RS, Nombela-Arrieta C, *et al. In vivo* generation of transplantable human hematopoietic cells from induced pluripotent stem cells. Blood 2013; 121(8): 1255-64.
[http://dx.doi.org/10.1182/blood-2012-06-434407] [PMID: 23212524]

[225] Ali A, Auvinen MK, Rautonen J. The aging population poses a global challenge for blood services. Transfusion 2010; 50(3): 584-8.
[http://dx.doi.org/10.1111/j.1537-2995.2009.02490.x] [PMID: 19912582]

[226] Daniels G, Castilho L, Flegel WA, *et al.* International society of blood transfusion committee on terminology for red blood cell surface antigens: Macao report. Vox Sang 2009; 96(2): 153-6.
[http://dx.doi.org/10.1111/j.1423-0410.2008.01133.x] [PMID: 19152607]

[227] Yazdanbakhsh K, Ware RE, Noizat-Pirenne F. Red blood cell alloimmunization in sickle cell disease: pathophysiology, risk factors, and transfusion management. Blood 2012; 120(3): 528-37.
[http://dx.doi.org/10.1182/blood-2011-11-327361] [PMID: 22563085]

[228] Reesink HW, Engelfriet CP, Schennach H, *et al.* Donors with a rare pheno (geno) type. Vox Sang 2008; 95(3): 236-53.
[http://dx.doi.org/10.1111/j.1423-0410.2008.01084.x] [PMID: 19121189]

[229] Shah S, Huang X, Cheng L. Concise review: stem cell-based approaches to red blood cell production for transfusion. Stem Cells Transl Med 2014; 3(3): 346-55.
[http://dx.doi.org/10.5966/sctm.2013-0054] [PMID: 24361925]

[230] Castro CI, Briceno JC. Perfluorocarbon-based oxygen carriers: review of products and trials. Artif Organs 2010; 34(8): 622-34.
[PMID: 20698841]

[231] Silverman TA, Weiskopf RB. Hemoglobin-based oxygen carriers: current status and future directions. Transfusion 2009; 49(11): 2495-515.
[http://dx.doi.org/10.1111/j.1537-2995.2009.02356.x] [PMID: 19903298]

[232] Bagnis C, Chiaroni J, Bailly P. Elimination of blood group antigens: hope and reality. Br J Haematol 2011; 152(4): 392-400.
[http://dx.doi.org/10.1111/j.1365-2141.2010.08561.x] [PMID: 21210778]

[233] Neildez-Nguyen TM, Wajcman H, Marden MC, *et al.* Human erythroid cells produced *ex vivo* at large scale differentiate into red blood cells *in vivo*. Nat Biotechnol 2002; 20(5): 467-72.
[http://dx.doi.org/10.1038/nbt0502-467] [PMID: 11981559]

[234] Giarratana M-C, Kobari L, Lapillonne H, *et al. Ex vivo* generation of fully mature human red blood cells from hematopoietic stem cells. Nat Biotechnol 2005; 23(1): 69-74.
[http://dx.doi.org/10.1038/nbt1047] [PMID: 15619619]

[235] Miharada K, Hiroyama T, Sudo K, Nagasawa T, Nakamura Y. Efficient enucleation of erythroblasts differentiated *in vitro* from hematopoietic stem and progenitor cells. Nat Biotechnol 2006; 24(10):

1255-6.
[http://dx.doi.org/10.1038/nbt1245] [PMID: 16980975]

[236] Lapillonne H, Kobari L, Mazurier C, *et al.* Red blood cell generation from human induced pluripotent stem cells: perspectives for transfusion medicine. Haematologica 2010; 95(10): 1651-9.
[http://dx.doi.org/10.3324/haematol.2010.023556] [PMID: 20494935]

[237] Dias J, Gumenyuk M, Kang H, *et al.* Generation of red blood cells from human induced pluripotent stem cells. Stem Cells Dev 2011; 20(9): 1639-47.
[http://dx.doi.org/10.1089/scd.2011.0078] [PMID: 21434814]

[238] Malik P, Fisher TC, Barsky LL, *et al.* An *in vitro* model of human red blood cell production from hematopoietic progenitor cells. Blood 1998; 91(8): 2664-71.
[PMID: 9531574]

[239] Timmins NE, Athanasas S, Günther M, Buntine P, Nielsen LK. Ultra-high-yield manufacture of red blood cells from hematopoietic stem cells. Tissue Eng Part C Methods 2011; 17(11): 1131-7.
[http://dx.doi.org/10.1089/ten.tec.2011.0207] [PMID: 21736478]

[240] Fujimi A, Matsunaga T, Kobune M, *et al. ex vivo* large-scale generation of human red blood cells from cord blood CD34+ cells by co-culturing with macrophages. Int J Hematol 2008; 87(4): 339-50.
[http://dx.doi.org/10.1007/s12185-008-0062-y] [PMID: 18369691]

[241] Hiroyama T, Miharada K, Sudo K, *et al.* Establishment of mouse embryonic stem cell-derived erythroid progenitor cell lines able to produce functional red blood cells. PLoS ONE 2008; 3(2): e1544-4.

[242] Lu S-J, Feng Q, Park JS, *et al.* Biologic properties and enucleation of red blood cells from human embryonic stem cells. Blood 2008; 112(12): 4475-84.
[http://dx.doi.org/10.1182/blood-2008-05-157198] [PMID: 18713948]

[243] Chang C-J, Mitra K, Koya M, *et al.* Production of embryonic and fetal-like red blood cells from human induced pluripotent stem cells. PLoS One 2011; 6(10): e25761.
[http://dx.doi.org/10.1371/journal.pone.0025761] [PMID: 22022444]

[244] Kobari L, Yates F, Oudrhiri N, *et al.* Human induced pluripotent stem cells can reach complete terminal maturation: *in vivo* and *in vitro* evidence in the erythropoietic differentiation model. Haematologica 2012; 97(12): 1795-803.
[http://dx.doi.org/10.3324/haematol.2011.055566] [PMID: 22733021]

[245] Ran D, Shia W-J, Lo M-C, *et al.* RUNX1a enhances hematopoietic lineage commitment from human embryonic stem cells and inducible pluripotent stem cells. Blood 2013; 121(15): 2882-90.
[http://dx.doi.org/10.1182/blood-2012-08-451641] [PMID: 23372166]

[246] Peyrard T, Bardiaux L, Krause C, *et al.* Banking of pluripotent adult stem cells as an unlimited source for red blood cell production: potential applications for alloimmunized patients and rare blood challenges. Transfus Med Rev 2011; 25(3): 206-16.
[http://dx.doi.org/10.1016/j.tmrv.2011.01.002] [PMID: 21377319]

[247] Szabo E, Rampalli S, Risueño RM, *et al.* Direct conversion of human fibroblasts to multilineage blood progenitors. Nature 2010; 468(7323): 521-6.
[http://dx.doi.org/10.1038/nature09591] [PMID: 21057492]

[248] Leberbauer C, Boulmé F, Unfried G, Huber J, Beug H, Müllner EW. Different steroids co-regulate long-term expansion *versus* terminal differentiation in primary human erythroid progenitors. Blood 2005; 105(1): 85-94.
[http://dx.doi.org/10.1182/blood-2004-03-1002] [PMID: 15358620]

[249] Migliaccio AR. Erythroblast enucleation. Haematologica 2010; 95(12): 1985-8.
[http://dx.doi.org/10.3324/haematol.2010.033225] [PMID: 21123437]

[250] Nakamura Y. Establishment of immortalized human erythroid progenitor cell lines able to produce enucleated red blood cells. ISBT Sci Ser 2012; 7(1): 202-6.
[http://dx.doi.org/10.1111/j.1751-2824.2012.01542.x]

[251] Kitamura T, Tange T, Terasawa T, *et al.* Establishment and characterization of a unique human cell line that proliferates dependently on GM-CSF, IL-3, or erythropoietin. J Cell Physiol 1989; 140(2): 323-34.
[http://dx.doi.org/10.1002/jcp.1041400219] [PMID: 2663885]

[252] Wong S, Keyvanfar K, Wan Z, Kajigaya S, Young NS, Zhi N. Establishment of an erythroid cell line from primary CD36+ erythroid progenitor cells. Exp Hematol 2010; 38(11): 994-1005.e1, 2.
[http://dx.doi.org/10.1016/j.exphem.2010.07.012] [PMID: 20696208]

[253] Carotta S, Pilat S, Mairhofer A, *et al.* Directed differentiation and mass cultivation of pure erythroid progenitors from mouse embryonic stem cells. Blood 2004; 104(6): 1873-80.
[http://dx.doi.org/10.1182/blood-2004-02-0570] [PMID: 15166028]

[254] Huang X, Shah S, Wang J, *et al.* Extensive *ex vivo* expansion of functional human erythroid precursors established from umbilical cord blood cells by defined factors. Mol Ther 2014; 22(2): 451-63.
[PMID: 24002691]

[255] Aiuti A, Cattaneo F, Galimberti S, *et al.* Gene therapy for immunodeficiency due to adenosine deaminase deficiency. N Engl J Med 2009; 360(5): 447-58.
[http://dx.doi.org/10.1056/NEJMoa0805817] [PMID: 19179314]

[256] Gaspar HB, Bjorkegren E, Parsley K, *et al.* Successful reconstitution of immunity in ADA-SCID by stem cell gene therapy following cessation of PEG-ADA and use of mild preconditioning. Mol Ther 2006; 14(4): 505-13.
[http://dx.doi.org/10.1016/j.ymthe.2006.06.007] [PMID: 16905365]

[257] Kohn DB, Candotti F. Gene therapy fulfilling its promise. N Engl J Med 2009; 360(5): 518-21.
[http://dx.doi.org/10.1056/NEJMe0809614] [PMID: 19179320]

[258] Ott MG, Schmidt M, Schwarzwaelder K, *et al.* Correction of X-linked chronic granulomatous disease by gene therapy, augmented by insertional activation of MDS1-EVI1, PRDM16 or SETBP1. Nat Med 2006; 12(4): 401-9.
[http://dx.doi.org/10.1038/nm1393] [PMID: 16582916]

[259] Stein S, Ott MG, Schultze-Strasser S, *et al.* Genomic instability and myelodysplasia with monosomy 7 consequent to EVI1 activation after gene therapy for chronic granulomatous disease. Nat Med 2010; 16(2): 198-204.
[http://dx.doi.org/10.1038/nm.2088] [PMID: 20098431]

[260] Kang EM, Choi U, Theobald N, *et al.* Retrovirus gene therapy for X-linked chronic granulomatous

disease can achieve stable long-term correction of oxidase activity in peripheral blood neutrophils. Blood 2010; 115(4): 783-91.
[http://dx.doi.org/10.1182/blood-2009-05-222760] [PMID: 19965657]

[261] Boztug K, Schmidt M, Schwarzer A, *et al.* Stem-cell gene therapy for the Wiskott-Aldrich syndrome. N Engl J Med 2010; 363(20): 1918-27.
[http://dx.doi.org/10.1056/NEJMoa1003548] [PMID: 21067383]

[262] Cartier N, Hacein-Bey-Abina S, Bartholomae CC, *et al.* Hematopoietic stem cell gene therapy with a lentiviral vector in X-linked adrenoleukodystrophy. Science 2009; 326(5954): 818-23.
[http://dx.doi.org/10.1126/science.1171242] [PMID: 19892975]

[263] Cartier N, Aubourg P. Hematopoietic stem cell transplantation and hematopoietic stem cell gene therapy in X-linked adrenoleukodystrophy. Brain Pathol 2010; 20(4): 857-62.
[http://dx.doi.org/10.1111/j.1750-3639.2010.00394.x] [PMID: 20626747]

[264] Borgna-Pignatti C, Rugolotto S, De Stefano P, *et al.* Survival and complications in patients with thalassemia major treated with transfusion and deferoxamine. Haematologica 2004; 89(10): 1187-93.
[PMID: 15477202]

[265] Modell B, Khan M, Darlison M. Survival in β-thalassaemia major in the UK: data from the UK Thalassaemia Register. Lancet 2000; 355(9220): 2051-2.
[http://dx.doi.org/10.1016/S0140-6736(00)02357-6] [PMID: 10885361]

[266] Angelucci E, Matthes-Martin S, Baronciani D, *et al.* Hematopoietic stem cell transplantation in thalassemia major and sickle cell disease: indications and management recommendations from an international expert panel. Haematologica 2014; 99(5): 811-20.
[http://dx.doi.org/10.3324/haematol.2013.099747] [PMID: 24790059]

[267] Thomas ED, Sanders JE, Buckner CD, Papayannopoulou T, Borgna-Pignatti C, De Stefano P, *et al.* Marrow transplantation for thalassaemia. The Lancet Elsevier. 1982; 320: pp. (8292)227-9.

[268] Issaragrisil S, Visuthisakchai S, Suvatte V, *et al.* Brief report: transplantation of cord-blood stem cells into a patient with severe thalassemia. N Engl J Med 1995; 332(6): 367-9.
[http://dx.doi.org/10.1056/NEJM199502093320605] [PMID: 7823998]

[269] Locatelli F, Rocha V, Reed W, *et al.* Related umbilical cord blood transplantation in patients with thalassemia and sickle cell disease. Blood 2003; 101(6): 2137-43.
[http://dx.doi.org/10.1182/blood-2002-07-2090] [PMID: 12424197]

[270] Raftopoulos H, Ward M, Bank A. High-level transfer and long-term expression of the human β-globin gene in a mouse transplant model. Ann N Y Acad Sci 1998; 850(1): 178-90.
[http://dx.doi.org/10.1111/j.1749-6632.1998.tb10474.x] [PMID: 9668539]

[271] Vunjak-Novakovic G, Lui KO, Tandon N, Chien KR. Bioengineering heart muscle: a paradigm for regenerative medicine. Annu Rev Biomed Eng 2011; 13(1): 245-67.
[http://dx.doi.org/10.1146/annurev-bioeng-071910-124701] [PMID: 21568715]

[272] Karlsson S, Bodine DM, Perry L. Expression of the human beta-globin gene following retroviral-mediated transfer into multipotential hematopoietic progenitors of mice.
[http://dx.doi.org/10.1073/pnas.85.16.6062]

[273] Rivella S, May C, Chadburn A, Rivière I, Sadelain M. A novel murine model of Cooley anemia and its

rescue by lentiviral-mediated human β-globin gene transfer. Blood 2003; 101(8): 2932-9. [http://dx.doi.org/10.1182/blood-2002-10-3305] [PMID: 12480689]

[274] Imren S, Payen E, Westerman KA, *et al.* Permanent and panerythroid correction of murine β thalassemia by multiple lentiviral integration in hematopoietic stem cells. Proc Natl Acad Sci USA 2002; 99(22): 14380-5. [http://dx.doi.org/10.1073/pnas.212507099] [PMID: 12391330]

[275] Lacerra G, Sierakowska H, Carestia C, *et al.* Restoration of hemoglobin A synthesis in erythroid cells from peripheral blood of thalassemic patients. Proc Natl Acad Sci USA 2000; 97(17): 9591-6. [http://dx.doi.org/10.1073/pnas.97.17.9591] [PMID: 10944225]

[276] Puthenveetil G, Scholes J, Carbonell D, *et al.* Successful correction of the human β-thalassemia major phenotype using a lentiviral vector. Blood 2004; 104(12): 3445-53. [http://dx.doi.org/10.1182/blood-2004-04-1427] [PMID: 15292064]

[277] Cavazzana M, Ribeil J-A, Payen E, *et al.* Outcomes of gene therapy for beta thalassemia major *via* transplantation of autologous hematopoietic stem cells transduced *ex vivo* with a lentiviral beta A T87Q globin vector. Blood 2014; 25(11): A22-.

[278] Ye L, Chang JC, Lin C, Sun X, Yu J, Kan YW. Induced pluripotent stem cells offer new approach to therapy in thalassemia and sickle cell anemia and option in prenatal diagnosis in genetic diseases. Proc Natl Acad Sci USA 2009; 106(24): 9826-30. [http://dx.doi.org/10.1073/pnas.0904689106] [PMID: 19482945]

[279] Wang Y, Jiang Y, Liu S, Sun X, Gao S. Generation of induced pluripotent stem cells from human β-thalassemia fibroblast cells. Cell Res 2009; 19(9): 1120-3. [http://dx.doi.org/10.1038/cr.2009.100] [PMID: 19690515]

[280] Wang Z, Vogel O, Kuhn G, Gassmann M, Vogel J. Decreased stability of erythroblastic islands in integrin β3-deficient mice. Physiol Rep 2013; 1(2): e00018. [http://dx.doi.org/10.1002/phy2.18] [PMID: 24303107]

[281] Hanna J, Wernig M, Markoulaki S, *et al.* Treatment of sickle cell anemia mouse model with iPS cells generated from autologous skin. Science 2007; 318(5858): 1920-3. [http://dx.doi.org/10.1126/science.1152092] [PMID: 18063756]

[282] Zou J, Mali P, Huang X, Dowey SN, Cheng L. Site-specific gene correction of a point mutation in human iPS cells derived from an adult patient with sickle cell disease. Blood 2011; 118(17): 4599-608. [http://dx.doi.org/10.1182/blood-2011-02-335554] [PMID: 21881051]

[283] Allen TD, Dexter TM. Ultrastructural aspects of erythropoietic differentiation in long-term bone marrow culture. Differentiation 1982; 21(2): 86-94. [http://dx.doi.org/10.1111/j.1432-0436.1982.tb01201.x] [PMID: 7084572]

[284] Bhatia SN, Ingber DE. Microfluidic organs-on-chips. Nat Biotechnol 2014; 32(8): 760-72. [http://dx.doi.org/10.1038/nbt.2989] [PMID: 25093883]

[285] Meissner P, Schröder B, Herfurth C, Biselli M. Development of a fixed bed bioreactor for the expansion of human hematopoietic progenitor cells. Cytotechnology 1999; 30(1-3): 227-34. [http://dx.doi.org/10.1023/A:1008085932764] [PMID: 19003372]

[286] Noll T, Jelinek N, Schmid S, Biselli M, Wandrey C. Cultivation of hematopoietic stem and progenitor

cells: biochemical engineering aspects. Adv Biochem Eng Biotechnol 2002; 74: 111-28. [http://dx.doi.org/10.1007/3-540-45736-4_6] [PMID: 11991176]

[287] Zandstra PW, Jervis E, Haynes CA, Kilburn DG, Eaves CJ, Piret JM. Concentration-dependent internalization of a cytokine/cytokine receptor complex in human hematopoietic cells. Biotechnol Bioeng 1999; 63(4): 493-501. [http://dx.doi.org/10.1002/(SICI)1097-0290(19990520)63:4<493::AID-BIT13>3.0.CO;2-0] [PMID: 10099630]

[288] Koller MR, Emerson SG, Palsson BO. Large-scale expansion of human stem and progenitor cells from bone marrow mononuclear cells in continuous perfusion cultures. Blood 1993; 82(2): 378-84. [PMID: 8329697]

[289] Koller MR, Palsson MA, Manchel I, Maher RJ, Palsson BO. Tissue culture surface characteristics influence the expansion of human bone marrow cells. Biomaterials 1998; 19(21): 1963-72. [http://dx.doi.org/10.1016/S0142-9612(98)00101-X] [PMID: 9863530]

[290] Jiang J, Papoutsakis ET. Stem-cell niche based comparative analysis of chemical and nano-mechanical material properties impacting *ex vivo* expansion and differentiation of hematopoietic and mesenchymal stem cells. Adv Healthc Mater 2013; 2(1): 25-42. [http://dx.doi.org/10.1002/adhm.201200169] [PMID: 23184458]

[291] Mantalaris A, Keng P, Bourne P, Chang AY, Wu JH. Engineering a human bone marrow model: a case study on *ex vivo* erythropoiesis. Biotechnol Prog 1998; 14(1): 126-33. [http://dx.doi.org/10.1021/bp970136+] [PMID: 9496677]

[292] Di Buduo CA, Wray LS, Tozzi L, *et al.* Programmable 3D silk bone marrow niche for platelet generation *ex vivo* and modeling of megakaryopoiesis pathologies. Blood 2015; 125(14): 2254-64. [http://dx.doi.org/10.1182/blood-2014-08-595561] [PMID: 25575540]

[293] Xiong F, Chen Z, Liu H, *et al. ex vivo* expansion of human umbilical cord blood hematopoietic progenitor cells in a novel three-dimensional culture system. Biotechnol Lett 2002; 24: 1421-6. [http://dx.doi.org/10.1023/A:1019802615710]

[294] Braccini A, Wendt D, Jaquiery C, *et al.* Three-dimensional perfusion culture of human bone marrow cells and generation of osteoinductive grafts. Stem Cells 2005; 23(8): 1066-72. [http://dx.doi.org/10.1634/stemcells.2005-0002] [PMID: 16002780]

[295] Scherberich A, Galli R, Jaquiery C, Farhadi J, Martin I. Three-dimensional perfusion culture of human adipose tissue-derived endothelial and osteoblastic progenitors generates osteogenic constructs with intrinsic vascularization capacity. Stem Cells 2007; 25(7): 1823-9. [http://dx.doi.org/10.1634/stemcells.2007-0124] [PMID: 17446558]

[296] Hull CW. Apparatus for production of three-dimensional objects by stereolithography. US Patent Office; 1986.

[297] Murphy SV, Atala A. 3D bioprinting of tissues and organs. Nat Biotechnol 2014; 32(8): 773-85. [http://dx.doi.org/10.1038/nbt.2958] [PMID: 25093879]

[298] Torisawa Y-S, Spina CS, Mammoto T, *et al.* Bone marrow-on-a-chip replicates hematopoietic niche physiology *in vitro*. Nat Methods 2014; 11(6): 663-9. [http://dx.doi.org/10.1038/nmeth.2938] [PMID: 24793454]

[299] Huh D, Hamilton GA, Ingber DE. From 3D cell culture to organs-on-chips. Trends Cell Biol 2011; 21(12): 745-54.
[http://dx.doi.org/10.1016/j.tcb.2011.09.005] [PMID: 22033488]

[300] Zhang C, Zhao Z, Abdul Rahim NA, van Noort D, Yu H. Towards a human-on-chip: culturing multiple cell types on a chip with compartmentalized microenvironments. Lab Chip 2009; 9(22): 3185-92.
[http://dx.doi.org/10.1039/b915147h] [PMID: 19865724]

[301] Krupnick AS, Shaaban A, Radu A, Flake AW. Bone marrow tissue engineering. Tissue Eng 2002; 8(1): 145-55.
[http://dx.doi.org/10.1089/107632702753503135] [PMID: 11886662]

[302] Thon JN, Mazutis L, Wu S, *et al.* Platelet bioreactor-on-a-chip. Blood 2014; 124(12): 1857-67.
[http://dx.doi.org/10.1182/blood-2014-05-574913] [PMID: 25606631]

[303] Zhang W, Lee WY, Siegel DS, Tolias P, Zilberberg J. Patient-specific 3D microfluidic tissue model for multiple myeloma. Tissue Eng Part C Methods 2014; 20(8): 663-70.
[http://dx.doi.org/10.1089/ten.tec.2013.0490] [PMID: 24294886]

[304] Green MD, Chen A, Nostro M-C, *et al.* Novel approaches for immune reconstitution and adaptive immune modeling with human pluripotent stem cells. Nat Biotechnol 2011; 29(3): 267-72.
[http://dx.doi.org/10.1038/nbt.1788] [PMID: 21358635]

[305] Davies EG. Immunodeficiency in DiGeorge syndrome and options for treating cases with complete athymia. 2013; 1–9.
[http://dx.doi.org/10.3389/fimmu.2013.00322]

[306] Tomlinson MJ. Tissue engineering the thymus and secondary lymphoid organs. Tissue Eng Journal of Postdoctoral Research 2013; 1(5): 12-27.

[307] Bennett AR, Farley A, Blair NF, Gordon J, Sharp L, Blackburn CC. Identification and characterization of thymic epithelial progenitor cells. Immunity 2002; 16(6): 803-14.
[http://dx.doi.org/10.1016/S1074-7613(02)00321-7] [PMID: 12121662]

[308] Blackburn CC, Manley NR, Palmer DB, Boyd RL, Anderson G, Ritter MA. One for all and all for one: thymic epithelial stem cells and regeneration. Trends Immunol 2002; 23(8): 391-5.
[http://dx.doi.org/10.1016/S1471-4906(02)02265-2] [PMID: 12133801]

[309] Montel-Hagen A, Chung B, Ge S, *et al.* Engineering the human thymic microenvironment to support thymopoiesis. Blood 2013; 122(21): 3494-4.

[310] Sun X, Xu J, Lu H, *et al.* Directed differentiation of human embryonic stem cells into thymic epithelial progenitor-like cells reconstitutes the thymic microenvironment *in vivo*. Cell Stem Cell 2013; 13(2): 230-6.
[http://dx.doi.org/10.1016/j.stem.2013.06.014] [PMID: 23910085]

[311] Tan JK, Watanabe T. Artificial engineering of secondary lymphoid organs. Adv Immunol 2010; 105: 131-57.
[http://dx.doi.org/10.1016/S0065-2776(10)05005-4] [PMID: 20510732]

[312] Kang JH, Super M, Yung CW, *et al.* An extracorporeal blood-cleansing device for sepsis therapy. Nat Med 2014; 20(10): 1211-6.

[http://dx.doi.org/10.1038/nm.3640] [PMID: 25216635]

[313] Rigat-Brugarolas LG, Elizalde-Torrent A, Bernabeu M, *et al.* A functional microengineered model of the human splenon-on-a-chip. Lab Chip 2014; 14(10): 1715-24. [http://dx.doi.org/10.1039/c3lc51449h] [PMID: 24663955]

[314] Cupedo T, Stroock A, Coles M. Application of tissue engineering to the immune system: development of artificial lymph nodes. Front Immunol 2012; 3: 343. [http://dx.doi.org/10.3389/fimmu.2012.00343] [PMID: 23162557]

[315] Giese C, Demmler CD, Ammer R, *et al.* A human lymph node *in vitro*--challenges and progress. Artif Organs 2006; 30(10): 803-8. [http://dx.doi.org/10.1111/j.1525-1594.2006.00303.x] [PMID: 17026580]

[316] Suematsu S, Watanabe T. Generation of a synthetic lymphoid tissue-like organoid in mice. Nat Biotechnol 2004; 22(12): 1539-45. [http://dx.doi.org/10.1038/nbt1039] [PMID: 15568019]

[317] Okamoto N, Chihara R, Shimizu C, Nishimoto S, Watanabe T. Artificial lymph nodes induce potent secondary immune responses in naive and immunodeficient mice. J Clin Invest 2007; 117(4): 997-1007. [http://dx.doi.org/10.1172/JCI30379] [PMID: 17364025]

[318] Choi RS, Vacanti JP. Preliminary studies of tissue-engineered intestine using isolated epithelial organoid units on tubular synthetic biodegradable scaffolds. Transplant Proc 1997; 29(1-2): 848-51. [http://dx.doi.org/10.1016/S0041-1345(96)00164-9] [PMID: 9123551]

[319] Perez A, Grikscheit TC, Blumberg RS, Ashley SW, Vacanti JP, Whang EE. Tissue-engineered small intestine: ontogeny of the immune system. Transplantation 2002; 74(5): 619-23. [http://dx.doi.org/10.1097/00007890-200209150-00006] [PMID: 12352876]

CHAPTER 4

Cell-based Therapy in Veterinary Medicine

Chad Maki, Thomas Ramos and **Fariborz Izadyar***

VetCell Therapeutics, 2917 Daimler Street, Santa Ana, CA 92705, USA

Abstract: Cell-based therapy is a growing field in veterinary medicine and has created a lot of hope and excitements for developing new therapies for degenerative diseases that otherwise cannot be treated by traditional medical approaches. Clinical studies in dogs, cats and horses show promising results indicating that stem cells and other cell-based products may facilitate tissue repair and improve quality of life in companion animals. In this review, different cell based therapies, their risks and benefits and their possible therapeutic use for veterinary medical application will be discussed.

Keywords: Companion animals, Degenerative diseases, Stem cells, Veterinary medicine.

RECENT ADVANCES AND APPLICATIONS OF CELL THERAPY

Cell therapy is the administration of live cells to the body of a recipient for treatment of a medical condition. It could be for replacement of absent cells like infusion of red blood cells to overcome anemia, administration of platelets for emergency thrombocytopenia, or application of T lymphocytes for regulation of the immune system and cancer therapy. This type of cell therapy is called cell replacement therapy and is commonly used in medical and veterinary medical protocols. Cell therapy can also be used for regenerative medical applications. In that case, stem cells and progenitor cells with the ability to produce other cells will be used. A good example is the administration of bone marrow cells, which contain hematopoietic stem cells for regeneration of the blood. This therapy can help reestablish hematopoiesis and regenerate blood in patients undergoing

* **Corresponding author Fariborz Izadyar:** VetCell Therapeutics, 2917 Daimler Street, Santa Ana, CA 92705, USA; Tel: 949-428-0511; E-mail: fizadyar@vetcelltherapeutics.com

Atta-ur-Rahman & Shazia Anjum (Eds.)

cytotoxic treatments such as chemotherapy or radiotherapy. Cell products with regenerative potential can regenerate cells, tissues and organs, to restore or establish normal function. In addition to live cells, regeneration could also be triggered by proteins and growth factors, extracellular matrices, scaffolds and small molecules, but these effects are generally transient, and repeated treatments are required.

The nomenclature to identify and describe various cells is to use the cluster of differentiation (abbreviated as CD) distinction. It refers to protein structures (or antigens) embedded on the outer membrane of cells. The system was first described to identify basic immune cells and since has been applied to many cell types relevant for the immune system. Over the past 30 years, data generated by the Human Leukocyte Differentiation Antigens Workshops have led to the characterization and formal designation of more than 400 CD molecules. These CD molecules are commonly used as cell markers, allowing the identification and isolation of leukocyte populations, subsets and differentiation stages. Some of these markers are commonly expressed by other cells including stem cells (Table **1**).

Table 1. Some of the important CD molecules and their expression profile.

CD1	The first-named CD; this complex glycoprotein is expressed in immature T-cells, some B cells and other, specialized immune cells in the skin.
CD3	A multimeric protein complex, known historically as the T3 complex, and is part of the T cell receptor.
CD4	A molecule on a mature “helper” T lymphocyte cell surface.
CD8	A molecule on a mature “cytotoxic” T lymphocyte cell surface.
CD19	A molecule on a mature B lymphocyte cell surface.
CD34	A monomeric cell surface antigen that is selectively expressed on hematopoietic progenitor cells.
CD90	Also known as Thy-1. CD90 is expressed on neuronal cells, a subset of CD34+ cells, activated endothelial cells and mesenchymal stem cells.

The types of cells used in cell therapy can be mature cells, such a T-cells or dendritic cells, to adult stem cells isolated from a fresh tissue source. An up and coming exciting use of mature cells is that of regulatory T-cells (Tregs) in therapeutic applications. CD4+Foxp3+ Tregs are long-lived cells that suppress immune responses *in vivo* in a dominant and antigen-specific manner. Tregs confer long-term protection against auto-inflammatory diseases in mouse models and have been shown to be effective in suppressing alloimmunity in models of graft-*versus*-host disease [1]. CD4+ Foxp3+ Treg therapy is now at the point to evaluate its safety and efficacy within preclinical testing in humans [2, 3].

Another interesting mature cell type being explored for cell therapy is the dendritic cell. Dendritic cells are highly specialized, bone marrow-derived antigen-presenting cells that induce or regulate innate and adaptive immunity. In general, dendritic cells express CD11b and CD11c, although there are many more markers to identify subsets of dendritic cells. Since the mid-1990s, dendritic cells have been used in clinical trials as cellular mediators for therapeutic vaccinations of patients with cancer. As sentinel members of the innate immune arm, dendritic cells secrete protective cytokines (Table **2**) in response to signals of inflammation [4]. Interleukins (ILs) are specific types of cytokines that are produced by leukocytes for regulating immune responses. IL-6 and IL-12 are particularly important because they play roles in establishing a local immune response. For example, IL-6 secreted from dendritic cells has immunosuppressive properties and IL-12 augments CD8 T cell activation. Dendritic cells capture process and present antigens *via* the major histocompatibility complex to naïve T-cells at lymphoid organs, thereby inducing adaptive CD4+ and CD8+ T-cell-mediated immune responses [5, 6]. Dendritic cell based therapeutic immunotherapies with oncogene inhibitors in patients appear to be the method of choice. Human clinical trials investigating targetable tumors are underway in renal cell carcinoma, prostate cancer, breast cancer and melanoma [7, 8].

Table 2. Some of the important Cytokines produced and their function.

Cytokine	Full Name	Characteristic
IL-2	Interleukin-2	Promotes T cell expansion, used in immunotherapy to support large numbers of tumor-infiltrating lymphocytes with anti-cancer activity.
IL-6	Interleukin-6	Multifunctional cytokine involved in modulating various physiological events, such as cell proliferation, differentiation, survival, and apoptosis.
IL-10	Interleukin-10	Produced by cell types that mediate anti-inflammatory activities, induce regulatory T-cells, and are involved in immunosuppression and tissue repair.

(Table 2) contd.....

Cytokine	Full Name	Characteristic
IL-12	Interleukin-12	Regulates inflammation by linking innate and adaptive immune responses. Most of the IL-12-induced effects are mediated by the secretion of interferon -γ.
VEGF	Vascular Endothelial Growth Factor	Angiogenic cytokine with immunosuppressive properties and a major mediator of angiogenesis.
INF- γ	Interferon-gamma	Known to enhance the immunosuppressive properties of MSCs.
TNF-α	Tumor Necrosis Factor-alpha	Multifunctional cytokine involved in apoptosis, cell survival, inflammation and immunity, used with T cell adaptive immunotherapies.

To improve the potency of cells with a therapeutic potential many ongoing trials are focusing on the combination of cytokines along with cells [9]. For instance, CD4+Foxp3+ Tregs have had positive results with the addition of IL-2 in clinical cancer and HIV trials [10 - 12]. Immune responses of Tregs with the clinical cytokines includes expression of inducible T-cell co-stimulator (ICOS) [13]; elevated levels of CD39, CD73 and transforming growth factor β1 (TGFβ1) [13]; synergistic immunostimulatory effects [14]; and establishment of Treg-dependent tolerance by intestinal macrophages [15].

A large proportion of cell therapy is in development with various types of stem cells. Recent advances in stem cell biology and cell culture techniques have facilitated the development of cell therapy for translational medicine. Stem cell therapies are a type of cell therapy in which the cells used are either stem cells, as in the case of bone marrow and adipose stem cells, or are derived from stem cells, as in the case with cardiomyocytes [16], retinal pigmented epithelial cells [17] and induced pluripotent stem cells [18, 19].

Adult stem cells have been used in pilot studies as potential cell-based therapy for various diseases. Certain types of adult stem cells are better candidates for therapeutic applications, as compared to other stem cell types. To date, the most common types of adult stem cells used are bone marrow stem cells, mobilized hematopoietic stem cells (*i.e.* CD34+) and adipose stem cells. These particular stem cell lineages are easily harvested from patients, have a high capacity of cell proliferation in culture, are able to migrate to host target tissues and have the ability to integrate into host tissues to interact with the surrounding environment [20, 21].

The motivation to derive therapies from BM, HSC and ASCs is chiefly due to their relative abundance and minimal ethical concerns. Embryonic stem cells (ESCs) are derived from blastocysts, which are part of 7-11 day old embryos. Although these have been touted as the most potent cell of choice [21], the therapeutic benefits are clouded by a controversy owing to the belief that the process of extraction of totipotent ESCs from an embryo destroys the embryo itself. Additionally, it is difficult to control the growth and differentiation of ESCs elevating the risk of tumorogenicity and teratoma formation. As such, ESCs are not viewed as therapeutic cells, but have fundamentally remained a strong platform for basic stem cell research.

The use of induced pluripotent stem cells (iPSCs) as a therapeutic resource comes with concerns from a safety perspective [22]. Similar to ESCs, they retain the ability to differentiate into other cell types, but without the ethical concerns since they are derived from somatic cells. Originally, iPSCs were generated from mouse fibroblasts by retroviral introduction of the transcription factors *Oct4, Sox2, c-Myc*, and *Klf4* [23]. The actions of these 4 transcription factors are thought to reprogram somatic cells, including fibroblasts, into embryonic stem cell (ESC)-like pluripotent cells [23]. By definition, pluripotent stem cells are an aggressive cell type and require tumor formation as a qualifying characteristic. Recent advances in iPS cell technology have generated different methods to generate these cells and each type harbors different biological properties [24 - 27]. With these new advances, the safety concerns must be mitigated through a comprehensive examination of the quality of both iPS cells and iPS cell-derived cells. The stability of genetic and epigenetic status of iPS cells needs to be determined, as well as the differentiation capability both *in vitro* and *in vivo*, and their tumorigenicity properties. All of these biologic safety points must be evaluated before any clinical application of iPSC-based cell therapy can be realized.

CELL THERAPY AND REGENERATIVE MEDICINE

Regenerative medicine is a game-changing area of medicine with the potential to heal and restore damaged tissues and organs, offering solutions and hope for patients who have conditions that today are beyond repair. Some animals can regenerate complex body parts with full function after amputation or injury. Invertebrates (animals without a spinal cord), such as the flatworm or planarian, can regenerate the head from a tail-piece, and the tail from a head-piece. Among vertebrates (animals with a spinal cord), fish can regenerate parts of the brain, eye, kidney, heart and fins. Frogs can regenerate the limb, tail, brain and eye tissue as tadpoles but not as adults. Salamanders can regenerate the limb, heart, tail, brain, eye tissues, kidney, brain and spinal cord throughout life. Recent research in different regenerating animals has shown that there are various stem cell programs for regenerating body parts built from multiple tissues, such as muscle, nerve and skin. Understanding how stem cell regeneration works in these animals could help in unraveling the principle mechanism of regeneration in mammals [28, 29].

Discoveries of stem cells in mammalian tissues have created considerable excitement in the field of medicine and veterinary medicine because they potentially could regenerate damaged tissues for which current therapeutic strategies have minimal effectiveness. Stem cells are defined as cells that have clonogenic, self-renewing capacities and the capability to differentiate into multiple cell lineages. Whereas embryonic stem cells are derived from mammalian embryos in the blastocyst stage and can generate terminally differentiated cells of all three embryonic germ layers, adult stem cells are capable of maintaining, generating and replacing terminally differentiated cells within their own specific tissue as a consequence of physiologic cell turnover or tissue injury. The traditional idea of organ-restricted stem-cell differentiation is now being challenged by the suggestion that adult stem cells retain developmental plasticity. Numerous *in vitro* studies revealed that hematopoietic and adipose-derived stem cells could not only make blood and adipocytes, they also can be differentiated into other cell types such as chondrocytes, osteocytes and myocytes [30, 31]. Preclinical and clinical studies provide evidence that in addition to tissue-specific stem cells, within the blood there are circulating progenitor cells that not only differentiate into hematopoietic cells, but also can participate in the repair and replacement of solid organ tissues. These circulating stem cells or so called multipotent adult progenitor cells [32, 33] are small cells (< 5 μm) and can easily migrate to every organ within the body through capillaries and repair damaged tissues. The mechanism by which adult stem cells repair different tissue types is not fully understood. In addition to the occurrence of cell fusion, there are 4 suggested mechanisms of adult stem cell differentiation into solid organ cells. Preclinical data support that transdifferentiation is the most likely model, allowing stem/progenitor cells to differentiate across lineage, tissue, and germ layer boundaries [34]. There is increasing evidence that manipulations to *in vivo* circulating adult stem cells can be made to repair or regenerate solid organ tissue, which offers potential clinical benefits in the treatment of many hereditary and acquired diseases [35, 36].

Preclinical studies in laboratory animals are leading us to believe that in the future stem cell therapy can become a routine treatment protocol for many debilitating diseases in companion animals but, before this becomes a reality, large scale clinical trials for safety and efficiency are needed. With the initiation of clinical trials such as those for osteoarthritis, tendon repair and chronic kidney failure, the journey has begun. It is important to realize that claims of stem cell treatments are not to be accepted without validated data from clinical trials and the approval of the regulatory agencies such as the Food and Drug Administration in the United States of America. At present, stem cell therapies in veterinary patients are not rigorously supervised by regulatory agencies in any country [37]. Unfortunately, this has resulted in the implementation of some therapies without rigorous examination of the efficacy in preclinical animal studies. In general, the therapeutic role of stem cells in regenerative medicine is not fully understood. It is unclear whether stem cells ultimately function once differentiated into a tissue-specific cell such as a tenocyte or whether they primarily improve tissue repair through secretion of immunomodulatory cytokines and growth factors or whether a combination of the two mechanisms occurs [38]. Most of the studies use autologous (cells from one individual to the same individual) cell products to avoid immune rejection. However, some stem cell products such as mesenchymal stem cells (MSCs) seem to have immune modulatory effects [39]. If stem cells are truly immunomodulatory, then allogeneic (from one individual to another individual from the same species) cell transplantations could be possible. Safe and efficacious applications of allogeneic stem cells would imply that off-the-shelf stem cell products could be developed for increased availability and rapid implementation of stem cell therapies early in a disease course and in the older patients where often autologous stem cells are greatly reduced in quantity and potency due to aging. Potentially allogeneic stem cell products could be more cost-effective than autologous stem cells and more pets could benefit from this advanced technology. With use of allogeneic cells, a patient could avoid a costly tissue harvesting procedure and labor intensive stem cell production and just incur a fee for the application of more potent and consistent stem cell products originating from young healthy donors that are free of microorganisms and infectious diseases.

FUNCTIONAL MODELS OF CELL THERAPY AND REGENERATIVE MEDICINE

The main concern for a successful cell therapy treatment lies in the ability to demonstrate a significant clinical improvement as compared to the standard treatment regimen. The drive to develop cell based therapies lies in the overall goal to reduce or outright replace the standard treatments based on pharmaceuticals or surgical procedures. Regenerative medicine, in this respect, may be viewed as an alternative to the standard treatment regimen; however, it can also be viewed as complimentary. For example, the application of cell therapy has been used in horses to repair bone fractures and damaged ligaments [40]. The horse offers itself as a good animal model of spontaneous joint disorders that are clinically relevant to similar human disorders or instances of joint disease. A benefit of developing a veterinary animal model is the safety and efficacy can be thoroughly evaluated beyond the laboratory research induced animal models [41]. The cell of choice for this application has been bone marrow derived mesenchymal stem cell, although some testing is starting to be conducted on using adipose derived stem cells. The results for these trials have been mixed, where some labs are reporting successful orthoscopic repairs [42] and other labs not determining any significant results [43].

Much of regenerative medicine and cell therapy is still in the research and early clinical trial stage. The cell types that are stable and relatively easy to isolate (*i.e.* T-cells, dendritic cells, CD34+) have been validated with clinical trials and have been part of therapeutic applications for a number of years. The more esoteric cell types (*i.e.* ADSC, iPS cells, glioma stem cells, dental pulp stem cells, *etc.*) require further clinical study to prove their effectiveness for different indications [44, 45].

The principal question that clinicians, scientists, regulators, and the general public is asking is: what are the cells actually doing? Each group in the above parties is very interested in the question of how the cells function, since it would provide understanding and the foundation of how to maximize the benefits for regenerative medicine. Clinicians are interested in how the cells can modulate a disease or contribute to a cure. Scientists are interested in the mechanisms of how the stem cells undertake their regenerative traits. Regulators are interested in the safety and effectiveness of cell therapies. Whereas the general public want reliable, easy to digest information about the general features of therapies and how they could be useful for treating diseases.

Although there is no absolute confirmation on what stem cells actually do, there has been a lot of research in this area with substantial data. The potentially confusing inkling here is that different types of stem cells used for therapy may have different mechanisms of action. For example, CD34 hematopoietic stem cells would function much differently than neuronal stem cells. This may feel quite obvious, but it has to be recognized that different cells have different mechanisms and could be used for different therapeutic applications. Overall, there are 5 generally agreed upon mechanisms of integration that have been reported for stem cell therapy:

1. Trophic Support: Stem cells release several factors and cytokines that support resident cells and they respond to an injury or tissue damage. These growth factors are secreted by stem cells and support cell survival and contribute to wound healing (Table **2**). Other cellular activities include complex mechanisms such as anti-scarring, angiogenesis, anti-apoptosis, nascent stem cell recruitment and cellular growth [46]. A couple of interesting cases of trophic support are described with the use of islet survival and function after transplantation and for neurodegenerative diseases [47, 48].
2. Anti-Inflammatory: In line with trophic support, stem cells and MSCs, in particular, secrete cytokines that are anti-inflammatory in nature which help to balance the production of inflammation brought on by T-cells. The immunosuppressive effects mediated by MSCs mainly involve interactions

with T-cells. When T-cells are activated, they secrete pro-inflammatory cytokines such as TNF-α, IFN-γ, and IL-12. MSCs have shown the ability to inhibit the secretion of TNF-α, IFN-γ, and IL-12, mediated by the type 1 helper T cell pathway (Fig. **1**), [49, 50]

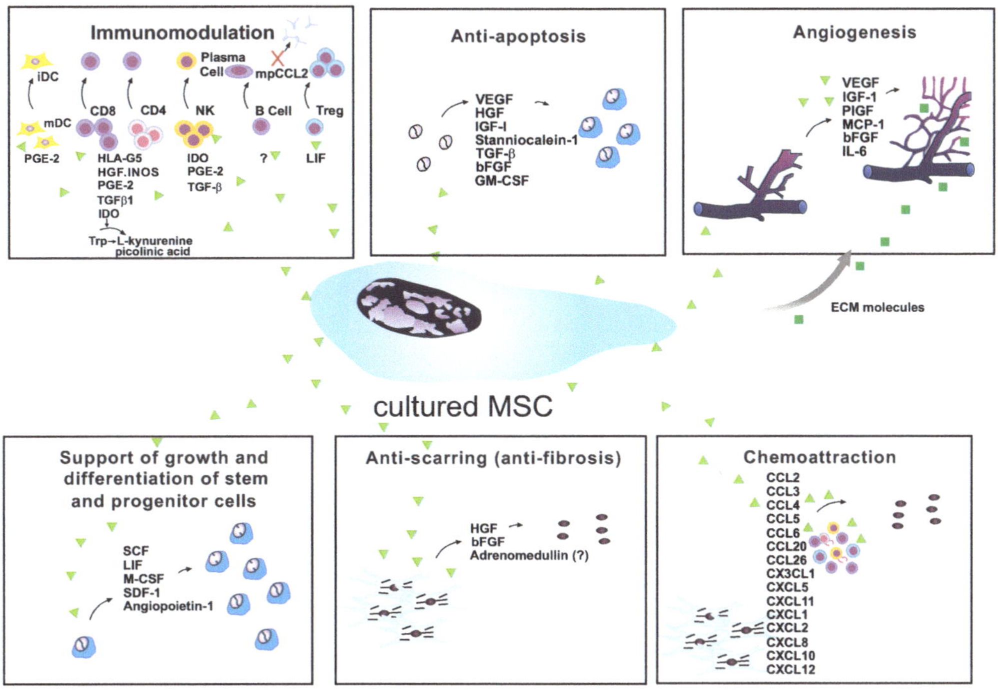

Fig. (1). Different mechanisms by which MSCs may exert their effects.
Copyright granted by Dr. Caplan. Annu. Rev. Pathol. Mech. Dis. 2011.6:457-78.

3. Differentiation into tissue specific cell types: The hope and promise of stem cells center on their potential capability of creating many different types of tissues. This particular function for the stem cells is quite significant and requires strong evidence that this process actually occurs *in vivo*. In an *in vitro* system, mesenchymal stem cells from adipose tissue in particular differentiate into chondrocytes (cartilage cells), osteocytes (bone cells) and adipocytes (fat cells). In fact, the International Society for Cell Therapy (ISCT) uses trilineage differentiation as a characteristic of human MSCs. The same criteria may be applied to growing and testing MSCs from any animal. In an *in vivo* system, the task is challenging to provide evidence to the scientific community regarding differentiation. To do this successfully depends on the ability to monitor

transplanted cells over a long period of time.

The only practical approach to monitor transplanted cells is to use some type of cell label. There are many physical factors to consider in choosing a label. Some examples are: what type of signal is generated, how long will the signal last, will the label diffuse out of cells, what is the labeling efficiency, and what happens to the label when cells die. It is equally important to demonstrate efficient long term labeling without deleterious biological effects on cell phenotype and differentiation capacity. Given these variables, cell-tracking experiments using a label must be heavily scrutinized. The typical categories of labels used in cell tracking studies include: radioactive (Indium-111), fluorescent (Green Fluorescent Protein), bioluminescence (Luciferase), and iron oxidized nanoparticles. Each option comes with its own set of challenges that each researcher has to overcome in order to properly optimize for an experiment or trial.

Fluorescent labeling is commonly used in laboratory animals to track cells. However, this method allows tracking cells only for a short period of time and requires a fluorescent microscope [51]. Using fluorescent labeling and whole body imaging, cell trafficking in the body of the laboratory mice can be tracked in live animals using an *in vivo* imaging system. Radio-labeling also has been successfully used to track stem cells, however it requires a specific radioactive laboratory and safety considerations [52]. Iron oxidized nanoparticles were recently used in clinical trials in human to track cells in patients. These particles can be up taken by the cells and be visualized by CT-scan for more than a year after transplantation [53]. While theoretically cells can be tracked by these systems, it is very unlikely that either of these systems can track few cells locating in an organ. Therefore, there is a limitation in detection level of transplanted cells and other technical hurdles that must be overcome.

4. Homing to an injury site: Stem cells are capable of locating injured sites by homing to them. Examples of this are seen while using MSCs for treatment. The therapeutic delivery of MSCs is typically performed *via* intravenous systemic injection into a major vessel. Some time is required to allow the MSCs to home and engraft within sites of injury. Not all the MSCs will actually end up at the injury site. Initially, intravenous injected MSCs are rapidly

cleared from the circulation and become entrapped within the lungs. After 1–5 days post injection, MSCs begin to exit the lungs and are found within the liver, spleen, kidneys and bone marrow.

The apprehension with this feature is that the MSCs do have the inherent ability to migrate. Although some of the MSCs will home into a site of injury, a large proportion of them will migrate to other tissues. An example of this was noted in dogs where repeated injections into the periocular and intra-articular regions showed migration and subsequent engraftment of allogenic MSCs in the thymus as well as the gastrointestinal tract [54]. Although in this situation the migrating MSCs did not contribute to any harm, some regulatory agencies could cite this as a safety concern. The burden of proof to show that the cells will not harm the patient is a difficult task and tends to require large clinical trials and an exhausting cell tracking long term study. As this group showed, after two weeks post injection, the allogenic MSCs were able to deliver favorable immunomodulatory support and demonstrated the potential for a therapeutic benefit using MSCs.

5. Immune system modulation: The most anticipated function of stem cells and widely studied are the effects that these cells have on the immune system. An immunomodulatory effect is well described with treatments that involve MSCs. This effect can be broad, from direct inhibition of lymphocyte proliferation, induction of regulatory T and B cells, to resetting the immune system. Although MSCs are not part of the immune system according to the established definitions [55], they interact with all immune cell types. They secrete a large range of anti-inflammatory as well as pro-inflammatory factors, among them cytokines, chemokines and prostaglandins, which target immune cells and affect their function [56]. In addition to paracrine interactions, MSCs express cell surface molecules that undergo interaction with various immune cells. The expression of certain adhesion markers, such as ICAM-1, increase the recruitment of activated immune cells to allow the immune cells to interact with the MSCs [57].

These features may have consequences in the effectiveness of stem cells used in cell therapy. Due to a variable number and quality of stem cells in different individuals, immediate use of freshly isolated cells from an individual most likely will result in a variable cell product with regard to the final composition of cell types and potency. These products have variable effectiveness ranging from working well to not being effective. Clinicians who use point of care systems (automated machines for preparation of cell products) to process stem cells, are more likely to experience this type of variability. Many reasons may cause these variabilities including differences between patients, their age and their health, differences in tissue of origin, the effects of time needed for the point of care system and the skill level of the technician performing the procedure.

Compared to freshly isolated cell products, autologous purified or culture expanded cell products could be used to minimize this type of variation, but patient variability is still a major consideration with the most notable factor being patient age and metabolic status. Allogenic purified or culture expanded cells offer the most controlled and least variable type of cellular therapeutic. Purified stem cells present a powerful tool for use in therapy. Systems which are based on magnetic sorting and fluorescent labeling sorting offer the best ways to collect purified stem cells. Cell sorting is a method by which cells are labeled with fluorescent antibodies directed against specific cell surface proteins, and then separated and isolated away from the total population of cells by targeting these specific antibody labels. Cell types such as CD34 hematopoietic stem cells or CD133 progenitor cells, are routinely sorted using a magnetic antibody based system. This system is slightly more favorable than sorting with fluorescent labels, since this requires expensive equipment that has to be arranged in a certified clean room. In these systems, stream-in-air flow cytometers, are not completely closed systems and require preventative measures to ensure the environment is qualified and validated for therapeutic sorting.

The method to culture expanded cells is becoming a common technique to generate a pure population of stem cells. Typically, these result in MSCs from bone marrow, adipose tissue, cord blood and other tissue types. The culture system yields large numbers of cells, which end up containing a common set of surface markers that are accepted as MSC identifiers. Variability still exists within this system and this is mainly due to tissue origin, culture conditions and expansion time. To minimize the introduction of variability requires strict protocols, highly trained personnel, and a closely monitored environment. MSCs created for therapy also are generated in a certified clean room and must undergo testing to ensure sterility and a lack of any contaminates or toxins.

IMPLICATIONS OF CELL THERAPEUTIC APPLICATIONS

Cell therapy holds the potential for future treatments of damaged tissues from both mature and stem cell lineages. Cell therapy may provide a means to one-day restore dysfunctional or damaged tissue without the need for immunosuppression [58]. Autologous treatments for joint repair and blood regeneration were natural starting points for using stem cells. This allows patients to use their own cells with minimal risk of graft *versus* host disease (GVHD) developing and also minimal use of immunosuppression drugs.When cells are transplanted from one individual (donor) into a patient (recipient), the immune system of the recipient triggers an immune response against the transplanted cells as it would have against any foreign material. This action sets off a chain of events that could disrupt or damage the transplanted cells. Most recipients are prescribed a combination of drugs after transplantation; for example, they may be given a combination of cyclosporine, azathioprine, and prednisolone. These drugs are extremely potent and can cause serious side effects such as high blood pressure, kidney problems and liver disorders. In addition, immunosuppressant drugs lower the resistance to infection and can result in infections that become difficult to treat.

Interferon γ (IFN- γ) is known to enhance the immunosuppressive properties of MSCs [59]. MSCs are immunogenic due to their inherent expression of major histocompatibility (MHC) molecules. Undifferentiated human MSCs express low to medium levels of human leukocyte antigen (HLA) Class I and low levels of HLA Class II to avoid recognition by the immune system [60]. IFN- γ pre-activated MSCs (MSC- γ) directly or indirectly modulates T cell responses by enhancing or inducing MSC inhibitory factors. These factors are known to downregulate T cell activation, enhance T cell negative signaling and alter T-cells from a pro-inflammatory to an anti-inflammatory phenotype [61]. MSC- γ also interacts with antigen-presenting cells to increase or induce regulatory cells [62]. MSCs sec rete an array of growth factors and anti-inflammatory proteins in response to inflammatory signals. The key immunomodulatory cytokines include prostaglandin 2, TGF-β1, HGF, SDF-1, nitrous oxide, indoleamine 2,3-

dioxygenase, IL-4, IL-6, IL-10, IL-1 receptor antagonist and soluble TNF-α receptor [63 - 65].

This property gives donor MSCs the ability to go undetected by a host immune system in allogeneic therapies. Demonstration of immune system evasion in multiple animal models has deduced the use of these cells for therapeutic use in non-hematopoietic indications [66 - 68]. Since then, MSCs are involved in a series of clinical trials in a wide range of major diseases including stroke [69], heart failure [70], COPD [71] and liver failure [72]. In the next 5 – 10 years with the completion of numerous clinical trials, the industry will have a better idea about the efficacy of MSCs and their use as therapeutic agents.

Quality of Life and Anti-aging

Research in anti-aging or rejuvenation is gaining considerable attention in the literature. In general, aging is related to loss of functional stem cells accompanying loss of tissue and organ regeneration potentials. The health of these tissues and organs ultimately are tied to functions at the cellular level, especially the stability of telomeres. Telomeres are repetitive DNA protein structures that cap the ends of chromosomes, protect chromosome ends from degradation and fusion and are essential for maintenance of genomic integrity [73]. Telomere length has been shown to gradually shorten over time as cells divide. When telomeres become critically short, the cells enter a state of senescence; as such, telomere shortening has been implicated in accelerated aging [74].

Stem cells and telomeres share an intimate relationship with respect to aging. Most somatic cells lack telomerase activity, but somatic stem and progenitor cells express low levels of telomerase. Telomere elongation is of great importance for the maintenance of multipotency in stem cells (also pluripotency in ESCs) during development. Telomerase-deficient mouse models have demonstrated that a malfunction of telomerase impairs renewal capacity and shortens lifespan at the cellular level. The result of such telomerase malfunctions is a shorter telomere with decreased DNA stability. Telomere shortening has a negative impact on cell proliferation by activating checkpoints that induce replicative senescence or apoptosis [75, 76]. Additionally, telomere erosion naturally occurs with age, and is proposed to be a fundamental factor in the aging process.

Stem cells also undergo the process of aging but it is delayed as compared to somatic cells. Some stem cell sources, such as adipose derived MSCs are available across all age groups. Infant derived MSCs are morphologically spindle-shaped, with long telomeres, and exhibit enhanced angiogenic and osteogenic capabilities compared with older cells [77]. Similarly, MSCs from bone marrow exhibit the same trend. MSCs obtained by marrow aspiration decline with age. In a paper discussing primate MSCs, age-related differences in gene expression profiles showed diminished processing of RNA transcripts [78]. This downward trend coupled with the negative regulation of gene expression suggests there is an age-related decline in overall MSC ability, which might lead to problems when using aged autologous MSCs for cell-based therapies.

An important part of a cell therapeutic treatment is to improve the quality of life and to support the ability to cope with stressors related to disease [79]. Concurrently, a cell based treatment requires evidence it is safe, efficacious, and also improves the quality of life of patients with severe disorders and injuries. In planning for clinical trials, a measurement for quality of life is a critical parameter to include. However, a potential weakness in the methodology of evaluating quality of life for a patient is that it is a qualitative survey and at times may be a self-assessment. Nonetheless, many clinical studies present results that the patient finds the overall experience to be a positive one [80 - 85].

MANUFACTURING CELL-BASED PRODUCTS

All manufacturing of cell-based products is highly dependent on fresh tissues and cells as its critical raw material. Much of the source of this biological raw material is relied on a steady normal healthy donor base, and to a lesser extent, individual patients for autologous therapy. A major challenge in cell therapy is to manufacture a consistent product with varying tissue based inputs. Donors who regularly supply raw material, such as blood or bone marrow, can have variations in the proportion of cellular and non-cellular fractions due to everyday circumstances.

Clinical and commercial consistency of manufacturing consistent cell-based products relies on a deep understanding of the product's critical quality attributes (CQAs). Controlled points that can be measured and evaluated during manufacturing are a key component to satisfying a product's CQAs. These are characteristics that should be within specified defined limits to ensure the product has the proper consistency, which could have an effect on its efficacy and safety. CQAs are a combination of safety, identity, purity and potency elements along with biochemical, physical, biological and microbiological attributes that can be defined, measured and continually monitored [86].

CQAs are defined by explicit manufacturing process parameters and critical raw material characteristics. The manufacturing process and critical raw materials must be sufficiently controlled and consistently performed to result in a product with CQAs within the specified limits [87]. To accomplish routine consistent CQAs, a manufacturer must set in place a stringent product development program, have devoted quality systems and a formal training structure for employees. It is important to also be open to advice and feedback for these systems. At times, a well experienced staff could offer points to improve the operational processes. These should be discussed and vetted. Ultimately is it vital to be clear on the expectations and to make the team's technical and interpersonal development a priority.

Cell therapy products are manufactured from either an origin of autologous or allogeneic cells or tissue. Autologous cell therapy products are manufactured from the patient's own cells or tissue, while allogeneic cell therapy products are derived from a donor other than the patient. Autologous cell therapy products have been in clinical use for over 30 years using cellular material from bone marrow, white blood cells and platelets. In the recent decade, a strong push has been made to develop autologous cell therapy derived from adipose derived stem cells. A search on the clinical trials website currently shows autologous adipose stem cells in 92 clinical trials for treatment of a variety of clinical indications including osteoarthritis, wound healing and Type 2 Diabetics [88]. The main driving point for autologous cell therapy products is the benefit of using stem cells in a transplant for the same patient is safer than using a source from a donor, because the body will not reject its own stem cells. As forthright as it sounds, it does not come without its challenges – especially with FDA regulation [89].

Where autologous cell therapy is personalized, allogeneic cell therapy is not and the manufacturing process differs in that it often involves more than minimally manipulation processes and extensive cell processing. Allogeneic cell therapy products hold great promise for addressing unmet medical needs but also present challenges related to product characterization, safety testing, and clinical trial design. The ideal business model for a commercial allogeneic cell therapy product is to have it be scalable, prepared in large batches, off the shelf inventory, favorable storage conditions and with efficient logistics.

Allogeneic cell therapy has the potential to be a disruptive force in regenerative medicine. Standard immunologic views hold that foreign tissue would elicit an immune reaction. This concept is marked in organ and tissue transplantation, in which aggressive immunosuppression is the norm to protect allografts from rejection [90]. The potential for this type of cell therapy is treating large patient numbers with a cell product obtained from a universal donor. Furthermore, from an economic perspective, one donor provides the necessary amount of raw material to manufacture a batch size suitable to spread the associated costs of production over the batch. Also as each product is sold, the per unit value is greater and generates overall more revenue as compared to a single batch run of an autologous single use product.

MSCs have been viewed as a choice cell type for allogeneic therapy due to its ability to evade or suppress the immune system [91, 92] Due to these properties, bone marrow derived MSCs have been involved in numerous clinical trials. The main objective of the clinical trials is to demonstrate that MSCs may be safely administered to humans without eliciting clinically relevant immune reactions. These early clinical trials of allogeneic MSC therapy focused on acute myocardial infarction [93 - 95]. The results of these trials have been promising, where MSCs did not produce an immune reaction and led to improved outcomes with regard to cardiac arrhythmias, pulmonary function, and left ventricular function [96]. In addition, MSC therapy improved indices of physical functional capacity and quality of life along with reduced tissue scaring [93, 97].

Cardiac-derived stem cells may have similar immunologic properties to MSCs as explored in the laboratory [98]. Although no clinical trial using allogeneic cardiac-derived stem cells is established, there has been an autologous trial using this cell type [99]. This study focused on CD117 c-kit-positive, lineage-negative cardiac stem cells as an autologous therapy for patients with ischemic heart disease. These initial results were encouraging. They suggested that intracoronary infusion of autologous CSCs was effective in improving systolic function and reducing the infarct size in patients [99]. In the laboratory, emerging data shows that cardiospheres have many properties of MSCs, as they express classic markers including CD105, CD90, and CD73 [100, 101]. Taken together, these findings with cardiac-derived stem cells suggest that allogeneic cell therapy may be applied to similar cell types that target particular medical conditions [97].

STEM CELLS, PROGENITOR CELLS AND BLOOD DERIVED PRODUCTS

Applications of cell therapy have benefited from the use of both somatic and stem cells over the past half-decade. The growth of cell therapy has evolved from the early days of whole organ transplantations. The successful development of whole organ bone marrow transplant in 1960s laid the foundation for future work in cell therapy. A key moment during the developing organ transplantation era, was demonstration that organ allografts could self-induce tolerance with the aid of immunosuppression [102].

The concepts of stem cell biology was in its infancy during this time and early marrow transplantations in the dog pointed to the existence of special primitive bone marrow cells that can restore hematopoiesis [103]. Roughly twenty years later, these primitive bone marrow cells were determined to be hematopoietic progenitor cells characterized by the surface glycoprotein CD34 [104, 105]. The discovery of CD34 as a hematopoietic cell surface antigen had transformed and accelerated studies involved in developmental hematopoiesis. Since those early days, CD34 hematopoietic stem cells have been isolated from bone marrow, umbilical cord blood and blood mobilized with granulocyte-colony stimulating factor (Neupogen). Current therapeutic uses of CD34 hematopoietic stem cells include transplantations in patients with myocardial infarcts [106], spinal cord injury [107] and retinal diseases [108].

Minimal criteria to define expanded multipotent human MSCs, as defined by the International Society for Cellular Therapy, include that they must be plastic-adherent when maintained in standard culture conditions, express CD105, CD73 and CD90, and lack of expression of CD45, CD34, CD14 or CD11b, CD79α or CD19 and HLA-DR surface molecules, and they must be capable of differentiating into osteoblasts, adipocytes and chondroblasts *in vitro* [109].

MSCs are broadly distributed in the body and can be isolated and expanded from a variety of sources, such as bone marrow, adipose tissue, synovial membrane, umbilical cord blood, peripheral blood, Wharton's jelly, trabecular bone, infrapatellar fat pad, dermis, periosteum and muscle [110]. The phenotypic characteristics of MSCs derived from different sources are similar, but the number of MSCs and their proliferation and differentiation potentials can differ [111]. As discussed above, cell therapy based on MSCs tend to be allogeneic and require more involved safety and efficacy testing in order to be approved by the FDA. The current regulatory landscape on allogeneic cell therapy is to treat is in a similar way as a new drug development. Laboratories and organizations developing MSCs for therapeutic treatment must conduct clinical studies that address safety concerns and the efficacy of the treatment. The FDA has is a zero-risk position on potential cell therapies for the recipients and this position could pose a challenge to pushing these cells through clinical trials.

A potent cell type that is gaining notoriety in this field is a genetically engineered T cell to produce special receptors on their surface called chimeric antigen receptors (CARs). These CAR T-cells are engineered to recognize specific antigens on tumor cells and kill them [112]. This method of immunotherapy builds upon the principle of adoptive cell transfer. In this model, T-cells are collected from the patient's own blood and are genetically engineered to produce the chimeric antigen receptors. The CAR T-cells are expanded in a cell culture system into the billions and are infused back into the patient. The CAR T-cells proliferate in the patient's body and, with guidance from their engineered receptor, recognize and kill cancer cells that harbor the specific antigen [113]. This new development of treatment is cell therapy, gene therapy, and immunotherapy all packaged together.

A search on clinicaltrials.gov website shows there are 35 planned or ongoing clinical trials focused on CAR T-cells [114]. Most of these studies are aimed at late-stage cancers for which other therapeutic options have failed. The anticipation of these trials is that these immunotherapies could replace standard cancer treatments in the future. The FDA will have its eye on the trials from the safety perspective since CAR T-cell therapy can cause several worrisome side effects. A particularly troublesome side effect is cytokine-release syndrome, where there is a massive release of cytokines into the bloodstream. This can lead to dangerously high fevers and quick drops in blood pressure [115]. These types of side effects can be managed with standard supportive therapies, including steroids. Scientists working with this immunotherapy caution that there is still much to learn about CAR T-cells, even though the early results from trials have generated considerable optimism.

THERAPEUTIC APPLICATIONS IN VETERINARY MEDICINE

The therapeutic application of stem cell-based technologies in veterinary medicine was first used by Douglas J. Hershel, to treat equine suspensory ligament desmitis [116]. This application involved direct injection of large volumes (20 to 60 mL) of bone marrow aspirate obtained from the sternum into an injured ligament (Fig. **2**). In this report of an uncontrolled, nonrandomized case study, the technique appeared to improve return to athletic function rates over conventional therapies. However, it is unlikely that the observed results were only due to the stem cells, as there are very few stem cells in the bone marrow aspirate. MSCs represent a very small fraction of the total population of nucleated cells from bone marrow from humans [117] and cats [118] and are presumed to be similar in other species, including the horse. These studies indicate that 0.001% to 0.01% of mononuclear cells isolated from a Ficoll density gradient of bone marrow aspirate are MSCs. The percentage of MSCs in raw bone marrow aspirate would be even smaller because the technique of Ficoll density gradient isolation removes several types of nucleated cells, including granulocytes and immature myeloid precursors. Any clinical effect of bone marrow aspirate might be attributed to the numerous bioactive substances in the acellular fraction such as growth factors produced by platelets or other cells. For example, bone marrow aspirate that lost the cells due to a freeze-thaw procedure has some stimulatory effects on matrix synthesis when applied *in vitro* to tendons and ligaments [119, 120].

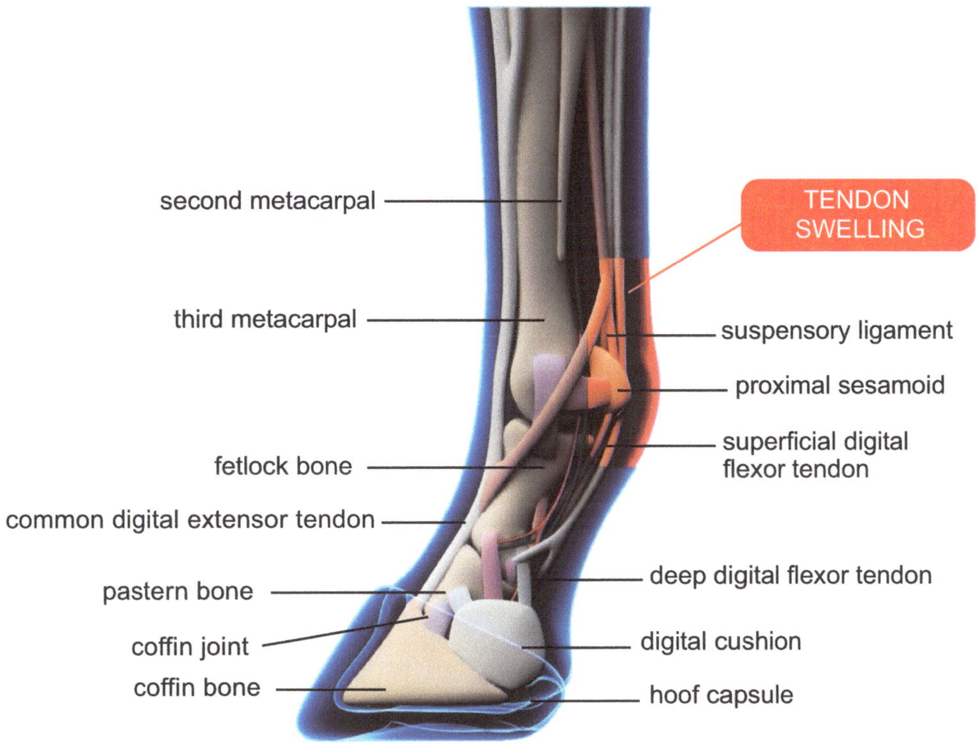

Fig. (2). Anatomical location of cell therapy for tendonitis.

In veterinary medicine stem cells have been predominantly used for orthopedic injury and, in terms of translational research for developing the technology for use in the clinic, the most advanced application has been in the horse. Stem cells have been used for studies of heart disease in the dog, although this has been mainly as a model for ischemic heart disease in human [121]. Unlike humans, most other mammals, including dogs, do not suffer naturally occurring clinical ischemic heart disease, so the relevance of this research for clinical veterinary medicine is more limited. However, other cardiac diseases, such as canine dilated cardiomyopathy, might be appropriate beneficiaries for this therapy in the future. Within the equine orthopedic field, MSCs have been used experimentally and in limited numbers of clinical cases for the surgical treatment of subchondral-bone cysts, bone-fracture repair and cartilage repair. However, by far the most frequent use clinically has been in the treatment of overstrain induced injuries of tendons in horses [122 - 124]. Commonly used cell products in veterinary practice are presented in Table **3**.

Table 3. Cell Products used for debilitating diseases of companion animals.

Cell Product	Cell Type	Species	Indication Used for	References
MSC	BM	Horse	Tendon repair	[122]
MSC+PRP	Adipose	Horse	Tendon repair	[123, 124]
SVF	Adipose	Horse	Tendon repair	[124]
PRP	Peripheral blood	Horse	Tendon	[125, 126]
SVF	Adipose	Dog	Osteoarthritis	[127]
MSC	Adipose	Dog	Osteoarthritis	[128]
PRP	Peripheral blood	Dog	Arthritis	[129]
MSC	BM	Dog	Ischemic heart, Myocardial infarction	[121, 130]
MSC	BM	Dog	Bone repair	[131]
MSC	BM	Dog	Intervertebral disc degeneration	[132]
MSC	Adipose	Cat	Chronic kidney disease	[133, 134]

MSC: mesenchymal stem cell; PRP: platelet rich plasma; SVF: stromal vascular fraction from adipose tissue.

Veterinary Clinical Studies

Currently there is an expanding list of major United States veterinary teaching institutions that are conducting clinical trials for various conditions using regenerative therapies. These include Tufts University, Colorado State University, UC Davis, University of Florida and The University of Pennsylvania to name a few. The majority of the studies are focusing on the use of MSCs, but there are other approaches including platelet rich plasma (PRP) and umbilical cord blood. There are multiple disease conditions that are being studied such as feline chronic kidney disease (CKD), feline chronic gingivostomatitis, canine inflammatory bowel disease, canine keratoconjunctivitis sicca, equine recurrent uveitis and many more. Clinical trial activity involving the use of stem cells will likely increase in the coming years and as we gain a better understanding about how stem cells are behaving and improving disease conditions *in vivo*, we will begin to see more specific therapies using more defined cell types for particular conditions.

There have been a few clinical studies performed on canines with osteoarthritis using autologous stromal vascular fraction from fat, which contains a small percentage of MSCs. Both studies showed improvement in patients by owner and veterinarian assessments, lameness scores and pressure plate analysis. A notable randomized, double-blinded, multicenter, controlled trial was published by Black *et al.* The study population included 21 dogs (14 females, 7 males), age range from 1 - 11 years, weight range from 25 – 110 lbs., a minimum 6 month duration of bilateral coxofemoral joint OA and a mix of many different breeds. This study tested autologous stromal vascular fraction collected from subcutaneous fat for intra-articular injection *vs.* saline solution in the control group in patients diagnosed with chronic coxofemoral degenerative joint disease. The SVF contains a mixture of cells including a small percentage of MSCs. There was a mixture of 30, 60 and 90 day significant improvements among the stem cell treated group for lameness, pain on manipulation, range of motion or owner improvement scores. There was also a purportedly overall large orthopedic effect size for stem cell treated animals [127].

Jose M Vilar *et al.* published a study using a mixture of autologous cultured MSCs derived from fat and platelet-rich plasma (PRGF) for intra-articular injections on 8 lame dogs with severe hip OA. This study focused on force platform analysis before and after treatment. Overall, they found improvement in peak vertical force (PVF) and vertical impulse (VI) over a 6-month follow up period. This is promising considering the animals used were severely arthritic and lame, although the sample size was small and there were no negative controls included in the study [135].

Antonio J. Villatoro *et al.* published a clinical case report describing allogeneic cultured adipose derived MSCs for the treatment of chronic keratoconjunctivitis sicca (KCS) in canines. 12 dogs were studied, all of which were affected by bilateral KCS for at least 6 months and were refractory to conventional treatments (corticosteroids, tacrolimus, cyclosporine and artificial tears), and had a schirmer tear test (STT) of <10mm / minute. MSCs were harvested from a donor dog, culture expanded and then utilized as an allogenic treatment *via* injection around the main lacrimal gland and the nictitans membrane (Fig. **3**). Patients were followed for up to 9 months. There were significant decreases in hyperemia, ocular discharge and corneal changes and significant STT increases. This study also highlights the safety of the use of allogenic MSCs [136].

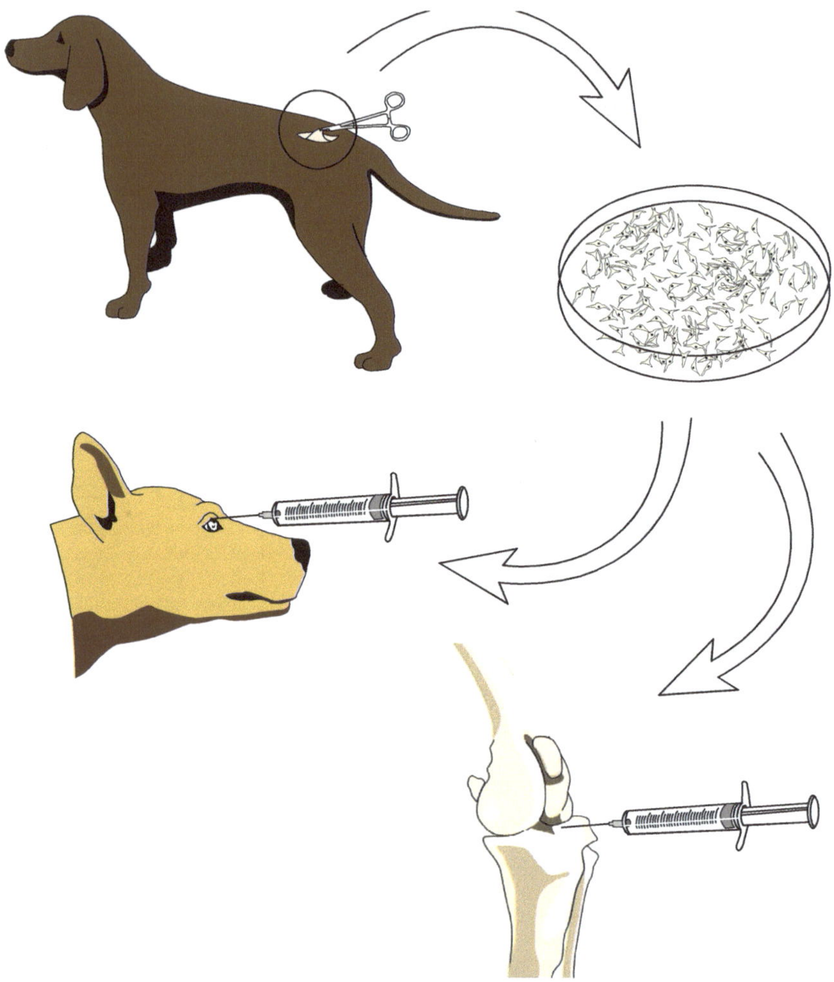

Fig. (3). Preparation of canine adipose derived MSCs for allogenic cell therapy.
Copyright granted by Dr. Murphy, Cytotherapy 2013 15:1488-1510.

Maria A. Fahie *et al.* published a randomized, controlled, 2-center clinical trial evaluating the efficacy of autologous platelet therapy for intra-articular injection for canine OA. With 3 month follow up times, they found significant improvements in lameness scores, pain scores and peak vertical force. There was no difference in control groups. This is the first known well controlled study demonstrating a reduction in clinical symptoms associated with osteoarthritis in one joint using platelet rich therapy [129].

Quimby *et al.*, studied safety and efficacy of intravenous infusion of allogeneic cryopreserved feline adipose derived MSCs for the treatment of chronic kidney disease in cats. They reported that frozen / thawed MSCs had an adverse effect on kidney function while cultured MSCs after freeze and thaw neither improved nor impaired kidney function of the diseased cats [137]. Stem cell therapy for chronic kidney disease will require more basic research and further clinical studies. This study also highlights that allogenic MSC applications can be utilized safely.

AGING AND REGENERATIVE POTENTIAL

It is generally accepted that younger individuals have greater regenerative potential than the old ones. While the exact mechanism of aging is not clear, it is believed that both regenerative potential and DNA repair mechanisms are impaired by aging. In the very early stages of embryonic development, cells have the ability to divide indefinitely and then differentiate into any type of cell in the body. Recent studies have revealed that some of this remarkable developmental potential of embryonic stem cells is retained by small populations of cells within most adult tissues defined as adult stem cells [34]. Intercellular signals that control the proliferation, differentiation and survival of stem cells are being identified and include a diverse array of growth factors, cytokines and cell adhesion molecules [138]. Intracellular mechanisms that regulate stem cell fate are also emerging and include established second messenger pathways, novel transcription factors and telomerase [139]. The possibility that a decline in the numbers or plasticity of stem cell populations contributes to aging and age-related disease is suggested by recent findings.

DNA repair maintains genomic stability and the loss of DNA repair capacity results in genetic instability that may lead to a decline of cellular function [140]. Adult stem cells are extremely important in the long-term maintenance of tissues throughout life. They regenerate and renew tissues in response to damage and replace senescent terminally differentiated cells that no longer function. Oxidative stress, toxic byproducts, reduced mitochondrial function and external exposures all damage DNA through base modification or mis-incorporation and result in DNA damage [141]. As in most cells, this damage may limit the survival of the stem cell population affecting tissue regeneration and even longevity. Normal stem cells appear to have strict control of gene expression and DNA replication whereas stem cells with loss of DNA repair may have altered patterns of proliferation, quiescence and differentiation [142]. Furthermore, stem cells with loss of DNA repair may be susceptible to malignant transformation either directly or through the emergence of cancer prone stem cells [143]. Human diseases and animal models of loss of DNA repair provide longitudinal analysis of DNA repair processes in stem cell populations and may provide links to the physiology of aging.

Somatic cells in adult tissues maintain a delicate balance between proliferation and apoptosis. Stem cells are important mediators of tissue maintenance and wound repair. Their longevity is dependent on careful control of gene expression, proliferation and cell cycle and differentiation signals. Of note, a number of human genetic abnormalities associated with aging, and those replicated in the mouse, suggest that loss of DNA repair may contribute to the aging process. Maintenance of the adult stem cell genome through robust DNA repair is fundamental in the prevention of aging and disease. Furthermore, failure of genomic maintenance is a leading cause of cancer, as well as senescence [143]. The global process of aging may in fact be linked to a global decline in tissue stem cell function and this in turn may be linked to changes in DNA damage repair processes.

Most of the data regarding the involvement of stem cells in aging has been obtained from human and mouse studies and very little is known in companion animals. The human body depends on adult stem cells in the bone marrow to repair damaged sites to remain healthy. The vital role of adult stem cells in both health and disease is one of the greatest health discoveries of modern times. As we age our stem cell production from our bone marrow declines. This is one of the causes of aging as well as exposure to various environmental (and sometimes dietary) toxins, stress, oxidation, inflammation and hormone decline. The decline of adult stem cells can lead to un-repaired damage in virtually any body part, leading to tissue breakdown, aging and loss of health. Without active, vigorous adult stem cells good health is compromised long-term. Scientific anti-aging research shows adult stem cell release rates from the bone marrow drop at an astonishing rate as we age, as per the following percentages: At 35 years the stem cell release rate drops by 45%. At 50 years the stem cell release rate drops by 50% and at 65 years the stem cell release rate drops by 90% leaving only 10% of adult stem cells circulating in the blood stream and this is when we need them the most to fight disease and sickness [96]. A baby has adult stem cell circulation of 100% in their blood stream. When a child fractures a leg or has injuries and illnesses, healing and repair occur quite rapidly because their system of adult stem cell release into the blood stream is functioning at optimal levels. With aging, everything slows down. It takes longer to get over coughs and colds; longer for cuts and scrapes to heal; eventually longer to get up in the morning. Part of the reason for this might be that adult stem cells are not released as efficiently from the bone marrow. Scientists believe that by increasing the release of circulating stem cells aging might be delayed [144]. Circulating stem cells can reach various organs and become cells of that organ, helping such organs to regain and maintain optimal health. Recent studies have suggested that the number of circulating stem cells is a key factor; the higher the number of circulating stem cells, the greater the ability of the body to heal itself. Scientific interest in adult stem cells has been focused on their ability to divide or self-renew, and generate all the cell types of the organ from which they originate, potentially regenerating damaged parts of an organ from a few cells.

DEBILITATING DISEASES IN COMPANION ANIMALS

Debilitating diseases are those, in which due to the severe damage, organ function is hampered and its gradual deterioration results into fibrosis and end stage organ failure. Arthritis and skeletal diseases are on the top of the list of all debilitating diseases in human and pets followed by lung disease, cardiovascular disease and neuromuscular diseases [145]. Degenerative joint disease (DJD), or osteoarthritis (OA), is chosen as major debilitating diseases in dogs and horses. A list of current use of stem cells within veterinary medicine with an emphasis on application of mesenchymal stem cells (MSCs), recovered from bone marrow, adipose tissue or umbilical cord is presented in Table **3**.

Degenerative joint disease (DJD) or Osteoarthritis (OA) is a degenerative disease of joints that occurs as a primary disease in aging patients but can also occur as a secondary disease in our younger patients. Primary OA is a result of normal or abnormal forces on an otherwise normal joint. This disease generally takes years of wear and tear before clinical symptoms manifest, but in young patients that have predisposing conditions such as hip dysplasia (HD) or elbow dysplasia symptoms may manifest as early as 1-2 years old. Secondary OA occurs due to some insult to the diseased joint such as joint trauma or infection followed by normal or abnormal forces on an abnormal joint. OA occurs in canine, feline and equine patients and likely other exotic species as well, but here we mainly focus on OA in canine patients with brief mention of equine and feline OA and feline chronic kidney disease (CKD), feline chronic gingivostomatitis (FCGS) and equine tendon and ligament injuries will also be briefly discussed. Additional clinical studies, including allogeneic MSC therapy for canine keratoconjunctivitis sicca (KCS) are mentioned in the "Veterinary Clinical Studies" section 8.1.

OA involves cartilage degeneration, fibrillation and loss, inflammation and hyperplasia of the synovial membrane, abnormal proliferation of bone (osteophyte production) and eventually exposure of subchondral bone. All of these changes can elicit varying degrees of pain and lameness. The classic finding in the patient's history is that the owner reports their dog seems to have difficulty rising after a nap or after a long walk. The patient may also be reported to have a low-grade lameness during these times. Additionally, the owner may notice that he / she seems a bit slower than normal, may be reluctant to run and play, and his overall demeanor may also be affected by clinical signs such as depression, lethargy and inappetence. Physical examination may reveal varying degrees of lameness, muscular asymmetry and joint enlargement. There may be a reduction in stability and range of motion and pain and crepitation may be noted. A diagnosis is reached based on a combination of patient history, physical exam, (including an orthopedic exam), and radiographs. Although rarely performed in general practice, advanced imaging techniques such as computerized tomography (CT), magnetic resonance imaging (MRI) and / or arthroscopy can be immensely beneficial. It is important to rule out other sources of pain or lameness such as fracture, cranial cruciate ligament disease, joint luxation / subluxation, bone fragmentation, intervertebral disc disease, infectious polyarthropathy or other causes that may show similar symptoms. This again highlights the importance of a thorough history and physical exam and warrants the investigation of a complete blood profile and chemistry, urinalysis, synovial fluid analysis and other imaging techniques such as thoracic / abdominal radiographs or abdominal ultrasound if warranted.

OA is an immensely important and debilitating disease as it is a progressive, degenerative, debilitating and painful and inevitably leads to joint failure. Clinical management of a patient with OA can depend on many factors. Orthopedic surgical corrective procedures should always be on the clinician's mind as it would be negligent not to recommend an orthopedic consult for conditions such as femoral head necrosis, cranial cruciate ligament rupture, fragmented coronoid process, and patella and hip luxation because these conditions can be greatly improved with a surgical approach, especially if dealt with early. In this section we will focus on patients that are not surgical candidates yet present with osteoarthritis of one or more joints.

ARTHRITIS AND JOINT FAILURE AS A MODEL FOR CHRONIC DISEASES

Arthritis is inflammation of one or more joints (Fig. **4**). The main symptoms of arthritis are joint pain and stiffness, which typically worsen with age. The most common types of arthritis are osteoarthritis and rheumatoid arthritis. Many pets develop some form of joint disease during their lives. It can be mild, even unnoticeable to the pet owner, or it can be debilitating, severely affecting the pet's quality of life, or even causing complete lameness. The majority of cases fall somewhere in between. While some pets may develop joint disease in the first half of their lives, signs usually do not appear until the latter half of life, which varies depending on pet's size and breed. Dogs seem to be more susceptible to arthritis than cats, and the larger dog breeds are more vulnerable than smaller breeds. The most common signs of joint disease include stiffness, limping, or favoring a limb - particularly after sleep or resting, inability to rise, reluctance to jump or even climb stairs, and noticeable pain. Common areas of bone and joint pain in the dog are illustrated in Fig. (**5**). There are many diseases that affect the joints of dogs that can be categorized in ten major classifications which are listed below. It is noticeable that cell therapy is most effective in degenerative and inflammatory arthritis.

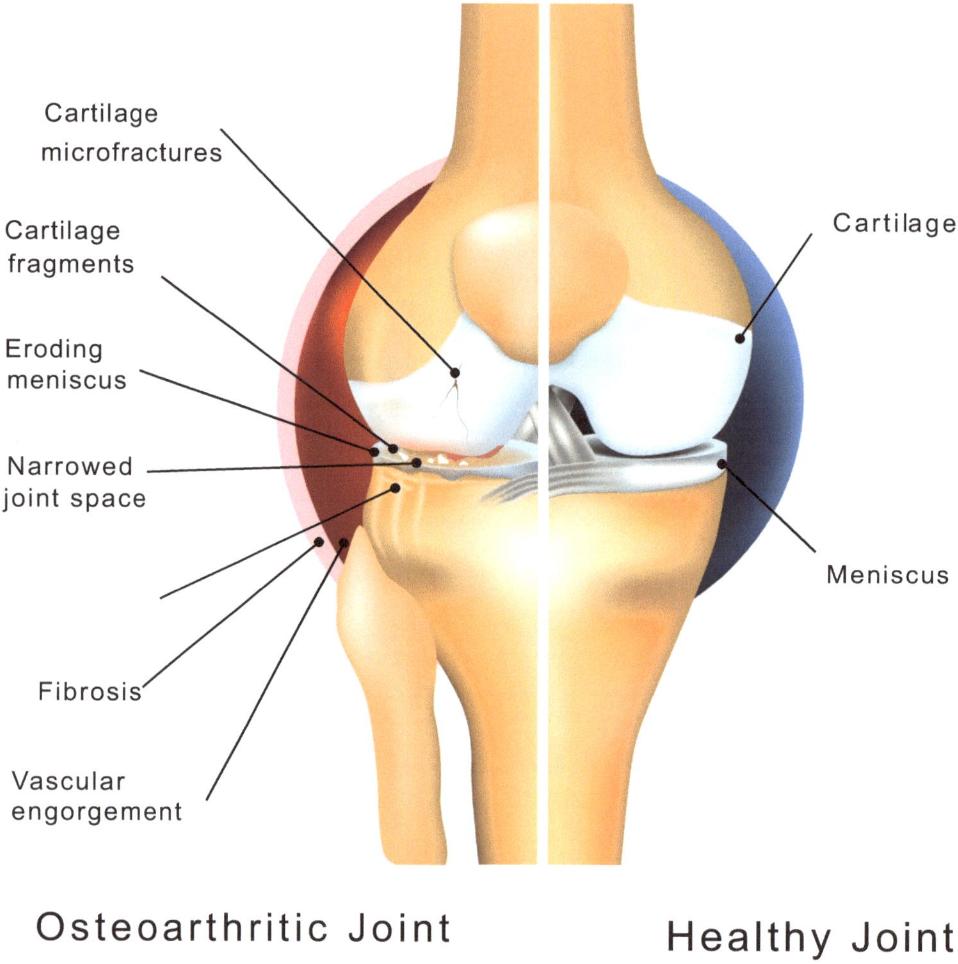

Fig. (4). Pathologic changes in the joint due to osteoarthritis.

1. Ligament, tendon, or muscle disease, *e.g.*, tiered or ruptured anterior cruciate ligament.
2. Fractures involving the joint.
3. Developmental disorders, *e.g.*, hip dysplasia, elbow dysplasia, osteochondritis dissecans, Legg-Perthes disease.

4. Congenital disorders, *e.g.*, Wobbler's syndrome (cervical spondylomyelopathy), luxated patella.
5. Dietary and hormonal disease, *e.g.*, hyperparathyroidism, obesity.
6. Metabolic disorders, *e.g.*, von Willebrand's disease (hemophilia) in dogs.
7. Cancer.
8. Degenerative joint disease (osteoarthritis).
9. Inflammatory joint disease, *e.g.*, Lyme disease, rheumatoid arthritis.
10. Degenerative spinal joint disease, *e.g.*, intervertebral disc disease, cauda equina syndrome.

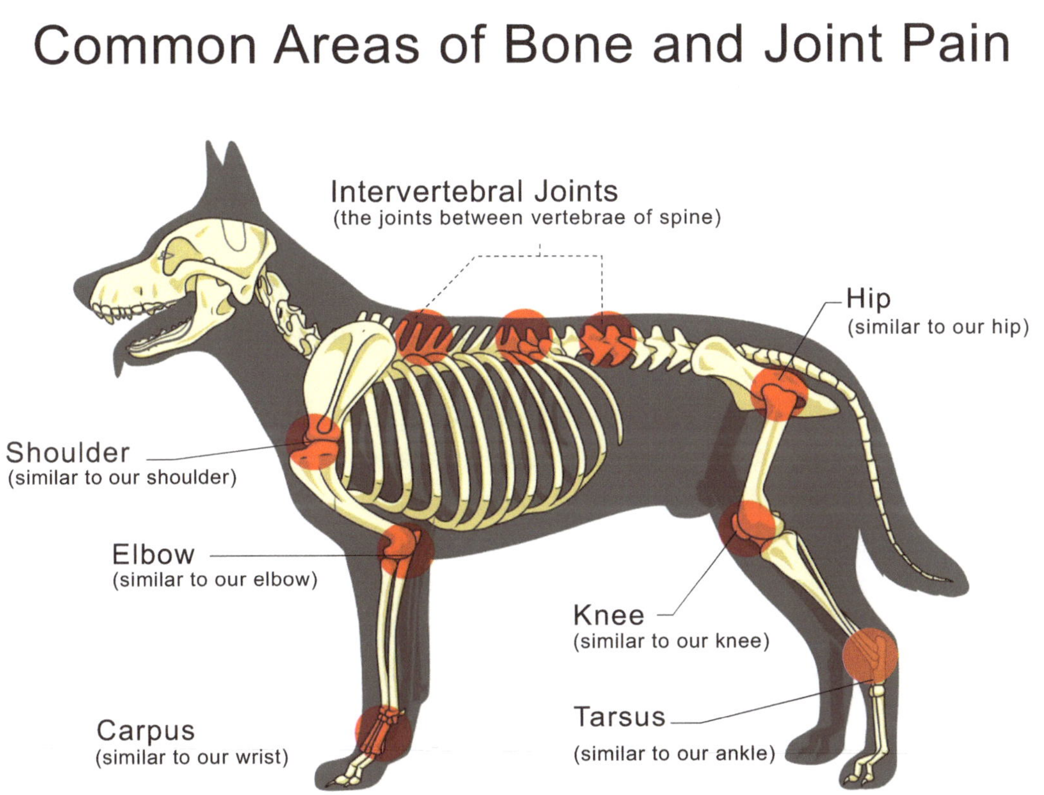

Fig. (5). Common areas of osteoarthritis in the dog.

Most of the information regarding unraveling the destructive mechanism of arthritis has been obtained from inflammatory and rheumatoid arthritis [146]. The synovial membrane in patients with inflammatory arthritis is characterized by hyperplasia, increased vascularity, and infiltration of inflammatory cells, primarily CD4+ T-cells. This is normally triggered by exogenous or endogenous antigens. Antigen-activated CD4+ T-cells stimulate monocytes, macrophages, and synovial fibroblasts to produce cytokines interlukin-1 (IL-1), interleukin-6 (IL-6), and TNF-α and to secrete matrix metalloproteinase and this results in erosion of cartilage and bone. Activated macrophages, lymphocytes and fibroblasts, as well as their products, can also stimulate angiogenesis, which may explain the increased vascularity found in the synovium of patients with arthritis (Fig. **4**).

CURRENT CLINICAL MANAGEMENT AND STEM CELL THERAPIES

Classically there are four main strategies that are utilized for medical management of canine OA: 1) weight management, 2) exercise moderation and physical rehabilitation therapy, 3) nutritional supplementation and 4) pharmacological management. Weight management seems fairly obvious but this is an important aspect for the patient. The aim should be to maintain a body condition score (BCS) of 4 / 9 to reduce the weight bearing forces on the joints and try to reduce the need for surgery and long term analgesics and anti-inflammatories. It is important for owners to continue exercising their pets but to do so in moderation. This helps to increase mobility and assists with weight management without inflicting excessive loads on the joints. Physical therapy is also beneficial to improve joint mobility and strengthen surrounding muscles and tendons. It has been shown that instituting a diet with a high N-3: N-6 omega fatty acid ratio aids patients affected by OA by reducing clinical symptoms. In theory the high N-3: N-6 ratio helps to reduce inflammation and pain by modifying cell walls and by blocking some of the genes that cause inflammation in arthritic joints [147 - 149].

Other alternative therapies are aimed at chondroprotection in the attempt to reduce cartilage degeneration and to increase cartilage matrix regeneration. Chondroitin sulfate and glucosamine hydrochloride supplements are used in this respect despite weak and conflicting clinical studies regarding their effectiveness at reducing clinical symptoms associated with OA. Adequan is an intramuscular injectable product that contains Polysulfated Glycosaminoglycans (PSGAGS). It is given twice weekly for up to 4 weeks. It has weak clinical evidence to support its use for OA by having anabolic effects on cartilage metabolism and reduction of lameness. Other products that have limited clinical data but may be useful are green-lipped mussel extract and Zeel. Although gabapentin, tramadol and amitriptyline are all prescribed for use as analgesics for OA, there is little clinical evidence to support their use [148].

Non-steroidal anti-inflammatories (NSAIDs) have been shown to have significant efficacy in multiple clinical trials for control of pain and improved function with OA, namely carprofen, firocoxib and meloxicam. NSAIDs are used for their anti-inflammatory and analgesic properties as they inhibit cyclooxygenase 1 and 2 (COX-1 and 2) pathways with preference for COX-2 inhibition. NSAIDs are widely used as a long term therapy to reduce symptoms associated with OA. They are generally started at the maximum daily dose and are reduced to a minimum daily effective dose. Although we do not have good evidence to support the use of NSAIDs at lower doses, we aim to reduce the potential for negative long term side effects such as gastrointestinal ulceration and hepatic and renal toxicity [148, 150].

The approach to the arthritic patient is multimodal and regardless of our best current efforts the disease progresses and eventually the animal suffers from total joint failure. In severe cases orthopedic arthrodesis surgeries or limb amputation can be performed in an attempt to remove the inciting cause of pain. With these techniques the animal completely loses normal joint function and success rates vary with arthrodesis procedures.

Mesenchymal stem cell (MSC) therapy for OA is intriguing for many reasons, but four properties in particular make them appealing: (1) they are a bioactive living system that can work and maintain a lasting and adaptable effect; (2) they are anti-inflammatory and therefore should slow or stop the progress of OA and improve clinical symptoms, (3) they can "home" to the site of injury within the joint to have a more targeted effect, and (4) they have the potential to create new cartilage either directly by differentiation into chondrocytes or by paracrine function to recruit nascent stem cell and progenitor cell populations. Currently, some clinicians are implementing autologous cellular therapies for joint disease including platelet rich plasma (PRP) preparations [151]. There are several point of care systems that prepare cell isolates from blood, adipose or bone marrow in the clinic for immediate use (Companion Regenerative Therapies, MediVet Biologics, Harvest Technologies). There are also cell therapy providers that work in a GMP environment for improved cleanliness and quality, and these providers can also offer fresh isolates such as the stromal vascular fraction (SVF) from adipose tissue (VetCell Therapeutics (VCT), VetStem Biopharma). They also have the capability of culture expansion and storage of cells for a greatly enhanced homogenous population of mesenchymal stem cells (MSCs) which are the cells of interest and multiple doses can be saved and stored for future use. Some clinicians promote intravenous injection of stem cell preparations for joint disease, but treatment *via* intraarticular injection seems to be a more logical approach and cells may not reach the joints or other injured sites through the vascular system [152]. There are many publications supporting the use of MSCs and SVF for OA, and currently groups are working on CVM (FDA) approval for allogenic cell based therapies that will be "off the shelf" cellular products when they are approved (see "Veterinary Clinical Studies" section 8.1 for examples of current clinical studies).

Equine patients suffer from the same disease process as canine patients with osteoarthritis, and like their canine counterparts NSAIDs are often the first palliative medical treatment modality that is implemented. There are a number of different NSAIDs on the market but the most widely used in the equine market are phenylbutazone, flunixin meglumine, carprofen, meloxicam and ketoprofen. Unlike the small animal industry, intra-articular injections are more widely practiced in equine medicine. This is likely because they have more numerous joints that are accessible, larger joint spaces, their high dependence on proper limb function to perform, and the fact that they are highly prized animals with narrow margins of competitive success and so have undergone more rigorous testing for medical therapies. Corticosteroids such as triamcinolone are injected intra-articular for anti-inflammatory effects and reduced nociception. Hyaluronan (HA) can be administered with corticosteroids to add a chondroprotective effect. HA is a large non-sulphated glycosaminoglycan component of synovial fluid that is responsible for viscoelasticity, boundary lubrication and steric hindrance. It also contributes compressive stiffness as part of the proteoglycan in the cartilage matrix. Although most of its support comes from *in vitro* studies, the belief is that

HA may contribute as a disease modifying agent [153].

Polysulphated glycosaminoglycans (PSGAGs) are also used in equine patients, although they can be injected either intra-articular or intra-muscular (Adequan use in canines is intra-muscular only). PSGAGs are also purportedly disease modifying agents with few studies proving their efficacy [153 - 155]. Oral nutraceuticals chondroitin sulphate and glucosamine are also administered as chondroprotectants but their efficacy is questionable. Tiludronate (Tildren) is a bisphosphonate that inhibits osteoclastic resorption of bone and is thought to reduce the activity of bone remodeling in active OA. One clinical study has shown efficacy of tiludronate for the treatment of distal tarsal OA [153, 156].

Interleukin-1 receptor antagonist protein (IRAP) or Autologous conditioned serum (ACS) is a new therapy that is aimed at anti-inflammation. ACS is generated from 50 cc of blood that is incubated overnight in a syringe with chromium coated glass spheres that stimulate white blood cells to produce anti-inflammatory cytokines. There is an increase in IRAP but the solution also contains many other blood derived proteins that make up the serum. The solution is injected intra-articular in 2 ml aliquots as needed. Although this therapy is widely adopted and it has been shown to reduce lameness in horses in a clinical study, clinical efficacy data is conflicting at this time and more clinical trials are needed [153, 157].

The equine industry has benefited from widely used stem cell and platelet rich injections into ligament injuries and affected joints for over ten years. The most widely used technique for stem cell use is to extract bone marrow or adipose tissue to obtain cellular isolates from these. At this point either the cell mixture is prepared for injection or it is cultured to gain a greatly enhanced population of mesenchymal stem cells (MSCs) which are then injected into the affected area. Concentrated platelets are obtained from blood and are prepared in a special manner in order to create platelet rich plasma (PRP) which has potent anti-inflammatory effects [157 - 159]. There is still a lot to be learned about cellular therapies but the treatments for equine patients remain promising. Some groups are also investigating tendon and ligament injury treatments in canine patients utilizing stem cell therapies [160]. While this is a novel and promising approach, more controlled studies and trials are needed to assess the effectiveness and understand dosing schedules for stem cell therapy applications in muscle, tendon and ligament injuries in canines.

Feline osteoarthritis (OA) has remained elusive for many years. This may be due to the difficulty with detecting pain and lameness in feline patients by either the owner or the veterinarian. Cats are prey animals and will hide any symptoms of pain or discomfort until end-stage disease. They are also more agile and carry less weight and therefore can manage and disguise arthritic pain more readily. More recently clinicians have begun investigating cats for arthritis and remarkably most cats over the age of 12 have significant OA of either the hips, stifles, shoulders or elbows. Feline patients may be difficult to diagnose, but clinical symptoms may include the inability to jump up to a previously common perching site (such as the countertop), reduced activity, reduced mobility or agility, inappetence, lethargy, pain or aggression. Many of these symptoms are non-specific thus highlighting the difficulty in detection of OA without radiographic studies. Feline patients will likely be screened, diagnosed and treated for OA in the near future, which heightens the need for effective treatments in our feline patients. As of now, there are not many viable options. One may institute a high omega-3:6 fatty acid diet, chondroitin / glucosamine supplements, and pain control as needed (such as trans-mucosal buprenorphine). NSAID use in cats is controversial due to the risk of acute kidney failure, but likely feline patients would benefit greatly from their use. In the USA there is an approved NSAID Onsior, but it is only licensed for use 3 days at a time following a spay procedure. In the UK metacam is licensed for use at 0.05 mg / kg per os once daily long term [161].

Chronic kidney disease (CKD) may affect canine and equine patients, but it is one of the most common aging conditions of feline patients. There are varying reports from 7.5% to over 50% of cats affected over the age of 10 years, but this only highlights that CKD is a commonly diagnosed disease in the feline [162]. Symptoms of CKD do not arise until 65-75% of nephron loss has been reached in both kidneys. The kidneys are not very plastic as other organs such as the liver are, but they can withstand widespread damage before a patient becomes ill. Azotemia is slow to develop because remaining functional nephrons compensate by excreting more solute, which leads to a solute osmotic diuresis, thus effectively increasing single nephron glomerular filtration rate (SNGFR) to maintain a normal glomerular filtration rate (GFR). CKD can arise due to many primary causes such as hereditary nephropathies, pyelonephritis, renal neoplasia, renal calculi, hydronephrosis, polycystic kidney disease, chronic tubulointerstitial inflammation, nephrotoxicity, and idiopathic. Clinical signs include polyuria / polydipsia (pu / pd), weight loss and BCS reduction, anorexia, emesis, halitosis, oral and gastrointestinal inflammation and ulceration, weakness, lethargy, depression, anemia, reduced immunity and poor hemostasis (platelet dysfunction). Secondary complications with CKD include renal secondary hyperparathyroidism, systemic hypertension, anemia, and various metabolic derangements such as dehydration, hypokalemia and metabolic acidosis [163, 164].

Diagnosis of CKD initially involves a combination of elevated or increasing blood urea nitrogen (BUN) and creatinine levels and a low urine specific gravity. The International Renal Interest Society (IRIS) provides much information regarding diagnosis and management of CKD and they offer recommendations about staging the disease, which involves further workup including urine protein: creatinine values and blood pressure. Further helpful diagnostics include ultrasonography, renal biopsy, and GFR.

CKD is generally considered irreversible and treatment is supportive rather than curative. The aim is to improve the quality of life and extend the duration of life by eliminating or reducing systemic signs associated with the disease and by attempting to slow progression of deteriorating renal function. Treatment includes dietary management mainly to decrease phosphate (protein) intake (Hills, Purina, Royal Canin prescription diets, *etc.*), hydration through utilizing wet foods, offering water at all times, and sometimes twice weekly subcutaneous (SQ) crystalloid fluids and periodic hospitalization with intravenous (IV) fluids. ACE inhibitors such as amlodipine may be instituted to treat associated hypertension, oral potassium supplements may be needed for hypokalemia, and blood transfusions and / or recombinant erythropoietin may be utilized for severe anemia of chronic disease. A questionable 'probiotic' product called Azodyl has entered the market, which supposedly can help the body to excrete excess nitrogenous waste through the gastrointestinal tract and thus reduce BUN. This may be the case but evidence is not strong to support this claim, although it does seem to have a positive effect by aiding hydration in the cats. Lastly, kidney transplantation can be an option, but there are only a handful of hospitals offering this in the United States and it has its own set of serious risks. CKD is a disease that would benefit immensely from a disease modifying agent to stop or even reverse glomerular and tubulointerstitial damage. It is a challenging discovery as the kidney is a complicated organ with many working parts and it is for the most part non-regenerative. Stem cells that can act as local anti-inflammatories and recruit nascent stem cells may have lasting benefit. With recent advances in 3-D printing, this may allow whole sections of functional kidneys to be grown in the lab and later engrafted into the patient [163, 164]. There are a few CKD clinical reports in the literature and one is mentioned in the "Veterinary Clinical Studies" section 8.1.

Feline chronic gingivostomatitis (FCGS) is a fairly common oral inflammatory disease of unknown etiology. It is reported to affect approximately 0.7% of the population and approximately 30% of these cases are non-responders to tooth extractions and require further medical management [165, 166]. It is very frustrating to treat and oftentimes requires full mouth tooth extractions and in some cases even this does not cure the disease. The condition is thought to be a chronic overactive immune response to low levels of plaque antigens and some associations may be feline viruses, gram negative anaerobes and high density cat populations. The lesions are submucosal and involve infiltrates of plasma cells, lymphocytes, macrophages and neutrophils. Clinical signs include mouth pain, head shyness, salivation, halitosis, reduced appetite, reduced grooming, sub-mandibular lymph node enlargement and personality changes. The interior of the mouth or at least the gum margins at affected teeth is red, angry and raw. Treatment ultimately includes removal of all cheek teeth, but medical therapy with corticosteroids, cyclosporine and antibiotics may be attempted. Stem cell therapy that can control the inflammatory and immune response in these cats could change the way they are treated in the future. A recent clinical study utilizing autologous MSCs *via* repeated intravenous administration has shown great improvement in feline patients affected by FCGS that were non-responders to tooth extractions and medical therapy. This is a neat example that highlights MSC treatment for a disease process that is different from that of osteoarthritis where there is no interest in tissue regeneration. This study shows great promise for utilizing MSCs for the treatment of a chronic immune-mediated and inflammatory disorder, where the immune-modulatory effect is most crucial [165]. It will be exciting to see the result of a similar study performed utilizing allogenic rather than autologous MSCs.

For many chronic disease condition, but osteoarthritis in particular, there is a great need for a therapy that helps reduce pain, inflammation and lameness, and increases range of motion and the patient's quality of life without having unwanted side effects. An even more desirable treatment would entail a disease modifying therapy that can stop progression of OA and even reverse its effects. The time may be upon us that this theoretical therapy may come to life. Stem cell research and therapies have been around with bone marrow transplants for almost 50 years, but recent embryonic and adult stem cell research has boomed since the early 2000s. Currently, significant efforts are underway to try to translate our knowledge about stem cells and their regenerative capabilities into clinical practice where both animal and human patients can benefit from their use. There are a number of clinical trials underway in both the human and veterinary medical fields. The majority of them focus on adult MSCs as, unlike embryonic stem cells (ESCs), they are deemed to be non-tumor forming and lack any ethical burdens, yet they are plastic cells with multipotent regenerative capabilities, immunepriviledged and immunomodulatory, anti-inflammatory, do not provoke an inflammatory response, home to sites of injury, integrate into host tissues, recruit nascent stem cells, and have been shown to reduce pain, inflammation and lameness, and to increase range of motion in canine patients with OA [39, 127, 135].

ORGAN FAILURE AND TISSUE ENGINEERING

The adult body undergoes continuous cell loss on a daily basis and these cells are replaced by new cells made from stem cells residing in different tissues. As an example skin will be completely renewed approximately every 14 days [167]. Therefore, there is a balance between the tissue degradation and regeneration in a dynamic manner. In chronic diseases, this balance is shifted towards tissue degradation and the body will produce an excess of fibroblasts and extra cellular matrix, mainly collagen to fill the gap. Infiltration of fibroblasts and collagen results in fibrosis and scar formation. Fibrotic tissue does not have the function of the original tissue thus the functionality of that organ will be affected. Chronic diseases result in fibrosis and scar formation and ultimately lead to end stage organ failure. Stem cell therapy can help reverse tissue damage before fibrosis develops and progresses. Therapies using low level lasers have been proven to be safe and effective for the reduction of fibrosis and pain management in companion animals [168]. Low level laser therapy could potentially prime chronic wounds or damaged tissues for stem cells to penetrate and initiate angiogenesis and tissue regeneration.

The adult body can also regenerate some organs using natural repair mechanism, such as what the liver can do. If part of the liver is lost by disease or injury, the liver grows back to its original size [169]. Unfortunately, many other adult tissues do not regenerate, and a goal in regenerative medicine is to find ways to kick-start tissue regeneration in the body, or to engineer replacement tissues. Tissue engineering is an interdisciplinary field that brings together the principles of the life sciences and medicine with those of engineering [170]. The increase in its development over the past decade has resulted from a variety of factors; advances in genomics and proteomics, the advent of new biomaterials as potential templates for tissue growth, improvements in bioreactor design, and increased understanding of healing processes. Possibly the greatest contribution has come from our increased knowledge and understanding of stem cell biology, which is paving the way for the generation of unlimited cells of specific phenotypes for incorporation into engineered tissue constructs. Thus, tissue engineering approaches for expanding and engrafting the differentiated progeny of pluripotent or adult stem cells have promising potential for tissue repair and will make a major contribution to medicine in the 21st century. Engineered bladder [171] and trachea [172] have already been reported in human medicine. In veterinary medicine, production of engineered tendons and cartilages could potentially help replace the damaged tissues and improve the quality of life and possibly increase athletic performance and the lifespan of our companion animals [173].

LIMITATIONS AND CHALLENGES OF CELL-BASED THERAPY

Reality *vs.* Fantasy

While cell therapy can theoretically be useful for tissue regeneration, it is not the answer to all degenerative diseases that conventional medical therapies fail to treat. There is still a lot to learn about cell-based therapy and the more we understand the mechanisms by which cells repair the tissue we can implement a better strategy for their proper applications. We still do not know if cells integrate into the damaged organ or if they release cytokines and growth factors to recruit endogenous progenitor cells to repair the damaged tissues or both. We also do not know if cells stay in the area of transplantation or they can migrate to other areas. Cell migration depends on the type of cell and location of cell delivery and body damage. Studies regarding MSCs show that intravenous transplantation of MSCs result in localization of more than 90% of these cells in the lung. However, over time, cells will migrate to other major organs including heart, kidney and liver [174]. Therefore, one might think targeted therapy would be more desirable than intravenous transplantation. Transplantation of radio-labeled canine MSCs showed that following intra-articular injection, MSCs have migrated to other organs [54]. Caplan *et al.* demonstrated that if MSCs are collected by surgical removal to repair a damaged tissue in other places and if the MSCs are returned to the body, they will most likely go back to the surgical area than to the pre-existing damaged area [61]. This highlights their homing capability, but also a potential difficulty in targeting specific disease sites.

There are many references indicating cytokines and growth factors produced by cell products may recruit patient's own progenitor cells [175]. A perfect example of such a product is platelet rich plasma (PRP). PRP is a cell based product derived from peripheral blood and has four to six times more platelet concentration than whole blood. PRP contains numerous cytokines including PDGF, VEGF, and TGF-B and has been shown to have regenerative potentials [176]. However, it is again wishful thinking that PRP is useful for all regenerative indications, as different sets of factors with different concentrations might be needed for specific therapeutic indications. As PRP is approved by the FDA, in medical practice it is being widely tested for numerous clinical indications from hair regeneration and intervertebral disc regeneration to osteoarthritis [177]. While there are indications that cell-based products have helped patients by reducing pain and improving their quality of life, most of the studies are anecdotal and double blinded placebo controlled studies are needed to prove the efficacy of the cell products specifically from the disease modifying perspective.

Technical Challenges

Like any new technology, cell-based therapy encounters numerous challenges on its way that needs to be resolved to make it more effective and acceptable for clinical implication. Some of these challenges are: 1) what indications can benefit from cell therapy and what type of cells can be used for what indications? Generally, degenerative diseases, immune mediated tissue damage and inflammatory diseases can potentially benefit from cell based therapy. However as what type of cell can improve what indication not much information is available. Pluripotent and multipotent stem cells can theoretically be differentiated to many cell types, while progenitor cells will differentiate only to limited cell types. Thus, the ability to differentiate faster, more efficiently and consistently depends on the stem cell sources. 2) Do we need to differentiate stem cells to the cell type we need prior to application or let them differentiate in the body? If we pre-differentiate MSCs, do we lose the other important paracrine, homing, anti-inflammatory and immunomodulatory effects that are so desired? Some stem cells can differentiate to other cell types when they are being placed in an appropriate environment [178]. For other stem cells or conditions, it might be needed to differentiate them to the desired cell type prior to clinical applications. 3) If pluripotent or multipotent stem cells are used, how can one produce only one cell type from multipotent/pluripotent stem cells and how can they be separated from differentiated cells? This is a big safety concern as unwanted tissues might be generated in unwanted areas. Producing differentiated cells from stem cells is a reality but having a consistent protocol producing one cell type, or separating stem cells from differentiated cells needs more time to fully develop. 4) Another important aspect in cell therapy is the durability of cells after implantation. This will determine how long the cells will survive and can exert their therapeutic effect. This area of cell therapy is in its infancy and still a lot needs to be learned. However, as a general rule of aging, stem cells also age and over time they lose their replication potential. Studies have shown that aged stem cells have shorter telomere length tend to have less replication potential [179]. Thus one can imagine, stem cell sources from young donors could last longer than those from older donors. Also within the body some stem cells have longer telomere length and can last longer. For instance, germline stem cells (GSCs) located in the gonads that are responsible for production of gametes have a high telomerase activity and long telomeres [180]. As these cells are also multipotent, they are an ideal candidate for cell therapy and tissue regeneration. The major question regarding GSCs is can they be culture expanded in a safe system for therapeutic application and do they produce any unwanted side effects such as producing sperm at the injection site?

As discussed throughout the paper, another area that cell therapy is going to make a difference is with tissue engineering. While simple organs like the bladder have been made and applied in treating patients [181], building complex organs like a kidney and heart are more like fantasy at this stage and need more time to develop. 3D printing has revolutionized tissue engineering and a combination of cell products and an extracellular matrix (ECM) using 3D printer have been used to produce complex structures such as blood vessels, tendons and cartilage. In the near future, by increasing our knowledge about cell products and tissue engineering, more complex organs such as kidneys and hearts may be produced.

Regulatory Encounters

In the United States, clinical trials designed to demonstrate the safety and efficacy of cell therapeutic products are regulated by the U.S. Food and Drug Administration (FDA). European Medicines Agency (EMA) is the organization responsible for regulation of biologics, drugs and cell products in Europe. Currently, regulations intended for cell based therapeutic products are based off of the Code of Federal Regulations (CFR) guidance originally written for the development and manufacturing of pharmaceuticals. This is primarily due to the FDA's position on cell therapy: "Stem cells, like other medical products that are intended to treat, cure or prevent disease, generally require FDA approval before they can be marketed" [182]. The role of the FDA is to protect the public from unapproved (illegal) stem cell therapies and their potential dangers. The FDA cautions the public to make sure that any stem cell treatment that is being considered has been approved by the FDA or is being studied under a clinical investigation. Presently, there are 10 FDA approved cell therapeutic products and related devices for human use [183]. However, there are no approved cell therapeutic products for use with food-producing and companion animals.

Cell therapeutic products intended to be used as a new animal drug are regulated by the FDA's Center for Veterinary Medicine (CVM). The center approves animal drugs and cell based products for companion (pet) animals, such as dogs, cats, and horses; and for food-producing animals, such as cattle, pigs, and chickens. Although not as stringent as regulations for human cell therapeutic products, the recent published guidelines provide awareness for developing, manufacturing, or marketing cell based products for animal use [184].

The CVM has two pathways to market a cell based product: the formal approval process and an application for low priority status. Both processes involve establishing a working relationship with the CVM to review manufacturing procedures, the clinical protocol and intended use of the product. The formal approval process is modeled after the traditional clinical trial pathway for human drugs. The purpose of the approval process is to prove efficacy of the cell based product and to show it is safe, *e.g.* it will not be tumorigenic after 6 – 9 months. This process is reserved for products that meet at least one of the following criteria: are more than minimally manipulated, used for non-homologous use, use in a food producing animal, the product is dependent on the cell's metabolism, involved combination with other cell types or articles during manufacturing and if the product is combined with another drug or device [184]. If none of these criteria are met, then the cell based product may be submitted for a low priority status. A safety and efficacy trial is not needed for this status. An information package justifying how the product is not more than minimally manipulated, a review of the manufacturing process, the indications of use, packaging inserts, proper labeling strategy, defined product composition, any summarized pilot data, and marketing materials would be submitted to the CVM for their review. It could take up to 60 days to make a determination for low priority status. Once approved, the cell based product may be freely marketed and sold in the U.S and its territories.

Over the past decade, a dramatic increase in the number of peer reviewed publications exploring the therapeutic use of various stem cell populations, and evaluating their use in different preclinical disease models has trended out. While some cell types have shown encouraging clinical success, only a minority of new cell therapeutic treatments have proven to be both safe and effective. This provides a challenge to bring cell therapies to the market. Regulatory oversight of the approval process is based on the existing framework applied to pharmaceuticals. Some experts contend that the regulatory framework is overly burdensome for cell therapies and it needs refining to become more predictable and efficient [185]. The main focal point of regulatory oversight is with allogeneic products, since products used for autologous treatments may be judged to be minimally manipulated, which earn a low priority status [89].

However even though the applications of certain stem cell treatments have been proven to be safe, the actual treatment may not be an improvement over standard therapy. Examples of this are documented in human applications of bone marrow derived mononuclear cells [186] and MSCs [187]. Use of these cells is fairly new in the clinics and their underperformance may be a function of insufficient understanding of the fate of cell products following infusion. Recently, some late stage clinical trials have failed to meet primary endpoints that were focused on repairing damaged or diseased connective tissue. The failure of these cells could relate to unknown mechanisms of MSCs following infusion [188]. To demonstrate the result of infused cell therapeutic products, the incorporation of cell-tracking techniques is critical during the safety and efficacy tests of a trial. As an example, one study reported that only a small percentage of infused MSCs (<1%) reach the target tissue. Cell entrapment is commonly observed in capillaries within the liver, spleen and lungs [189]. From the FDA's perspective, if the efficacy trial does not prove significant then the cell therapeutic product will not receive approval.

The quality systems and manufacturing operations, consistent with cGMP standards, are also components that weigh heavily on the FDA approval process. The production knowledge must include an understanding of variability in raw materials, the relationship between a process and product's critical quality attributes (CQA), and the connection between CQAs and a product's clinical properties. An autologous cell based product, due to its custom nature, will always present a variable starting material. These products also tend to have minimal refinement as they may contain multiple cell types (stem and non-stem cells) at which concentrations will change from patient to patient. To compensate for this occurrence, the other components in manufacturing require thoughtful and rigorous controls. An allogeneic product offers the opportunity to have more control over raw material and entails well-controlled comparative clinical trials as well as pre-approval evaluation followed by post-approval evaluation [190]. As such, the FDA is intricately involved in the approval process for a veterinary cell based allogeneic product over the course of 2 - 3 years.

Current CVM (FDA) regulations balance protecting public health by requiring data demonstrating that certain uses of cells are safe and effective. The regulatory and clinical trial process is intended to ensure public safety during this time of scientific advancement. Whether it is for human use or for veterinary use, the FDA must be convinced that cell based products perform as they are claimed and with significant outcomes.

FUTURE IMPLICATIONS

While cell therapy has been recently introduced to the veterinary industry and is being tested for conditions like osteoarthritis, feline chronic gingivostomatitis (FCGS) and tendon repair, there is hope that in the future more complicated degenerative diseases could also benefit from this technology. Diabetes mellitus is a metabolic/endocrine disease, which depending on pathogenesis is classified as two types. Type 1 diabetes or juvenile diabetes, which is more common in dogs than cats, is due to insufficient production of insulin by beta cells in the endocrine pancreas. Type 2 diabetes is more common in cats and is due to the metabolic and nutritional imbalance causing desensitization of cells to insulin. Current medical management for diabetes includes low carbohydrate diet, exercise and daily insulin administration. Transplantation of stem cells or beta cells produced from stem cells into the pancreas could potentially cure type 1 diabetic patients. Studies in the laboratory animals have shown that stem cells could benefit mice that have lost their beta cells in a diabetic induced model [191].

Neurodegenerative diseases also might benefit from stem cells. Spinal cord injury in particular has created a lot of attention, as transplantation of stem cells and regeneration of neurons in the damaged area might alleviate some of the problems associated with the disease and help patients reduce urinary incontinence and increase some of their motor function. Due to their anatomy and long lumbar region, some breeds of dogs are predisposed for intervertebral disc degeneration and spinal cord injury including Dachshund and Pekingese. Studies in laboratory animals showed that stem cells injected into mice can repair damaged spinal cords and help partially paralyzed mice to walk again [192]. A variety of stem cells including olfactory nerve stem cells, embryonic stem cells and induced pluripotent stem cells have been used for spinal cord regeneration each have their own ethical and practical considerations [193].

Chronic kidney disease (CKD) and congestive heart failure are two other degenerative diseases that could potentially benefit from cell therapy. In the case of the heart, stem cells could be injected into heart muscle using catheterization. Numerous studies have shown that autologous bone marrow stem cells [194] or allogeneic mesenchymal stem cells [195], or cardiac muscle stem cells [196] have improved cardiac output and exercise tolerance of human patients. Applications of stem cells in human medicine for cardiac regeneration are very promising and a lot of clinical trials have been started using a variety of stem cells. But in veterinary medicine, especially in small animals, stem cells might not be as effective as there are always underlining factors contributing to heart failure that have to be addressed. As such, without fixing those issues all the effort to repairing a damaged heart will be negated. Chronic kidney failure is the most important debilitating disease in cats. Current medical management of the disease is diet change (low protein), anti-uremic medications, and fluid therapy. Due to the complex anatomy of the kidney it is very difficult to introduce stem cells to the damaged area. Two methods have been tried in cats with CKD; transplantation into the kidney artery and intravenous infusion. Transplantation into the kidney artery has been very challenging as it requires extensive sedations which itself cause kidney damage [134]. While there is a lot of hope for these patients, the right type and the correct dose of cells and the best time and location of injection needs to be further investigated.

Two other important areas that stem cell therapy could potentially benefit companion animals are immune mediated diseases (anemia, systemic lupus erythematous, atopy, *etc.*) and immune suppression in conjunction with organ transplant. Mesenchymal stem cells are shown to have both immune modulatory [38] and immune suppression [9] effects. In the future, cell products or their derivatives could improve the quality and prolong the life of many patients that otherwise could not have been treated with conventional medical therapy.

CONFLICT OF INTEREST

The authors confirm that they have no conflict of interest to declare for this publication.

ACKNOWLEDGEMENTS

The authors would like to thank Dr. Steven Budsberg for general discussions regarding osteoarthritis and cell therapy applications; Dr. Jeff Mayo for discussion about the use of PRP in small animal orthopedic conditions; Dr. Dori Borjesson for general discussions related to cell therapy for immune-mediated diseases (feline chronic gingivostomatitis in particular); Dr. Jessica Quimby for general discussions about the application of cell therapy in chronic kidney disease; and Dr. Sherman and Dr. Debra Canapp for general discussions about cell therapy applications for tendon, ligament, and joint injuries and diseases in dogs. We also acknowledge Dr. James Schafer and Dr. Guy Wouters for valuable discussions about regulatory considerations of veterinary cell therapy in the USA and EU. We would like to give our utmost appreciation to Mr. Thomas C. K. Yuen for his continuous support and inspiration to prepare this manuscript.

REFERENCES

[1] Tang Q, Bluestone JA, Kang SM. CD4(+)Foxp3(+) regulatory T cell therapy in transplantation. J Mol Cell Biol 2012; 4(1): 11-21.
[http://dx.doi.org/10.1093/jmcb/mjr047] [PMID: 22170955]

[2] Nishikawa H, Sakaguchi S. Regulatory T cells in cancer immunotherapy. Curr Opin Immunol 2014; 27(April): 1-7.
[http://dx.doi.org/10.1016/j.coi.2013.12.005] [PMID: 24413387]

[3] Whiteside TL. Regulatory T cell subsets in human cancer: are they regulating for or against tumor progression? Cancer Immunol Immunother 2014; 63(1): 67-72.

[http://dx.doi.org/10.1007/s00262-013-1490-y] [PMID: 24213679]

[4] Palucka K, Banchereau J. Cancer immunotherapy *via* dendritic cells. Nat Rev Cancer 2012; 12(4): 265-77.
[http://dx.doi.org/10.1038/nrc3258] [PMID: 22437871]

[5] Mehta-Damani A, Markowicz S, Engleman EG. Generation of antigen-specific CD8+ CTLs from naive precursors. J Immunol 1994; 153(3): 996-1003.
[PMID: 8027569]

[6] Mehta-Damani A, Markowicz S, Engleman EG. Generation of antigen-specific CD4+ T cell lines from naive precursors. Eur J Immunol 1995; 25(5): 1206-11.10..

[7] Datta J, Berk E, Cintolo JA, Xu S, Roses RE, Czerniecki BJ. Rationale for a multimodality strategy to enhance the efficacy of dendritic cell-based cancer immunotherapy. Front Immunol 2015; 6: 271-81.
[http://dx.doi.org/10.3389/fimmu.2015.00271] [PMID: 26082780]

[8] Anguille S, Smits EL, Lion E, van Tendeloo VF, Berneman ZN. Clinical use of dendritic cells for cancer therapy. Lancet Oncol 2014; 15(7): e257-67.
[http://dx.doi.org/10.1016/S1470-2045(13)70585-0] [PMID: 24872109]

[9] Vacchelli E, Aranda F, Obrist F, *et al.* Immunostimulatory cytokines in cancer therapy(2014) Trial watch,OncoImmunology, 2014; 3:6(e29030)
[http://dx.doi.org/10.4161/onci.29030.]

[10] Brandenburg S, Takahashi T, de la Rosa M, *et al.* IL-2 induces *in vivo* suppression by CD4(+)CD25(+)Foxp3(+) regulatory T cells. Eur J Immunol 2008; 38(6): 1643-53.
[http://dx.doi.org/10.1002/eji.200737791] [PMID: 18493984]

[11] Camisaschi C, Filipazzi P, Tazzari M, *et al.* Effects of cyclophosphamide and IL-2 on regulatory CD4+ T cell frequency and function in melanoma patients vaccinated with HLA-class I peptides: impact on the antigen-specific T cell response. Cancer Immunol Immunother 2013; 62(5): 897-908.
[http://dx.doi.org/10.1007/s00262-013-1397-7] [PMID: 23589107]

[12] Weiss L, Letimier FA, Carriere M, *et al. In vivo* expansion of naive and activated CD4+CD25+FOXP3+ regulatory T cell populations in interleukin-2-treated HIV patients. Proc Natl Acad Sci USA 2010; 107(23): 10632-7.
[http://dx.doi.org/10.1073/pnas.1000027107] [PMID: 20498045]

[13] Sim GC, Martin-Orozco N, Jin L, *et al.* IL-2 therapy promotes suppressive ICOS+ Treg expansion in melanoma patients. J Clin Invest 2014; 124(1): 99-110.
[http://dx.doi.org/10.1172/JCI46266] [PMID: 24292706]

[14] West EE, Jin HT, Rasheed AU, *et al.* PD-L1 blockade synergizes with IL-2 therapy in reinvigorating exhausted T cells. J Clin Invest 2013; 123(6): 2604-15.
[http://dx.doi.org/10.1172/JCI67008] [PMID: 23676462]

[15] Mortha A, Chudnovskiy A, Hashimoto D, *et al.* Microbiota-dependent crosstalk between macrophages and ILC3 promotes intestinal homeostasis. Science 2014; 343(6178): 1249288.
[http://dx.doi.org/10.1126/science.1249288] [PMID: 24625929]

[16] Sousonis V, Malliaras K, Terrovitis J, Nanas J. Cell Therapy in Cardiac Diseases. Introduction to Translational Cardiovascular Research 2015; pp. 565-82.

[17] Sun J, Mandai M, Kamao H, *et al.* Protective effects of human iPS-derived retinal pigmented epithelial cells in comparison with human mesenchymal stromal cells and human neural stem cells on the degenerating retina in rd1 mice. Stem Cells 2015; 33(5): 1543-53. [http://dx.doi.org/10.1002/stem.1960] [PMID: 25728228]

[18] Raikwar SP, Kim EM, Sivitz WI, Allamargot C, Thedens DR, Zavazava N. Human iPS cell-derived insulin producing cells form vascularized organoids under the kidney capsules of diabetic mice. PLoS One 2015; 10(1): e0116582. [http://dx.doi.org/10.1371/journal.pone.0116582] [PMID: 25629318]

[19] Hess D. Introduction to cell therapy in brain injury. Cell Therapy for Brain Injury 2015; pp. 1-9. [http://dx.doi.org/10.1007/978-3-319-15063-5_1]

[20] Barker RA, Jain M, Armstrong RJ, Caldwell MA. Stem cells and neurological disease. J Neurol Neurosurg Psychiatry 2003; 74(5): 553-7. [http://dx.doi.org/10.1136/jnnp.74.5.553] [PMID: 12700287]

[21] Nadig RR. Stem cell therapy - Hype or hope? A review. J Conserv Dent 2009; 12(4): 131-8. [http://dx.doi.org/10.4103/0972-0707.58329] [PMID: 20543921]

[22] Okano H, Nakamura M, Yoshida K, *et al.* Steps toward safe cell therapy using induced pluripotent stem cells. Circ Res 2013; 112(3): 523-33. [http://dx.doi.org/10.1161/CIRCRESAHA.111.256149] [PMID: 23371901]

[23] Takahashi K, Yamanaka S. Induction of pluripotent stem cells from mouse embryonic and adult fibroblast cultures by defined factors. Cell 2006; 126(4): 663-76. [http://dx.doi.org/10.1016/j.cell.2006.07.024] [PMID: 16904174]

[24] Stadtfeld M, Nagaya M, Utikal J, Weir G, Hochedlinger K. Induced pluripotent stem cells generated without viral integration. Science 2008; 322(5903): 945-9. [http://dx.doi.org/10.1126/science.1162494] [PMID: 18818365]

[25] Kaji K, Norrby K, Paca A, Mileikovsky M, Mohseni P, Woltjen K. Virus-free induction of pluripotency and subsequent excision of reprogramming factors. Nature 2009; 458(7239): 771-5. [http://dx.doi.org/10.1038/nature07864] [PMID: 19252477]

[26] Okita K, Matsumura Y, Sato Y, *et al.* A more efficient method to generate integration-free human iPS cells. Nat Methods 2011; 8(5): 409-12. [http://dx.doi.org/10.1038/nmeth.1591] [PMID: 21460823]

[27] Miyoshi N, Ishii H, Nagano H, *et al.* Reprogramming of mouse and human cells to pluripotency using mature microRNAs. Cell Stem Cell 2011; 8(6): 633-8. [http://dx.doi.org/10.1016/j.stem.2011.05.001] [PMID: 21620789]

[28] Sánchez Alvarado A, Tsonis PA. Bridging the regeneration gap: genetic insights from diverse animal models. Nat Rev Genet 2006; 7(11): 873-84. [http://dx.doi.org/10.1038/nrg1923] [PMID: 17047686]

[29] Brockes JP, Kumar A, Velloso CP. Regeneration as an evolutionary variable. J Anat 2001; 199(Pt 1-2): 3-11. [http://dx.doi.org/10.1046/j.1469-7580.2001.19910003.x] [PMID: 11523827]

[30] Kern S, Eichler H, Stoeve J, Klüter H, Bieback K. Comparative analysis of mesenchymal stem cells from bone marrow, umbilical cord blood, or adipose tissue. Stem Cells 2006; 24(5): 1294-301. [http://dx.doi.org/10.1634/stemcells.2005-0342] [PMID: 16410387]

[31] Vieira NM, Brandalise V, Zucconi E, Secco M, Strauss BE, Zatz M. Isolation, characterization, and differentiation potential of canine adipose-derived stem cells. Cell Transplant 2010; 19(3): 279-89. [http://dx.doi.org/10.3727/096368909X481764] [PMID: 19995482]

[32] Reyes M, Verfaillie CM. Characterization of multipotent adult progenitor cells, a subpopulation of mesenchymal stem cells. Ann N Y Acad Sci 2001; 938: 231-3. [http://dx.doi.org/10.1111/j.1749-6632.2001.tb03593.x] [PMID: 11458512]

[33] Schwartz RE, Reyes M, Koodie L, *et al.* Multipotent adult progenitor cells from bone marrow differentiate into functional hepatocyte-like cells. J Clin Invest 2002; 109(10): 1291-302. [http://dx.doi.org/10.1172/JCI0215182] [PMID: 12021244]

[34] Wagers AJ, Weissman IL. Plasticity of adult stem cells. Cell 2004; 116(5): 639-48. [http://dx.doi.org/10.1016/S0092-8674(04)00208-9] [PMID: 15006347]

[35] Petit I, Szyper-Kravitz M, Nagler A, *et al.* G-CSF induces stem cell mobilization by decreasing bone marrow SDF-1 and up-regulating CXCR4. Nat Immunol 2002; 3(7): 687-94. [http://dx.doi.org/10.1038/ni813] [PMID: 12068293]

[36] Kucia M, Ratajczak J, Reca R, Janowska-Wieczorek A, Ratajczak MZ. Tissue-specific muscle, neural and liver stem/progenitor cells reside in the bone marrow, respond to an SDF-1 gradient and are mobilized into peripheral blood during stress and tissue injury. Blood Cells Mol Dis 2004; 32(1): 52-7. [http://dx.doi.org/10.1016/j.bcmd.2003.09.025] [PMID: 14757413]

[37] Yingling GL, Nobert KM. Regulatory considerations related to stem cell treatment in horses. J Am Vet Med Assoc 2008; 232(11): 1657-61. [http://dx.doi.org/10.2460/javma.232.11.1657] [PMID: 18518806]

[38] Yagi H, Soto-Gutierrez A, Parekkadan B, *et al.* Mesenchymal stem cells: Mechanisms of immunomodulation and homing. Cell Transplant 2010; 19(6): 667-79. [http://dx.doi.org/10.3727/096368910X508762] [PMID: 20525442]

[39] Meirelles LdaS, Fontes AM, Covas DT, Caplan AI. Mechanisms involved in the therapeutic properties of mesenchymal stem cells. Cytokine Growth Factor Rev 2009; 20(5-6): 419-27. [http://dx.doi.org/10.1016/j.cytogfr.2009.10.002] [PMID: 19926330]

[40] Milner PI, Clegg PD, Stewart MC. Stem cell-based therapies for bone repair. Vet Clin North Am Equine Pract 2011; 27(2): 299-314. [http://dx.doi.org/10.1016/j.cveq.2011.05.002] [PMID: 21872760]

[41] Koch TG, Betts DH. Stem cell therapy for joint problems using the horse as a clinically relevant animal model. Expert Opin Biol Ther 2007; 7(11): 1621-6. [http://dx.doi.org/10.1517/14712598.7.11.1621] [PMID: 17961087]

[42] Atesok K, Li R, Stewart DJ, Schemitsch EH. Endothelial progenitor cells promote fracture healing in a segmental bone defect model. J Orthop Res 2010; 28(8): 1007-14. [PMID: 20135674]

[43] McDuffee LA, Pack L, Lores M, Wright GM, Esparza-Gonzalez B, Masaoud E. Osteoprogenitor cell therapy in an equine fracture model. Vet Surg 2012; 41(7): 773-83. [http://dx.doi.org/10.1111/j.1532-950X.2012.01024.x] [PMID: 22804243]

[44] Jung Y, Bauer G, Nolta JA. Concise review: Induced pluripotent stem cell-derived mesenchymal stem cells: progress toward safe clinical products. Stem Cells 2012; 30(1): 42-7. [http://dx.doi.org/10.1002/stem.727] [PMID: 21898694]

[45] Pranke PHL, Chagastelles PC. Potentialities of adipose-derived mesenchymal stem cells collected from liposuction for use in cellular therapy 2014; 151-7.

[46] Caplan AI, Dennis JE. Mesenchymal stem cells as trophic mediators. J Cell Biochem 2006; 98(5): 1076-84. [http://dx.doi.org/10.1002/jcb.20886] [PMID: 16619257]

[47] Joyce N, Annett G, Wirthlin L, Olson S, Bauer G, Nolta JA. Mesenchymal stem cells for the treatment of neurodegenerative disease. Regen Med 2010; 5(6): 933-46. [http://dx.doi.org/10.2217/rme.10.72] [PMID: 21082892]

[48] Park KS, Kim YS, Kim JH, *et al.* Trophic molecules derived from human mesenchymal stem cells enhance survival, function, and angiogenesis of isolated islets after transplantation. Transplantation 2010; 89(5): 509-17. [PMID: 20125064]

[49] Abreu S, Antunes M, Branco V, Cruz F, *et al.* Bone marrow, adipose and lung tissue-derived mesenchymal stromal cells yield distinct effects on lung inflammation and remodeling in experimental allergic asthma. A109 Remodeling and the Matrix 2016; A2828-8.

[50] Manning CN, Martel C, Sakiyama-Elbert SE, *et al.* Adipose-derived mesenchymal stromal cells modulate tendon fibroblast responses to macrophage-induced inflammation *in vitro*. Stem Cell Res Ther 2015; 6: 74. [http://dx.doi.org/10.1186/s13287-015-0059-4] [PMID: 25889287]

[51] Lanzkron SM, Collector MI, Sharkis SJ. Hematopoietic stem cell tracking *in vivo* : a comparison of short-term and long-term repopulating cells. Blood 1999; 93(6): 1916-21. [PMID: 10068664]

[52] Gholamrezanezhad A, Mirpour S, Bagheri M, *et al. In vivo* tracking of 111In-oxine labeled mesenchymal stem cells following infusion in patients with advanced cirrhosis. Nucl Med Biol 2011; 38(7): 961-7. [http://dx.doi.org/10.1016/j.nucmedbio.2011.03.008] [PMID: 21810549]

[53] Jing XH, Yang L, Duan XJ, *et al. In vivo* MR imaging tracking of magnetic iron oxide nanoparticle labeled, engineered, autologous bone marrow mesenchymal stem cells following intra-articular injection. Joint Bone Spine 2008; 75(4): 432-8. [http://dx.doi.org/10.1016/j.jbspin.2007.09.013] [PMID: 18448377]

[54] Wood JA, Chung DJ, Park SA, *et al.* Periocular and intra-articular injection of canine adipose-derived mesenchymal stem cells: an *in vivo* imaging and migration study. J Ocul Pharmacol Ther 2012; 28(3): 307-17. [http://dx.doi.org/10.1089/jop.2011.0166] [PMID: 22175793]

[55] Hoogduijn MJ. Are mesenchymal stromal cells immune cells? Arthritis Res Ther 2015; 17: 88. [http://dx.doi.org/10.1186/s13075-015-0596-3] [PMID: 25880839]

[56] Soleymaninejadian E, Pramanik K, Samadian E. Immunomodulatory properties of mesenchymal stem cells: cytokines and factors. Am J Reprod Immunol 2012; 67(1): 1-8. [http://dx.doi.org/10.1111/j.1600-0897.2011.01069.x] [PMID: 21951555]

[57] Augello A, Tasso R, Negrini SM, *et al.* Bone marrow mesenchymal progenitor cells inhibit lymphocyte proliferation by activation of the programmed death 1 pathway. Eur J Immunol 2005; 35(5): 1482-90. [http://dx.doi.org/10.1002/eji.200425405] [PMID: 15827960]

[58] Geissler EK, Hutchinson JA. Cell therapy as a strategy to minimize maintenance immunosuppression in solid organ transplant recipients. Curr Opin Organ Transplant 2013; 18(4): 408-15. [PMID: 23838645]

[59] Polchert D, Sobinsky J, Douglas G, *et al.* IFN-gamma activation of mesenchymal stem cells for treatment and prevention of graft *versus* host disease. Eur J Immunol 2008; 38(6): 1745-55. [http://dx.doi.org/10.1002/eji.200738129] [PMID: 18493986]

[60] Siegel G, Schäfer R, Dazzi F. The immunosuppressive properties of mesenchymal stem cells. Transplantation 2009; 87(9) (Suppl.): S45-9. [http://dx.doi.org/10.1097/TP.0b013e3181a285b0] [PMID: 19424005]

[61] Murphy MB, Moncivais K, Caplan AI. Mesenchymal stem cells: environmentally responsive therapeutics for regenerative medicine. Exp Mol Med 2013; 45: e54. [http://dx.doi.org/10.1038/emm.2013.94] [PMID: 24232253]

[62] Aggarwal S, Pittenger MF. Human mesenchymal stem cells modulate allogeneic immune cell responses. Blood 2005; 105(4): 1815-22. [http://dx.doi.org/10.1182/blood-2004-04-1559] [PMID: 15494428]

[63] Iyer SS, Rojas M. Anti-inflammatory effects of mesenchymal stem cells: novel concept for future therapies. Expert Opin Biol Ther 2008; 8(5): 569-81. [http://dx.doi.org/10.1517/14712598.8.5.569] [PMID: 18407762]

[64] Weiss DJ, Bertoncello I, Borok Z, *et al.* Stem cells and cell therapies in lung biology and lung diseases. Proc Am Thorac Soc 2011; 8(3): 223-72. [http://dx.doi.org/10.1513/pats.201012-071DW] [PMID: 21653527]

[65] Yagi H, Soto-Gutierrez A, Kitagawa Y, Tilles AW, Tompkins RG, Yarmush ML. Bone marrow mesenchymal stromal cells attenuate organ injury induced by LPS and burn. Cell Transplant 2010; 19(6): 823-30. [http://dx.doi.org/10.3727/096368910X508942] [PMID: 20573305]

[66] Zhou Y, Yuan J, Zhou B, *et al.* The therapeutic efficacy of human adipose tissue-derived mesenchymal stem cells on experimental autoimmune hearing loss in mice. Immunology 2011; 133(1): 133-40. [http://dx.doi.org/10.1111/j.1365-2567.2011.03421.x] [PMID: 21366561]

[67] Kavanagh H, Mahon BP. Allogeneic mesenchymal stem cells prevent allergic airway inflammation by inducing murine regulatory T cells. Allergy 2011; 66(4): 523-31.

[http://dx.doi.org/10.1111/j.1398-9995.2010.02509.x] [PMID: 21091718]

[68] González MA, Gonzalez-Rey E, Rico L, Büscher D, Delgado M. Adipose-derived mesenchymal stem cells alleviate experimental colitis by inhibiting inflammatory and autoimmune responses. Gastroenterology 2009; 136(3): 978-89.
[http://dx.doi.org/10.1053/j.gastro.2008.11.041] [PMID: 19135996]

[69] Bhasin A, Srivastava MV, Mohanty S, Bhatia R, Kumaran SS, Bose S. Stem cell therapy: a clinical trial of stroke. Clin Neurol Neurosurg 2013; 115(7): 1003-8.
[http://dx.doi.org/10.1016/j.clineuro.2012.10.015] [PMID: 23183251]

[70] Bartunek J, Behfar A, Dolatabadi D, *et al.* Cardiopoietic stem cell therapy in heart failure: the C-CURE (Cardiopoietic stem Cell therapy in heart failURE) multicenter randomized trial with lineage-specified biologics. J Am Coll Cardiol 2013; 61(23): 2329-38.
[http://dx.doi.org/10.1016/j.jacc.2013.02.071] [PMID: 23583246]

[71] Weiss DJ, Casaburi R, Flannery R, LeRoux-Williams M, Tashkin DP. A placebo-controlled, randomized trial of mesenchymal stem cells in COPD. Chest 2013; 143(6): 1590-8.
[http://dx.doi.org/10.1378/chest.12-2094] [PMID: 23172272]

[72] Shi M, Zhang Z, Xu R, *et al.* Human mesenchymal stem cell transfusion is safe and improves liver function in acute-on-chronic liver failure patients. Stem Cells Transl Med 2012; 1(10): 725-31.
[http://dx.doi.org/10.5966/sctm.2012-0034] [PMID: 23197664]

[73] Hoffmeyer K, Raggioli A, Rudloff S, *et al.* Wnt/β-catenin signaling regulates telomerase in stem cells and cancer cells. Science 2012; 336(6088): 1549-54.
[http://dx.doi.org/10.1126/science.1218370] [PMID: 22723415]

[74] Xi H, Li C, Ren F, Zhang H, Zhang L. Telomere, aging and age-related diseases. Aging Clin Exp Res 2013; 25(2): 139-46.
[http://dx.doi.org/10.1007/s40520-013-0021-1] [PMID: 23739898]

[75] Sperka T, Wang J, Rudolph KL. DNA damage checkpoints in stem cells, ageing and cancer. Nat Rev Mol Cell Biol 2012; 13(9): 579-90.
[http://dx.doi.org/10.1038/nrm3420] [PMID: 22914294]

[76] Tümpel S, Rudolph KL. The role of telomere shortening in somatic stem cells and tissue aging: lessons from telomerase model systems. Ann N Y Acad Sci 2012; 1266: 28-39.
[http://dx.doi.org/10.1111/j.1749-6632.2012.06547.x] [PMID: 22901253]

[77] Wu W, Niklason L, Steinbacher DM. The effect of age on human adipose-derived stem cells. Plast Reconstr Surg 2013; 131(1): 27-37.
[http://dx.doi.org/10.1097/PRS.0b013e3182729cfc] [PMID: 22965240]

[78] Hacia JG, Lee CC, Jimenez DF, *et al.* Age-related gene expression profiles of rhesus monkey bone marrow-derived mesenchymal stem cells. J Cell Biochem 2008; 103(4): 1198-210.
[http://dx.doi.org/10.1002/jcb.21498] [PMID: 17685434]

[79] Lazebnik LB, Golovanova EV, Slupskaia VA, Trubitsina IE, Kniazev OV, Shaposhnikova NA. [Realities and prospects for the use of cell technologies for the treatment of chronic diffuse liver diseases]. Eksp Klin Gastroenterol 2011; 6(6): 3-10.
[PMID: 22168071]

[80] Salama H, Zekri AR, Ahmed R, *et al.* Assessment of health-related quality of life in patients receiving stem cell therapy for end-stage liver disease: an Egyptian study. Stem Cell Res Ther 2012; 3(6): 49. [http://dx.doi.org/10.1186/scrt140] [PMID: 23206927]

[81] Leonhartsberger N, Ramoner R, Falkensammer C, *et al.* Quality of life during dendritic cell vaccination against metastatic renal cell carcinoma. Cancer Immunol Immunother 2012; 61(9): 1407-13. [http://dx.doi.org/10.1007/s00262-012-1207-7] [PMID: 22278360]

[82] Sharma A, Gokulchandran N, Chopra G, *et al.* Administration of autologous bone marrow-derived mononuclear cells in children with incurable neurological disorders and injury is safe and improves their quality of life. Cell Transplant 2012; 21(S1) (Suppl. 1): S79-90. [http://dx.doi.org/10.3727/096368912X633798] [PMID: 22507683]

[83] Zhu H, Yang X, Li J, *et al.* Immune response, safety, and survival and quality of life outcomes for advanced colorectal cancer patients treated with dendritic cell vaccine and cytokine-induced killer cell therapy. Article, ID: BioMed Research International Vol 2014; p. 603871.

[84] Fitzpatrick E, DeHart M, Tercero J, *et al.* Stem cell therapy to promote sensory and motor nerve function recovery in a limb transplant model (TRAN3P.863). J Immunology 2014; vol.192

[85] Siqueira RC, Messias A, Messias K, *et al.* Quality of life in patients with retinitis pigmentosa submitted to intravitreal use of bone marrow-derived stem cells (Reticell -clinical trial). Stem Cell Res Ther 2015; 6: 29. [http://dx.doi.org/10.1186/s13287-015-0020-6] [PMID: 25890251]

[86] Critical Quality Attributes definition [Wikipedia homepage on internet]. Cited 6th July 2015. Available from: https://en.wikipedia.org/wiki/Critical_Quality_Attributes. 2015.

[87] Ramos T, Wang W, Okhovat J, McPherson G, Burger S. Human-derived raw materials: Controlled, consistent collections enable successful manufacturing of cell-based regenerative medicine products. J Immunology 2014; 192: 69.15..

[88] SVF clinical trial search update [Clinical trial homepage on internet]. Cited 8th July 2015. Available from: https://clinicaltrials.gov/. 2015.

[89] McAllister TN, Audley D, LHeureux N. Autologous cell therapies: challenges in US FDA regulation. Regen Med 2012; 7(6) (Suppl.): 94-7. [http://dx.doi.org/10.2217/rme.12.83] [PMID: 23210819]

[90] Chinen J, Buckley RH. Transplantation immunology: solid organ and bone marrow. J Allergy Clin Immunol 2010; 125(2) (Suppl. 2): S324-35. [http://dx.doi.org/10.1016/j.jaci.2009.11.014] [PMID: 20176267]

[91] Ankrum JA, Ong JF, Karp JM. Mesenchymal stem cells: immune evasive, not immune privileged. Nat Biotechnol 2014; 32(3): 252-60. [http://dx.doi.org/10.1038/nbt.2816] [PMID: 24561556]

[92] van den Akker F, de Jager SC, Sluijter JP. Mesenchymal stem cell therapy for cardiac inflammation: immunomodulatory properties and the influence of toll-like receptors. Mediators of Inflammation 2013; 2013 [http://dx.doi.org/10.1155/2013/181020]

[93] Hare JM, Fishman JE, Gerstenblith G, *et al.* Comparison of allogeneic *vs* autologous bone marrow–derived mesenchymal stem cells delivered by transendocardial injection in patients with ischemic cardiomyopathy: the POSEIDON randomized trial. JAMA 2012; 308(22): 2369-79. [http://dx.doi.org/10.1001/jama.2012.25321] [PMID: 23117550]

[94] Hare JM, Traverse JH, Henry TD, *et al.* A randomized, double-blind, placebo-controlled, dose-escalation study of intravenous adult human mesenchymal stem cells (prochymal) after acute myocardial infarction. J Am Coll Cardiol 2009; 54(24): 2277-86. [http://dx.doi.org/10.1016/j.jacc.2009.06.055] [PMID: 19958962]

[95] Heldman AW, DiFede DL, Fishman JE, *et al.* Transendocardial mesenchymal stem cells and mononuclear bone marrow cells for ischemic cardiomyopathy: the TAC-HFT randomized trial. JAMA 2014; 311(1): 62-73. [http://dx.doi.org/10.1001/jama.2013.282909] [PMID: 24247587]

[96] Scheubel RJ, Zorn H, Silber RE, *et al.* Age-dependent depression in circulating endothelial progenitor cells in patients undergoing coronary artery bypass grafting. J Am Coll Cardiol 2003; 42(12): 2073-80. [http://dx.doi.org/10.1016/j.jacc.2003.07.025] [PMID: 14680729]

[97] Karantalis V, Schulman IH, Balkan W, Hare JM. Allogeneic cell therapy: a new paradigm in therapeutics. Circ Res 2015; 116(1): 12-5. [http://dx.doi.org/10.1161/CIRCRESAHA.114.305495] [PMID: 25552688]

[98] Lauden L, Boukouaci W, Borlado LR, *et al.* Allogenicity of human cardiac stem/progenitor cells orchestrated by programmed death ligand 1. Circ Res 2013; 112(3): 451-64. [http://dx.doi.org/10.1161/CIRCRESAHA.112.276501] [PMID: 23243206]

[99] Bolli R, Chugh AR, DAmario D, *et al.* Cardiac stem cells in patients with ischaemic cardiomyopathy (SCIPIO): initial results of a randomised phase 1 trial. Lancet 2011; 378(9806): 1847-57. [http://dx.doi.org/10.1016/S0140-6736(11)61590-0] [PMID: 22088800]

[100] Tateishi K, Ashihara E, Honsho S, *et al.* Human cardiac stem cells exhibit mesenchymal features and are maintained through Akt/GSK-3beta signaling. Biochem Biophys Res Commun 2007; 352(3): 635-41. [http://dx.doi.org/10.1016/j.bbrc.2006.11.096] [PMID: 17150190]

[101] Pouly J, Bruneval P, Mandet C, *et al.* Cardiac stem cells in the real world. J Thorac Cardiovasc Surg 2008; 135(3): 673-8. [http://dx.doi.org/10.1016/j.jtcvs.2007.10.024] [PMID: 18329492]

[102] Starzl TE, Marchioro TL, Waddell WR. The reversal of rejection in human renal homografts with subsequent development of homograft tolerance. Surg Gynecol Obstet 1963; 117: 385-95. [PMID: 14065716]

[103] Thomas ED, Lochte HL Jr, Lu WC, Ferrebee JW. Intravenous infusion of bone marrow in patients receiving radiation and chemotherapy. N Engl J Med 1957; 257(11): 491-6. [http://dx.doi.org/10.1056/NEJM195709122571102] [PMID: 13464965]

[104] Civin CI, Strauss LC, Brovall C, Fackler MJ, Schwartz JF, Shaper JH. Antigenic analysis of hematopoiesis. III. A hematopoietic progenitor cell surface antigen defined by a monoclonal antibody raised against KG-1a cells. J Immunol 1984; 133(1): 157-65.

[PMID: 6586833]

[105] Katz FE, Tindle R, Sutherland DR, Greaves MF. Identification of a membrane glycoprotein associated with haemopoietic progenitor cells. Leuk Res 1985; 9(2): 191-8. [http://dx.doi.org/10.1016/0145-2126(85)90082-7] [PMID: 3857408]

[106] Lezaic L, Socan A, Poglajen G, *et al.* Intracoronary transplantation of CD34(+) cells is associated with improved myocardial perfusion in patients with nonischemic dilated cardiomyopathy. J Card Fail 2015; 21(2): 145-52. [http://dx.doi.org/10.1016/j.cardfail.2014.11.005] [PMID: 25459687]

[107] Al-Zoubi A, Jafar E, Jamous M, *et al.* Transplantation of purified autologous leukapheresis-derived CD34+ and CD133+ stem cells for patients with chronic spinal cord injuries: long-term evaluation of safety and efficacy. Cell Transplant 2014; 23 (Suppl. 1): S25-34. [http://dx.doi.org/10.3727/096368914X684899] [PMID: 25372344]

[108] Siqueira J. Clinical Trials Currently Being Conducted with the Use of Adult Stem Cells Transplantation to Treat Retinal Diseases. Transplant Technol Res 2014; 4: 2.

[109] Dominici M, Le Blanc K, Mueller I, *et al.* Minimal criteria for defining multipotent mesenchymal stromal cells. The International Society for Cellular Therapy position statement. Cytotherapy 2006; 8(4): 315-7. [http://dx.doi.org/10.1080/14653240600855905] [PMID: 16923606]

[110] Vonk LA, de Windt TS, Slaper-Cortenbach IC, Saris DB. Autologous, allogeneic, induced pluripotent stem cell or a combination stem cell therapy? Where are we headed in cartilage repair and why: a concise review. Stem Cell Res Ther 2015; 6(1): 94. [http://dx.doi.org/10.1186/s13287-015-0086-1] [PMID: 25976213]

[111] Bourin P, Bunnell BA, Casteilla L, *et al.* Stromal cells from the adipose tissue-derived stromal vascular fraction and culture expanded adipose tissue-derived stromal/stem cells: a joint statement of the International Federation for Adipose Therapeutics and Science (IFATS) and the International Society for Cellular Therapy (ISCT). Cytotherapy 2013; 15(6): 641-8. [http://dx.doi.org/10.1016/j.jcyt.2013.02.006] [PMID: 23570660]

[112] CAR T-Cell Therapy [Cancer homepage on internet]. Cited 17th July 2015. Available from: http: //www.cancer.gov/about-cancer/treatment/research/car-t-cells. 2015.

[113] Srivastava S, Riddell SR. Engineering CAR-T Cells: Design Concepts. Trends Immunol 2015. [http://dx.doi.org/S1471-4906(15)00151-9.]

[114] CAR T cells clinical trial search update [Clinical trial homepage on internet]. Cited 17th July 2015. Available from: https://clinicaltrials.gov/. 2015.

[115] Kochenderfer JN, Rosenberg SA. Treating B-cell cancer with T cells expressing anti-CD19 chimeric antigen receptors. Nat Rev Clin Oncol 2013; 10(5): 267-76. [http://dx.doi.org/10.1038/nrclinonc.2013.46] [PMID: 23546520]

[116] Herthel DJ. Suspensory desmitis therapies. Proc 12th ACVS Symp. 165-67.

[117] Pittenger MF, Mackay AM, Beck SC, *et al.* Multilineage potential of adult human mesenchymal stem cells. Science 1999; 284(5411): 143-7. [http://dx.doi.org/10.1126/science.284.5411.143] [PMID: 10102814]

[118] Martin DR, Cox NR, Hathcock TL, Niemeyer GP, Baker HJ. Isolation and characterization of multipotential mesenchymal stem cells from feline bone marrow. Exp Hematol 2002; 30(8): 879-86. [http://dx.doi.org/10.1016/S0301-472X(02)00864-0] [PMID: 12160839]

[119] Schnabel LV, Mohammed HO, Miller BJ, *et al.* Platelet rich plasma (PRP) enhances anabolic gene expression patterns in flexor digitorum superficialis tendons. J Orthop Res 2007; 25(2): 230-40. [http://dx.doi.org/10.1002/jor.20278] [PMID: 17106885]

[120] Schnabel LV, Sonea HO, Jacobson MS, Fortier LA. Effects of platelet rich plasma and acellular bone marrow on gene expression patterns and DNA content of equine suspensory ligament explant cultures. Equine Vet J 2008; 40(3): 260-5. [http://dx.doi.org/10.2746/042516408X278030] [PMID: 18267879]

[121] Silva GV, Litovsky S, Assad JA, *et al.* Mesenchymal stem cells differentiate into an endothelial phenotype, enhance vascular density, and improve heart function in a canine chronic ischemia model. Circulation 2005; 111(2): 150-6. [http://dx.doi.org/10.1161/01.CIR.0000151812.86142.45] [PMID: 15642764]

[122] Smith RK, Korda M, Blunn GW, Goodship AE. Isolation and implantation of autologous equine mesenchymal stem cells from bone marrow into the superficial digital flexor tendon as a potential novel treatment. Equine Vet J 2003; 35(1): 99-102. [http://dx.doi.org/10.2746/042516403775467388] [PMID: 12553472]

[123] Carvalho AdeM, Badial PR, Álvarez LE, *et al.* Equine tendonitis therapy using mesenchymal stem cells and platelet concentrates: a randomized controlled trial. Stem Cell Res Ther 2013; 4(4): 85-98. [http://dx.doi.org/10.1186/scrt236] [PMID: 23876512]

[124] Richardson LE, Dudhia J, Clegg PD, Smith R. Stem cells in veterinary medicineattempts at regenerating equine tendon after injury. Trends Biotechnol 2007; 25(9): 409-16. [http://dx.doi.org/10.1016/j.tibtech.2007.07.009] [PMID: 17692415]

[125] Rich FR. Single-center study of 83 horses with suspensory injuries treated with adipose-derived stem and regenerative cells. Stem Cell Discovery 2014; 4: 44-53. [http://dx.doi.org/10.4236/scd.2014.42005]

[126] Bosch G, van Schie HT, de Groot MW, *et al.* Effects of platelet-rich plasma on the quality of repair of mechanically induced core lesions in equine superficial digital flexor tendons: A placebo-controlled experimental study. J Orthop Res 2010; 28(2): 211-7. [PMID: 19714688]

[127] Black LL, Gaynor J, Gahring D, *et al.* Effect of adipose-derived mesenchymal stem and regenerative cells on lameness in dogs with chronic osteoarthritis of the coxofemoral joints: a randomized, double-blinded, multicenter, controlled trial. Vet Ther 2007; 8(4): 272-84. [PMID: 18183546]

[128] Vilar JM, Batista M, Morales M, *et al.* Assessment of the effect of intraarticular injection of autologous adipose-derived mesenchymal stem cells in osteoarthritic dogs using a double blinded force platform analysis. BMC Vet Res 2014; 10: 143-50. [http://dx.doi.org/10.1186/1746-6148-10-143] [PMID: 24984756]

[129] Fahie MA, Ortolano GA, Guercio V, *et al.* A randomized controlled trial of the efficacy of autologous

platelet therapy for the treatment of osteoarthritis in dogs. J Am Vet Med Assoc 2013; 243(9): 1291-7. [http://dx.doi.org/10.2460/javma.243.9.1291] [PMID: 24134578]

[130] Kraitchman DL, Tatsumi M, Gilson WD, *et al.* Dynamic imaging of allogeneic mesenchymal stem cells trafficking to myocardial infarction. Circulation 2005; 112(10): 1451-61. [http://dx.doi.org/10.1161/CIRCULATIONAHA.105.537480] [PMID: 16129797]

[131] Bruder SP, Kraus KH, Goldberg VM, Kadiyala S. The effect of implants loaded with autologous mesenchymal stem cells on the healing of canine segmental bone defects. J Bone Joint Surg Am 1998; 80(7): 985-96. [http://dx.doi.org/10.2106/00004623-199807000-00007] [PMID: 9698003]

[132] Hiyama A, Mochida J, Iwashina T, *et al.* Transplantation of mesenchymal stem cells in a canine disc degeneration model. J Orthop Res 2008; 26(5): 589-600. [http://dx.doi.org/10.1002/jor.20584] [PMID: 18203202]

[133] Boyd LM, Langston C, Thompson K, Zivin K, Imanishi M. Survival in cats with naturally occurring chronic kidney disease (20002002). J Vet Intern Med 2008; 22(5): 1111-7. [http://dx.doi.org/10.1111/j.1939-1676.2008.0163.x] [PMID: 18691369]

[134] Quimby JM, Webb TL, Gibbons DS, Dow SW. Evaluation of intrarenal mesenchymal stem cell injection for treatment of chronic kidney disease in cats: a pilot study. J Feline Med Surg 2011; 13(6): 418-26. [http://dx.doi.org/10.1016/j.jfms.2011.01.005] [PMID: 21334237]

[135] Vilar JM, Morales M, Santana A, *et al.* Controlled, blinded force platform analysis of the effect of intraarticular injection of autologous adipose-derived mesenchymal stem cells associated to PRGF-Endoret in osteoarthritic dogs. BMC Vet Res 2013; 9: 131-40. [http://dx.doi.org/10.1186/1746-6148-9-131] [PMID: 23819757]

[136] Villatoro AJ, Fernandez V, Claros S, Rico-Llanos GA, Becerra J, Andrades JA. Use of adipose derived mesenchymal Stem Cells in keratoconjunctivitis sicca in canine model. BioMed Research International. 2015.

[137] Quimby JM, Webb TL, Habenicht LM, Dow SW. Safety and efficacy of intravenous infusion of allogeneic cryopreserved mesenchymal stem cells for treatment of chronic kidney disease in cats: results of three sequential pilot studies. Stem Cell Res Ther 2013; 4(2): 48-55. [http://dx.doi.org/10.1186/scrt198] [PMID: 23632128]

[138] Lapidot T, Petit I. Current understanding of stem cell mobilization: the roles of chemokines, proteolytic enzymes, adhesion molecules, cytokines, and stromal cells. Exp Hematol 2002; 30(9): 973-81. [http://dx.doi.org/10.1016/S0301-472X(02)00883-4] [PMID: 12225788]

[139] Baxter MA, Wynn RF, Jowitt SN, Wraith JE, Fairbairn LJ, Bellantuono I. Study of telomere length reveals rapid aging of human marrow stromal cells following *in vitro* expansion. Stem Cells 2004; 22(5): 675-82. [http://dx.doi.org/10.1634/stemcells.22-5-675] [PMID: 15342932]

[140] Kenyon J, Gerson SL. The role of DNA damage repair in aging of adult stem cells. Nucleic Acids Res 2007; 35(22): 7557-65. [http://dx.doi.org/10.1093/nar/gkm1064] [PMID: 18160407]

[141] Cooke MS, Evans MD, Dizdaroglu M, Lunec J. Oxidative DNA damage: mechanisms, mutation, and disease. FASEB J 2003; 17(10): 1195-214.
[http://dx.doi.org/10.1096/fj.02-0752rev] [PMID: 12832285]

[142] Cheung TH, Rando TA. Molecular regulation of stem cell quiescence. Nat Rev Mol Cell Biol 2013; 14(6): 329-40.
[http://dx.doi.org/10.1038/nrm3591] [PMID: 23698583]

[143] Gerson SL, Reese J, Kenyon J. DNA repair in stem cell maintenance and conversion to cancer stem cells. Ernst Schering Found Symp Proc. 231-44.

[144] Rauscher FM, Goldschmidt-Clermont PJ, Davis BH, *et al.* Aging, progenitor cell exhaustion, and atherosclerosis. Circulation 2003; 108(4): 457-63.
[http://dx.doi.org/10.1161/01.CIR.0000082924.75945.48] [PMID: 12860902]

[145] Fortier LA, Travis AJ. Stem cells in veterinary medicine. Stem Cell Res Ther 2011; 2(1): 9-15.
[http://dx.doi.org/10.1186/scrt50] [PMID: 21371354]

[146] McInnes IB, Schett G. The pathogenesis of rheumatoid arthritis. N Engl J Med 2011; 365(23): 2205-19.
[http://dx.doi.org/10.1056/NEJMra1004965] [PMID: 22150039]

[147] Fossum TW, Hedlund CS, Johnson AL, *et al.* Small Animal Surgery. 3rd ed., Mosby Elsevier 2007.

[148] Budsberg SC. Managing the Chronic Pain and Dysfunction of Osteoarthritis in the Dog Part 1+2. Pacific Veterinary Conference. Long Beach. 2015.

[149] Budsberg SC. Evidence for use of neutraceuticals in osteoarthritis – Is "Joint juice" insane quackery or is the internet right? Pacific Veterinary Conference, Long Beach July 2015.

[150] Budsberg SC. NSAIDS in the management of chronic pain – Are good drugs getting a bad name? Pacific Veterinary Conference, Long Beach July 2015.

[151] Carr BJ, Canapp SO. Regenerative medicine for soft tissue injury and osteoarthritis. Practical Techniques From The NAVC Institute, July/August 2016, Today's Veterinary Practice 2016.

[152] Harting MT, Jimenez F, Xue H, *et al.* Intravenous mesenchymal stem cell therapy for traumatic brain injury. J Neurosurg 2009; 110(6): 1189-97.
[http://dx.doi.org/10.3171/2008.9.JNS08158] [PMID: 19301973]

[153] Goodrich LR, Nixon AJ. Medical treatment of osteoarthritis in the horse – A review. Vet J 2006; 171: 51-69.

[154] Frisbie DD, Kawcak CE, McIlwraith CW, Werpy NM. Evaluation of polysulfated glycosaminoglycan or sodium hyaluronan administered intra-articularly for treatment of horses with experimentally induced osteoarthritis. Am J Vet Res 2009; 70(2): 203-9.
[http://dx.doi.org/10.2460/ajvr.70.2.203] [PMID: 19231952]

[155] Kristiansen KK, Kold SE. Multivariable analysis of factors influencing outcome of 2 treatment protocols in 128 cases of horses responding positively to intra-articular analgesia of the distal interphalangeal joint. Equine Vet J 2007; 39(2): 150-6.
[http://dx.doi.org/10.2746/042516407X170094] [PMID: 17378444]

[156] Gough MR, Thibaud D, Smith RK. Tiludronate infusion in the treatment of bone spavin: a double

blind placebo-controlled trial. Equine Vet J 2010; 42(5): 381-7.
[http://dx.doi.org/10.1111/j.2042-3306.2010.00120.x] [PMID: 20636772]

[157] Textor J. Autologous biologic treatment for equine musculoskeletal injuries: platelet-rich plasma and IL-1 receptor antagonist protein. Vet Clin North Am Equine Pract 2011; 27(2): 275-98.
[http://dx.doi.org/10.1016/j.cveq.2011.05.001] [PMID: 21872759]

[158] Koch TG, Berg LC, Betts DH. Current and future regenerative medicine - principles, concepts, and therapeutic use of stem cell therapy and tissue engineering in equine medicine. Can Vet J 2009; 50(2): 155-65.
[PMID: 19412395]

[159] Smith RK. Mesenchymal stem cell therapy for equine tendinopathy. Disabil Rehabil 2008; 30(20-22): 1752-8.
[http://dx.doi.org/10.1080/09638280701788241] [PMID: 18608378]

[160] Canapp SO, Canapp DA, Carr BJ, Cox C, Barrett JG. Supraspinatus Tendinopathy in 327 Dogs: A retrospective Study. Veterinary Evidence Online 2016; 1(3): 2396-9776.

[161] Budsberg SC. Feline osteoarthritis – What we know and don't know. Pacific Veterinary Conference. Long Beach. 2008.

[162] Polzin DJ, *et al.* Medical management of feline chronic renal failure. In: Bonagura JD, Kirk RW, Eds. Current Veterinary Therapy XI. Philadelphia, PA: Saunders 1992; pp. 848-53.

[163] Birchard and Sherding *et al.* Saunders Manual of Small Animal Practice. W.B. Saunders Company 1994.

[164] Sparkes AH, Caney S, Chalhoub S, *et al.* ISFM consensus guidelines on the diagnosis and management of feline chronic kidney disease. J Feline Med Sur 2016; 18: 219-39.

[165] Arzi B, Mills-Ko E, Verstraete FJ, *et al.* Therapeutic efficacy of fresh, autologous mesenchymal stem cells for severe refractory Gingivostomatitis in cats. Stem Cells Transl Med 2016; 5(1): 75-86.
[http://dx.doi.org/10.5966/sctm.2015-0127] [PMID: 26582907]

[166] Healey KA, Dawson S, Burrow R, *et al.* Prevalence of feline chronic gingivo-stomatitis in first opinion veterinary practice. J Feline Med Surg 2007; 9(5): 373-81.
[http://dx.doi.org/10.1016/j.jfms.2007.03.003] [PMID: 17507275]

[167] Simons BD. Stem cell renewal theory turns 60. Nat Rev Mol Cell Biol 2013; 14(12): 754.
[http://dx.doi.org/10.1038/nrm3706] [PMID: 24263358]

[168] Godine RL. Low level laser therapy (LLLT) in veterinary medicine. Photomed Laser Surg 2014; 32(1): 1-2.
[http://dx.doi.org/10.1089/pho.2013.9867] [PMID: 24359266]

[169] Taub R. Liver regeneration: from myth to mechanism. Nat Rev Mol Cell Biol 2004; 5(10): 836-47.
[http://dx.doi.org/10.1038/nrm1489] [PMID: 15459664]

[170] Howard D, Buttery LD, Shakesheff KM, Roberts SJ. Tissue engineering: strategies, stem cells and scaffolds. J Anat 2008; 213(1): 66-72.
[http://dx.doi.org/10.1111/j.1469-7580.2008.00878.x] [PMID: 18422523]

[171] Atala A. Tissue engineering of human bladder. Br Med Bull 2011; 97: 81-104.

[http://dx.doi.org/10.1093/bmb/ldr003] [PMID: 21324973]

[172] Gonfiotti A, Jaus MO, Barale D, *et al.* The first tissue-engineered airway transplantation: 5-year follow-up results. Lancet 2014; 383(9913): 238-44. [http://dx.doi.org/10.1016/S0140-6736(13)62033-4] [PMID: 24161821]

[173] Rodrigues MT, Reis RL, Gomes ME. Engineering tendon and ligament tissues: present developments towards successful clinical products. J Tissue Eng Regen Med 2013; 7(9): 673-86. [http://dx.doi.org/10.1002/term.1459] [PMID: 22499564]

[174] Sohni A, Verfaillie CM. Mesenchymal stem cells migration homing and tracking. Stem Cells International 2013; 2013(2013): 1-8.

[175] Gnecchi M, Zhang Z, Ni A, Dzau VJ. Paracrine mechanisms in adult stem cell signaling and therapy. Circ Res 2008; 103(11): 1204-19. [http://dx.doi.org/10.1161/CIRCRESAHA.108.176826] [PMID: 19028920]

[176] El-Sharkawy H, Kantarci A, Deady J, *et al.* Platelet-rich plasma: growth factors and pro- and anti-inflammatory properties. J Periodontol 2007; 78(4): 661-9. [http://dx.doi.org/10.1902/jop.2007.060302] [PMID: 17397313]

[177] Yazawa M, Ogata H, Nakajima T, Mori T, Watanabe N, Handa M. Basic studies on the clinical applications of platelet-rich plasma. Cell Transplant 2003; 12(5): 509-18. [http://dx.doi.org/10.3727/000000003108747073] [PMID: 12953925]

[178] Ning L, Goossens E, Geens M, *et al.* Mouse spermatogonial stem cells obtain morphologic and functional characteristics of hematopoietic cells *in vivo* . Hum Reprod 2010; 25(12): 3101-9. [http://dx.doi.org/10.1093/humrep/deq269] [PMID: 20940138]

[179] Blasco MA. Telomere length, stem cells and aging. Nat Chem Biol 2007; 3(10): 640-9. [http://dx.doi.org/10.1038/nchembio.2007.38] [PMID: 17876321]

[180] Wright WE, Piatyszek MA, Rainey WE, Byrd W, Shay JW. Telomerase activity in human germline and embryonic tissues and cells. Dev Genet 1996; 18(2): 173-9. [http://dx.doi.org/10.1002/(SICI)1520-6408(1996)18:2<173::AID-DVG10>3.0.CO;2-3] [PMID: 8934879]

[181] Atala A, Bauer SB, Soker S, Yoo JJ, Retik AB. Tissue-engineered autologous bladders for patients needing cystoplasty. Lancet 2006; 367(9518): 1241-6. [http://dx.doi.org/10.1016/S0140-6736(06)68438-9] [PMID: 16631879]

[182] Vaccines, blood and biologics [FDA homepage on internet]. Cited 22th July 2015. Available from: http://www.fda.gov/BiologicsBloodVaccines/CellularGeneTherapyProducts/ApprovedProducts/ 2015.

[183] FDA.gov. FDA.gov [homepage on internet]. Cited 22nd July 2015. FDA approved Marketed Products [homepage on internet]. Cited 19th July 2015. Available from: http://www.fda.gov/AboutFDA/Transparency/Basics/ucm194655.htm. 2015.

[184] Guidance for Industry Cell-based Products for Animal Use. U.S. Department of Health and Human Services. Food and Drug Administration. Cited June 2015. Available from: http://www.fda.gov/downloads/AnimalVeterinary/GuidanceComplianceEnforcement/GuidanceforIndustry/UCM405679.pdf.

[185] Werner M, Mayleben T, Van Bokkelen G. Autologous cell therapies: the importance of regulatory oversight. Regen Med 2012; 7(6) (Suppl.): 100-3.
[http://dx.doi.org/10.2217/rme.12.90] [PMID: 23210820]

[186] Geach T. Stem cells: BMMC treatment safe but not an improvement on standard therapy. Nat Rev Cardiol 2014; 11(12): 686.
[http://dx.doi.org/10.1038/nrcardio.2014.175] [PMID: 25367650]

[187] Ankrum J, Karp JM. Mesenchymal stem cell therapy: Two steps forward, one step back. Trends Mol Med 2010; 16(5): 203-9.
[http://dx.doi.org/10.1016/j.molmed.2010.02.005] [PMID: 20335067]

[188] Karp JM, Leng Teo GS. Mesenchymal stem cell homing: the devil is in the details. Cell Stem Cell 2009; 4(3): 206-16.
[http://dx.doi.org/10.1016/j.stem.2009.02.001] [PMID: 19265660]

[189] Phinney DG, Prockop DJ. Concise review: mesenchymal stem/multipotent stromal cells: the state of transdifferentiation and modes of tissue repaircurrent views. Stem Cells 2007; 25(11): 2896-902.
[http://dx.doi.org/10.1634/stemcells.2007-0637] [PMID: 17901396]

[190] Fink DW Jr. FDA regulation of stem cell-based products. Science 2009; 324(5935): 1662-3.
[http://dx.doi.org/10.1126/science.1173712] [PMID: 19556496]

[191] Soria B, Roche E, Berná G, León-Quinto T, Reig JA, Martín F. Insulin-secreting cells derived from embryonic stem cells normalize glycemia in streptozotocin-induced diabetic mice. Diabetes 2000; 49(2): 157-62.
[http://dx.doi.org/10.2337/diabetes.49.2.157] [PMID: 10868930]

[192] Lee TH. Functional effect of mouse embryonic stem cell implantation after spinal cord injury. J Exerc Rehabil 2013; 9(2): 230-3.
[http://dx.doi.org/10.12965/jer.130004] [PMID: 24278865]

[193] Nandoe Tewarie RS, Hurtado A, Bartels RH, Grotenhuis A, Oudega M. Stem cell-based therapies for spinal cord injury. J Spinal Cord Med 2009; 32(2): 105-14.
[http://dx.doi.org/10.1080/10790268.2009.11760761] [PMID: 19569457]

[194] Stamm C, Westphal B, Kleine HD, *et al.* Autologous bone-marrow stem-cell transplantation for myocardial regeneration. Lancet 2003; 361(9351): 45-6.
[http://dx.doi.org/10.1016/S0140-6736(03)12110-1] [PMID: 12517467]

[195] Amado LC, Saliaris AP, Schuleri KH, *et al.* Cardiac repair with intramyocardial injection of allogeneic mesenchymal stem cells after myocardial infarction. Proc Natl Acad Sci USA 2005; 102(32): 11474-9.
[http://dx.doi.org/10.1073/pnas.0504388102] [PMID: 16061805]

[196] Dawn B, Stein AB, Urbanek K, *et al.* Cardiac stem cells delivered intravascularly traverse the vessel barrier, regenerate infarcted myocardium, and improve cardiac function. Proc Natl Acad Sci USA 2005; 102(10): 3766-71.
[http://dx.doi.org/10.1073/pnas.0405957102] [PMID: 15734798]

CHAPTER 5

From Trials to Therapeutics: Are Mesenchymal Stem Cells Promising in Cellular Therapy?

Ganapathi M. Bhat[1,*], **Pooja Shahapurkar**[1] and **Rasika G. Bhat**[2]

[1] *Jaslok Hospital and Research centre, 15, Dr. G. Deshmukh Marg, Mumbai 400026, Maharashtra, India*

[2] *Department of Medical Genetics, College of Nursing, Bombay Hospital, Mumbai 400020, Maharashtra, India*

Abstract: Mesenchymal stem cells (MSCs) are today, the most favoured cellular candidates for regenerative therapeutics. Though discovered early in the 1960s, only recent decades have witnessed extensive research involving MSCs. MSCs, termed as multipotent mesenchymal stromal cells in 2006, by the International Society of Cellular Therapy have gained greater acceptance in view of their ubiquitous presence in tissues, exemption from ethical concerns, clonogenic potential, trilineage differentiation, versatile plasticity and ability to orchestrate host tissue interactions. Biological properties of MSCs that contribute to therapeutic efficiency include facilitating secretion of bioactive factors, induction of cellular recruitment and retention of progenitor faculties. Researchers, however continue to be intrigued by variability in the *in vivo* identity of MSCs which is influenced by various factors that include tissue of origin, age of MSCs, number of isolates and isolation efficiency, associated metabolic disorders, foetal or adult status, gene expression, protein and transcription factors and allogenic or autologous extract . Although early results in clinical studies are promising, transformation of MSCs into a mature clinically viable option would mean a patient wait.

Keywords: Cellular therapy, Critical limb ischemia, Crohn's disease, Graft *versus* host disease, Mesenchymal stem cells, Osteogenesis imperfecta, Telomere, Tissue repair, Wharton's jelly, Wound healing.

*Corresponding author Ganapathi M. Bhat:** Jaslok Hospital and Research centre, 15, Dr. G. Deshmukh Marg, Mumbai 400026, Maharashtra, India; Tel: +91-022-66573333; Fax: +91- 022-23526833; E-mail: dr_bhat2@yahoo.com

Atta-ur-Rahman & Shazia Anjum (Eds.)

INTRODUCTION

Successive failures of embryonic stem cells and induced pluripotent stem cells have potentiated the research utilization of mesenchymal stem cells (MSCs). As defined by the International Society of Cellular Therapy, these multipotent, plastic adherent spindle-shaped cells can be propagated through multiple passages in cell culture and differentiated into osteogenic, adipogenic and chondrogenic lineages under permissive conditions [1]. Conventionally isolated from the bone marrow, over the past couple of decades, the isolation of MSCs has been possible from virtually every tissue [2] exhibiting similarity in morphology and, to a certain extent in surface marker profile [3]. Culture-expanded MSCs may lose some of these markers but continue to remain multipotential [3].

By virtue of a variety of biological properties exerted either individually or in combination MSCs have contributed to therapeutic effects in a variety of disease conditions (Fig. **1**). The distinguishable properties include capability to: differentiate into different cell lineages, homing and migration to sites of inflammation and injury, secrete bioactive molecules and promote repair, produce immunomodulatory effects by interacting with the immune cells [4].

Though all MSCs were previously regarded as having low immunogenicity, in view of the low expression of MHC class I and II along with other co-stimulatory molecules such as CD40, CD80, and CD86, recent studies have suggested that allo MSCs may not be as immune-privileged [5] as previously reported except for a few foetal origin MSCs [6], such as those from Wharton's jelly [7].

Owing to the ease of isolation and expansion, MSCs have been studied and have undergone trials in a diverse range of clinical conditions ranging from graft *versus* host disease, autoimmune diseases such as Crohn's disease, cardiovascular diseases such as acute myocardial infarction, stroke and orthopaedic applications including bone and cartilage repair [8]. Primitive MSCs isolated from foetal and perinatal tissues, umbilical cord, placenta, and amniotic fluid have been found to possess longer telomeres [9], higher proliferative potential [10], greater colony-forming capacity, and ability to readily differentiate into bone and muscle [11] and in addition, into non-mesenchymal cells such as neural [12] and hepatic cells

[13]. Increasing age of the MSC may limit the ability to be expanded, leading to rapid senescing in culture and restricted differentiation capacity. In addition, differentiation capabilities, growth kinetics, and yield vary significantly between MSC cell populations from different tissue sources, impacting their clinical utility [14].

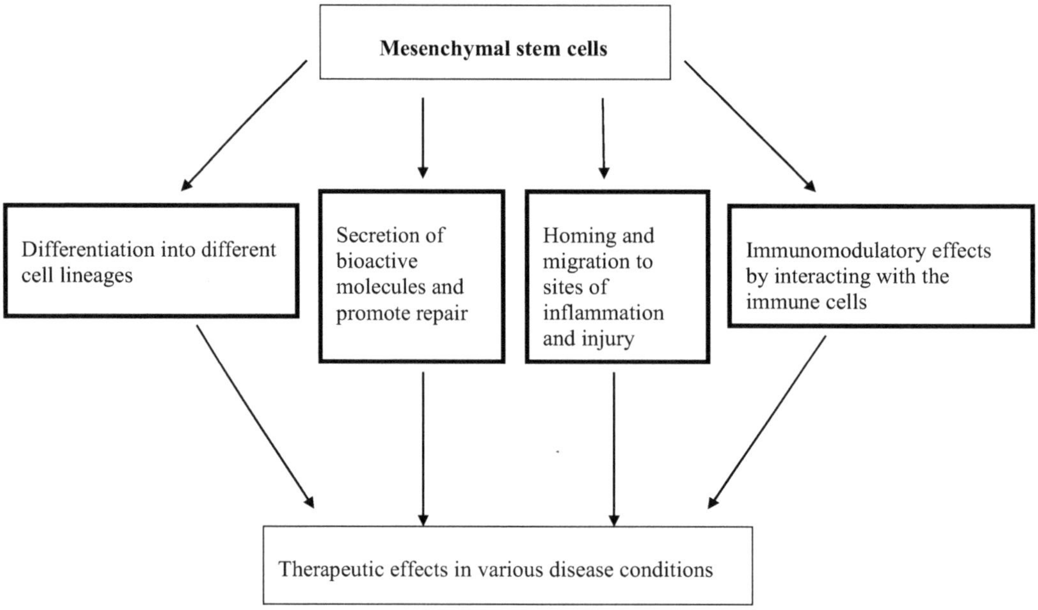

Fig. (1). A schematic diagram of the biological properties of MSCs.

Safety profile of MSCs has been encouraging with no reports of any major health concern in the various human disease settings and clinical trials conducted [15]. Further studies to evaluate tumor formation and genomic integrity are though necessary, the results from current studies have demonstrated no solid evidence of malignant growth or cellular transformation induced by chromosomal inconsistencies [16]. However there is concern about acquired mutations that may induce transformation as a result of prolonged culture.

The lack of standard validated protocols for cell isolation, expansion, and quality control hinder the monitoring of the clinical effectiveness of MSC-based therapy [14]. The absence of governmental regulatory policies at both local and global levels may impact the advance of MSC-based therapeutics into the clinic [17].

This review attempts to summarize the potential uses of MSCs in the treatment of various diseases and discuss the plausible factors influencing *in vivo* therapeutic efficacy.

THERAPEUTIC APPLICATIONS

Therapeutic Debut: Taking the First Step

The first phase 1 trial that evaluated the feasibility of *ex vivo* expansion and infusion in 15 bone marrow transplant patients marked the therapeutic debut of MSCs in 1995 [18].

However, subsequent attempts by Horwitz *et al.*, extended the therapeutic capability of allogeneic bone marrow-derived MSCs (BM-MSCs) to enhance the therapeutic effects of conventional bone marrow transplantation treatment in osteogenesis imperfecta (OI) [19]. The study demonstrated objective clinical responses indicated by new bone formation and increase in total bone mineral content associated with increases in growth velocity and reduced frequencies of bone fracture in 3 children suffering from a severe form of OI. Repeated attempts by Horwitz *et al.* documented a similar clinical response in children with severe OI and treated with BMT [20].

This early success was adopted by Le Blanc *et al.* taking it one step further with a transfusion of allogeneic HLA-mismatched male foetal MSC of hepatic source into a 32-week female foetus *in-utero* diagnosed with multiple fractures and severe OI [21]. The study established the functional contribution of the differentiated MSC to the osteogenic environment even when the recipient is immunocompetent and HLA-incompatible.

Migratory Homing

Role in Wound Healing

MSCs possess the characteristic ability to migrate *in vivo* to sites of injury in response to chemotactic signals and directly influence anti-immunomodulative activities, paracrine signalling, and release of various mediators of tissue repair

[22] and [23] antimicrobial factors such as LL-37 inflammatory response. The use of MSCs isolated from multiple sources as facilitators of healing acute and chronic wounds is being studied extensively. MSCs may be administered at the site of injury *via* injection of topical application or, alternatively, through systemic administration. A significantly faster response time coupled with maximized therapeutic benefit has been noted with direct application of MSCs to the affected region.

BM-MScs facilitated faster healing of acute surgical wounds in about 5 patients when locally applied as a fibrin spray in a study by Falanga *et al.* Further, in another study, improved wound closure, decrease in wound size and angiogenesis was observed with use of topical MSCs in patients with chronic venous and diabetic foot ulcers [24, 25].

The results of the first randomized controlled study by Dash *et al.* in 24 patients, which investigated the clinical effects of both topical (direct into the ulcer) and intramuscularly administered cultured autologous BM-MSCs, exhibited significant improvement in pain and decrease in wound size *versus* control group (72% *versus* 25%) [26]. A study by Yoshikawa *et al.* used a collagen sponge impregnated with autologous BM-MSCs, which when applied topically resulted in complete closure in 13 out of 20 cases of chronic non healing wounds [27].

From current observations, BM-MSCs and adipose-derived Stromal Cells can prove to be beneficial cell sources for the treatment of chronic wounds. However several studies continue to evaluate the critical factors that influence success in wound healing.

Paracrine Effects

Role in Critical Limb Ischemia (CLI)

A number of *in vivo* studies in critical limb ischemia have provided sufficient evidence of the safety and utility of MSCs, both autologous and allogenic, administered intra-arterially, intramuscularly and intradermally in pain relief, wound healing and limb salvage [28]. The efficacy of MSCs in CLI may be attributed to the paracrine effects exerted directly at the local site or indirectly through the recruitment of pro-angiogenic factors [28]. The contributory role of whole cells and their multipotent capabilities remains unclear and has not demonstrated any specific benefit in studies.

Gupta and colleagues have successfully investigated the role of MSCs in patients with CLI and have also extended the application of MSCs in peripheral arterial disease (PAD) [29]. What clinicians would find appealing includes the relative ease of administration, no risk of allo-immunization with allogenic MSCs when administered in multiple doses (vaccination scheme) and therapeutic benefit, irrespective of the source of the MSCs. The future holds the promise of off -the - shelf, alginate-coated, refrigerated MSCs for CLI patients [30].

Regenerative Prowess: Role in Tissue Repair and Organ Rescue

Kidney Disease

MSCs, by virtue of their tropism for the compromised kidney, elicit repair through paracrine and/or endocrine mechanisms, providing a local prosurvival environment by preserving podocyte viability and lowering glomerular inflammation and sclerosis [31].

A study that evaluated the therapeutic value of MSCs (intravenously administered 0.7-1.0 $\times$ 10^6 MSC/kg body weight) in the early stages of chronic kidney disease (CKD) in two divided doses 1 week apart resulted in a statistically significant difference between each of the serum creatinine and creatinine clearance levels pre and post MSC injection at 1, 3 and 6 months post-infusion [32].

A safety and clinical feasibility study by Guiseppe Remuzzi *et al.*, demonstrated the role of autologous MSCs in facilitating enlargement of Treg in the peripheral blood and controls memory CD8 T-cell function in two recipients of kidneys from living related donors, with preserved stable graft function after 1 year [33].

A randomised controlled trial by Jianming Tan *et al.* involving 159 patients who used autologous, intravenously given MSCs (1–2 $\times 10^6$/kg) as novel immunosuppressive induction therapy in living-related kidney transplant patients is worth taking note of [34]. The trial demonstrated that use of autologous MSCs resulted in reduced incidence of acute rejection and risk of opportunistic infection, and better estimated renal function at 1 year in comparison with anti-IL-2 receptor antibody induction therapy. No adverse events or any compromise in kidney transplant survival was observed.

Supra-renal aortic injection of allogeneic BM-MSCs reduced acute kidney ischaemia (AKI) post cardiac surgery by 20% and length of stay and readmission by ~40% compared to historical controls in a phase-1 clinical trial [35].

Currently recruiting trial NCT02266394 due for completion by 2019 is designed to evaluate the efficacy of MSC infusion prior to percutaneous transluminal renal angioplasty with stenting (PTRA) in restoring kidney function. The study also aims to assess the dose related impact of MSC on measures of organ function [36].

Liver Disease

The therapeutic potential of MSCs with regard to liver disease has been studied in a limited number of patients. Autologous BM-MSCs were used in the majority of the studies [37, 38]. Particularly, in four studies, allogeneic MSCs (one BM-MSCs and three UC-MSCs) were transplanted into patients with decompensated liver cirrhosis (LC), ursodeoxycholic acid (UDCA)-resistant primary biliary cirrhosis (PBC), and acute-on-chronic liver failure (ACLF) [39 - 42]. Allogeneic MSCs infusions were not associated with transplantation-related side effects and have demonstrated improvement in liver function. In another phase I-II trial, 1 alcoholic, 1 hepatitis C, 4 hepatitis B and 2 cryptogenic) with end-stage liver disease tolerated autologous MSCs infusion and their liver function improved [43]. Follow-up of the patients at 3 months and 6 months post-transplant revealed partial improvement of liver function tests with elevation of prothrombin concentration and serum albumin levels and decline of elevated bilirubin and Model for End-Stage Liver Disease score (MELD) [44].

Cardiac Disease

Application of MSCs for myocardial infarction (MI) and ischemic cardiomyopathy was initiated based on the safety and feasibility findings in clinical trials. Chen *et al.* demonstrated significant and sustained improvement in global left-ventricular ejection fraction, suggesting that MSCs influence the formation of new cardiomyocytes and neoangiogenesis in the human heart [45].

BM-MSCs are able to differentiate into endothelial cells *in vitro* [46], although cells from adipose tissue seem to be more prone to this differentiation [47]. Bioengineered MSCs implanted into vascular prostheses have been shown to contribute to the restoration of the vascular wall. In a pilot study, 69 patients who underwent primary percutaneous coronary intervention within 12 hours after onset of acute MI were randomized to receive an intracoronary injection of autologous BM-MSC or standard saline (MSCs = 34 and saline = 35) [48]. Several imaging techniques demonstrated that MSCs significantly improved left ventricular function. At 3 months after MSCs transplantation, left ventricular ejection fraction (LVEF) in the MSCs group increased significantly compared with that of pre-implantation and the control group. In ischemic cardiomyopathy trans-endocardial, intra-myocardial injection of MSCs demonstrated reverse remodelling and improved regional contractility of the treated scar 3 months after injection, which persisted for 12 months [49]. The advancement also reflected diastolic volume (EDV) and end systolic volume (ESV) in end.

Another study suggested the importance of IL-6 secretion and activation of Janus kinase transducers and activators of transcription (JAK-STAT) in cardiac repair by transplanted MSCs [50].

Orthopaedic and Bone Disorders

The application of MSCs in orthopedics has been evaluated, in majority, to address cartilage defects and to correct large bone defects [51, 52]. Increasingly innovative matrices which include poly (vinyl alcohol)/polycaprolactone (PVA/PCL) nanofibers [53], and alginate-embedded MSCs have [54] been tested *in vivo* animal models to facilitate MSC proliferation and chondrogenic differentiation. Use of BM-MSC s supported by scaffolds to repair osteochondral lesions of the talus and femur condyle in a study by Buda *et al.* resulted in significant clinical improvement, and integration of the graft, which was maintained over time [55]. In a study, locally directed intra-articular injection of

BM-MSCs in a patient with early osteoarthritis of the knee has shown, on MRI, to decrease the volume of the cartilage defect at 6 months [56].

While systemically administered MSCs exhibited migration, homing and differentiation at fracture site in a mouse model, *in vivo* there have been instances of erroneous accumulation in some off- target tissues [57], thus making local application more practicable for bone regeneration. Autologous BM-MSCs or autologous MSCs has been implanted in a number of patients to fill bone defects, improve healing of fractures and strengthen the bone [58]. Application of MSCs in tendon repair has demonstrated homogenous results in preclinical studies, while there is lack of data from clinical studies and also no consensus on an ideal carrier construct [59].

As opposed to focal articular cartilage defects, delivery of neocartilage constructs to inflamed joints (in osteoarthritis or rheumatoid arthritis) may result in rapid degradation [60]. Intra-articular delivery of *ex vivo* expanded bone marrow stromal cell populations in patients with osteoarthritis (OA) of the knee was safe and feasible with MRI-documented increased cartilage and meniscus thickening at 1 year post-procedure in 1 patient [61]. On the whole, application of MSC in preclinical studies is promising. However, the results of clinical studies remain discordant as a result of variability in cell sources, biomaterial carrier, study procedures and follow-up times.

Battling Immune Assaults: Role in Immune Mediated Disorders

Diabetes

The retention of endogenous insulin secretion is a significant clinical goal to enable long-term restoration of glucose metabolism in type 1 diabetes. Autologous human umbilical cord blood (hUCB) transfusion evaluated in 23 children with type 1 diabetes showed significant improvement in glucose control in addition retention of endogenous insulin production as assessed by stimulated C-peptide secretion [62].

In a noteworthy case report by Dave *et al.*, autologous adipose tissue-derived MSCs were infused with bone marrow-derived hematopoietic stem cells (HSCs) in portal, thymic circulation by mini-laparotomy and in subcutaneous circulation in 10 patients with insulin-dependent diabetes mellitus [63]. Significant reduction in HbA1C of 6% and sustained decrease in insulin requirement of 60% was noted at 18 months. In another study, *ex vivo* generation of insulin-secreting mesenchymal stem cells (IS-MSC) generated from human adipose tissue (h-AD) devoid of xenogenic material was found to be safe and effective in generating *in vitro* glucose responsive alteration in insulin and C-peptide secretion levels [64]. Intravenous delivery of Wharton's jelly MSCs (WJ-MSC) in new-onset type 1 diabetes resulted in a reduction in insulin supplementation or discontinuation in some patients and was not associated with any serious adverse events [65].

In type 2 diabetes, delivery of the first dose of WJ-MSC by the intravenous route and a repeated dose directly to the pancreas *via* the splenic artery demonstrated freedom from insulin in 7/17 patients and considerable reduction in insulin requirement in 5/15 [66] as a result of decrease in pro-inflammatory IL-β and IL-6 plasma level. In contrast, BM-MSC from aged patients did not exhibit similar response. However, in another attempt, adult aged BM-MSC co-cultured with activated T-cells was found to secrete more IL-6 than younger cells [67].

Systemic Lupus Erythematosus

Application of allogeneic umbilical cord (UC- MSCs) have demonstrated significant impact on disease activity, serologic changes, and stabilization of pro-inflammatory cytokines and no reported treatment-related mortality or other post or peri-procedural adverse event in 16 patients with severe and treatment-refractory systemic lupus erythematosus (SLE) [68]. In another non-placebo controlled study, intravenously administered WJ-MSC (two doses of 10^6/kg of WJ-MSC at 1-week intervals) resulted in significant improvement in renal function, decrease in the SLEDAI (Systemic Lupus Erythematosus Disease Activity Index) and BILAG (British Isles Lupus Assessment Group) scoring in at least half of the total patients studied (n=40). A repeated infusion was suggested for the patients who experienced disease relapse (n=7) [69].

Crohn's Disease

MSC treatment was attempted for use with standard Crohn's Disease therapy to enable more rapid tapering of immunosuppressive therapy. Onken *et al.* demonstrated a decrease in the Crohn's disease activity index (CDAI) score by day 28 post systemic administration of allogeneic BM-MSCs in Crohn's Disease, in patients who had failed treatment with steroids and immunosuppressant's and had active disease [70]. Irrespective of whether autologous or allogenic, systemically administered MSCs were found to be safe, with no toxic effects or generation of ectopic tissue.

Local application of autologous MSCs obtained from a lipoaspirate for the treatment of fistulae was first attempted in 2005 [71]. MSCs were found to be more effective than fibrin glue alone in patients with a suprasphincteric fistulous tract [71]. Considerable therapeutic benefit of local injection of autologous BM-MSCs in fistulising lesions was demonstrated in another study [72]. Selective expression of adhesion molecules and chemokine receptors present on the MSCs and, also, achievement of cell density at the inflamed site could explain the inconsistency between the efficacies of local injection of MSCs for the treatment of fistulas compared with systemic administration for treatment of luminal Disease [73]. A study by Duijvestein *et al.* further demonstrated that the possibility of patients benefitting from the synergism between MSC's and immunosuppressive therapy [74].

Neurological Disorders

The role of MSCs in four major neurological diseases; Parkinson's disease, Huntington's disease, amyotrophic lateral sclerosis and stroke is being investigated [75]. Adipose-tissue-derived MSCs display greater neuronal potential in addition to advantages in accessibility, proliferation potential *in vitro* and a lower senescence ratio than BM-MSCs *in vitro* and after co-culture with schwann cells or under the culture conditions used for neural stem cells [76].

In a study that investigated *ex vivo* expanded BM-MSCs in advanced Multiple Sclerosis (MS), 50% of cases showed beneficial outcomes suggesting that MSCs are safe and feasible for use in the treatment of MS patients [77]. In a phase I/II study, patients with MS and ALS treatment with intravenous MSC infusion led to increase in the proportion of CD4+CD45+Tregcells in the peripheral blood of the patient [78]. Despite positive preclinical results, clear evidence of the beneficial effect of MSCs for the treatment of neurodegenerative disorders is lacking [78].

In patients with advanced Parkinson's disease, transplant of autologous bone-marrow-derived stem cells into the sublateral ventricular zone was found to be a safe procedure; however was not powered to demonstrate efficacy [79]. MSC-DP cells are likely to be non-tumorigenic and able to stay by restoring dopaminergic function and motor behaviours *in vivo* reasonably in the transplanted region up to at least 9 months after engraftment, although not in large quantities [80].

Graft-versus-Host Disease (GVHD)

Application of haploidentical mesenchymal stem cells resulted in clinical response maintained for over a year in a patient with severe treatment-resistant grade IV in acute GVHD of the gut and liver [81]. In another study, MSCs demonstrated complete resolution and improved survival in patients with severe steroid refractory grade III-IV GVHD [82]. Infusion of *in vitro* expanded MSCs, regardless of the donor, may be an efficacious treatment choice in steroid-resistant, acute GVHD [83]. A Pre-manufactured formulation of hMSCs from an universal donor has been well tolerated and resulted in improved survival when used as rescue therapy in severe steroid- refractory, acute GVHD in preliminary studies [84].

Support of Hematopoietic Reconstitution

In two separate studies with patients receiving MSCs for graft failure, co-transplantation of MSCs resulted in early engraftment and 100% donor chimerism [85]. Co-transplantation of haploidentical MSCs facilitated the engraftment in paediatric recipients of unrelated donor UCB in a phase I/II study [86]. Another case report documented the evidence of recruitment of residual MSCs to repopulate the stroma promoted by allogeneic donor MSCs when engrafted in a patient with refractory, severe aplastic anaemia [87].

Solid Organ Transplant (SOT)

The applicability of MSCs has shown potential in inducing transplant tolerance in SOT. Despite limited experience with MSCs in clinical transplantation, available data supports the ability of MSCs to control proliferation of effector and memory CD8 T-cells and maintenance of donor-specific CD8 T-cell function in a kidney transplant setting [88].

DISCUSSION

By definition, human MSCs are a heterogeneous cell population residing in various tissues and possessing the capacity to proliferate and the potential to differentiate. Although the identification and isolation of MSCs was first carried out from bone marrow, they are now isolated from adult tissues that include adipose tissue [89], peripheral blood [90], skin, dental pulp [91], foetal lung [92], endometrium [93], synovial membrane, bone and muscle [94], and birth-associated tissues that include placenta [95], amnion, umbilical cord and cord blood [96]. Though, morphologically, MSCs derived from different tissues revealed no differences in gross appearance or in the expression of surface markers, discrepancies in clinical effectiveness among certain subpopulations of MSCs within the same primary culture have been observed [97]. Some interesting findings from the studies that have been noted include a higher proliferation potential and greater immunomodulatory capacity in foetal MSCs compared to adult MSCs. Also MSCs derived from the same tissue have better tissue specific differentiation capability, for example, BM-MSCs have better osteogenesis capability compared to MSCs from other sources [98] and, similarly, islet-derived MSCs for islet differentiation [99]. This may be attributed to peculiar biological properties that include morphological identity, immunophenotype, clonogenic potential, proliferation rate, cellular ageing, differentiation capacity, angiogenesis potential, and immunosuppression [4]. Moreover, non-specific degradation of certain surface markers as a result of disparate isolation methods used is also contributory [100].

MSCs derived from amniotic fluid have shown the capability to retain morphological identity up to over 29 passages in comparison to BM-MSCs, which exhibited morphological ageing cells as early as with 12 passages [101]. Similarly differences in yield have been noted ranging from very low (0.001-0.01% in mononuclear cells) [102] from bone marrow, to >500 fold greater from adipose tissue [103] to 0.9 - 1.5% from amniotic fluid (AF-MSCs). In contrast, the replication time of AF-MSCs was lower (1.6 days) than that of BM-MSCs (3.75 days) [104].

Potential to differentiate to other lineages was shown to be influenced by the tissue of origin of the MSC. For example, lower osteogenic and chondrogenic potential has been noted with MSCs isolated from the adipose tissue compared to those from bone marrow [98]. Further, MSCs from bone marrow and adipose tissue have lower adipogenic capacity compared with cord blood (CB-MSCs) and placenta derived (PL-MSCs) [105]. Osteogenic potential of foetal origin MSCs owing to the presence of osteogenic-specific gene expression than adult in BM-MSCs [106]. On the other hand, potential to differentiate to functional cardiomyocytes was demonstrated by foetal MSCs from umbilical cord although MSCs from bone marrow and adipose tissue did not [10].

In terms of proliferation, MSCs derived from cord blood demonstrated higher proliferation with every passage compared with BM-MSCs [107]. Similarly amniotic fluid MSCs exhibited greater proliferation capacity and expression of markers of neuronal stemness [108]. In addition to the source, cell culture media also impacted the expression of desmin by adipose tissue derived MSCs [109].

The source and cytokine profile may account for the varying immunomodulatory function of MSCs [110]. When observed in a study on anti-CD3/ CD28-induced peripheral blood mononuclear cells (PBMCs) proliferation, adipose tissue derived MSCs exhibited more potent immunosuppressive effects than BM-MSCs [111]. Cord blood derived MSCs demonstrated a higher expression of upstream angiopoietin-1 and lower levels of pro-inflammatory cytokines [112]. MSCs isolated from some specific adult organ tissues that included olfactory ecto-MSCs (OEMSCs), heart MSCs (H-MSCs), and lung MSCs (L-MSCs) have demonstrated to have *in vitro* immunosuppressive functions [113].

Appropriate homing of MSCs to relevant site is critical following their administration. WJ-MSCs may be a better choice to rescue and repair fibrosis and scarring owing to higher levels of angiogenic chemokines and angiogenic factors [114].

Autologous and allogenic MSCs though found to demonstrate comparable efficacy and safety in clinical trials may differ in their immunogenicity [115]. Nevertheless, application of the specific type of MSCs for specific disease conditions is recommended. On the other hand, autologous BM-MSCs were found to have superior in improving acute kidney injury [116]. Foetal tissues like amniotic fluid and umbilical cord are generally sources of allogenic MSCs. Allogenic MSCs when compared to their autologous counterparts appear to have the potential to be packaged as an off the shelf therapeutic option in future [117]. Another point worth consideration would be that though allogeneic UC-MSCs can be expanded and cryopreserved in a cell bank, at present it would not be possible to ascertain the future health status of the donor at time of donation [118].

Availability of CD45−/CD34+/CD133+ and angiogenic properties in adipose tissue derived MSCs is altered with ageing [119]. Matured adipose tissue derived MSC are not suitable to rescue age-associated impairments in cutaneous wound healing as a result of a significantly compromised ability to assist with angiogenesis. On the other hand, MSCs of foetal origin such as Wharton's jelly are spared from influence of pro-ageing factors [120]. Failure of adult BM-MSC to undergo neuroectodermal differentiation *in vitro*, makes them unsuitable as cell replacement options for neurologic diseases in elderly patients in autologous setting [121]. Researchers have also implicated down regulation of certain molecular pathways (SITR1, NAD+-dependent protein deacetylase) [122] in human MSCs with increasing passages, which promotes the age-dependent decline of stem cell function [123].

Autologous stem cell therapy in diabetic patients is not suitable since metabolic disorders such as diabetes influence stem cell niche and endogenous MSC properties [124]. Bone marrow stem cells from diabetic patients appeared phenotypically similar to healthy human BM-MSC but did not, however, express insulin, C-peptide, and other pancreatic markers in control healthy cells [124].

Gene expression profiles have also been shown to influence the properties of MSCs. For example, higher expression of gene E-cadherin and lower expression of gene Snail was noted with MSCs derived from dental pulp, favouring their superior ability to facilitate tissue repair [125]. Further, these cells were also found to highly express Pdx1 [126] (critical role in glucose-induced insulin gene transcription in adult β-cells), suggesting a probable role in diabetes, and Sox2 gene and Sox17 [127] (transcriptional regulators), suggesting a unique possibility to be induced *in vitro* to undergo neuronal differentiation.

Identifying the appropriate methods to improve survival and sustain therapeutic efficiency is critical for the success of MSC transplants. Nongenetic methods that do not alter tissue integrity are among the emerging methods being investigated [128]. A variety of microenvironmental preconditioning methods which include hypoxic exposure [129], heat shock treatment [130], exposure to oxidative stress, aggregate formation [131], and hydrogel encapsulation [132] have been tested in feasibility studies and have proven useful in facilitating stable and sustained engraftment.

It is critical to coordinate the clinicians' needs with the researchers' options in close interaction with the manufacturing laboratories, who overview the margin framed by the regulatory authorities [133]. Any significant changes in the manufacturing protocols require validation *in vitro* and in appropriate *in vivo* animal models to assure safety and efficacy as well as absence of toxicity (related to de-differentiation or unwanted differentiation, migration to unwanted sites) [133]. The translation of research-based protocols into good manufacturing practices (GMP) complaint procedures for large-scale production of clinical grade MSCs requires careful analysis of all the risks and benefits to identify and control all the critical aspects [134].

With regard to safety aspects, in general MSCs as mentioned earlier in this chapter have demonstrated no alarming adverse effects with use in limited settings and limited population [15]. Infusion related toxicity may be one of the manageable effects while tumorigenic promotion, induced cellular transformation arthritis [135] may be reason for concern. Unlike the embryonic stem cells autologous MSCs are less likely to trigger immunogenic response or result in teratoma formation. In a long term study that evaluated culture-expanded, autologous BMMSCs for various orthopedic conditions in 339 patients, no evidence of neoplastic complications was detected at implanted sites [61]. MSCs carrying chromosomal aberrations revealed no malignant transition *in vitro* and *in vivo* studies [16]. Despite no major safety concerns, further studies in larger number of patients may be required to allay doubts.

Currently registered clinical trials (www.clinicaltrials.gov) are investigating the use of MSCs in tissue repair (cardiac ischaemia, limb ischaemia, amyotrophic lateral sclerosis, diabetes, ischaemic stroke, osteoarthritis, liver cirrhosis, and liver failure) while a majority of them have been conducted for immune-mediated diseases including (rheumatoid arthritis, Crohn's disease, MS, GVHD) [136]. Further validation of the effectiveness of MSCs would require larger randomized controlled trials to be performed.

CONCLUSION

MSCs have emerged as cellular candidates that have kindled hope of harnessing the promise of regenerative medicine. With the unexpected therapeutic success in the first human study for treatment of osteogenesis imperfecta, MSCs further demonstrated usefulness in reconstitution of hematopoiesis following myeloablative chemotherapy. MSCs are today the most preferred over the other stem cells for regenerative therapy as a result of their effectiveness, safety and exemption from ethical issues. They have been increasingly utilized in therapeutic trials for chronic wound healing, critical limb ischemia, organ repair and autoimmune mediated disorders. Further studies elucidating the optimal source, quantity, expansion and administration suitable for a specific clinical condition would decide the therapeutic fate of these cells. The manufacturing, production and clinical use of MSC's necessitates the close involvement of fundamental research teams, research and development sectors, production laboratories and clinical groups, to finally catapult MSC s from the lab to life.

FINANCIAL SUPPORT

The authors confirm that they have not received any financial support for this publication.

CONFLICT OF INTEREST

The authors confirm that they have no conflict of interest to declare for this publication.

ACKNOWLEDGEMENTS

"We would like to thank Dr. Cheryl Travasso for editing of the manuscript."

REFERENCES

[1] Dominici M, Le Blanc K, Mueller I, *et al.* Minimal criteria for defining multipotent mesenchymal stromal cells. The International Society for Cellular Therapy position statement. Cytotherapy 2006; 8(4): 315-7.
[http://dx.doi.org/10.1080/14653240600855905] [PMID: 16923606]

[2] da Silva Meirelles L, Chagastelles PC, Nardi NB. Mesenchymal stem cells reside in virtually all post-natal organs and tissues. J Cell Sci 2006; 119(Pt 11): 2204-13.
[http://dx.doi.org/10.1242/jcs.02932] [PMID: 16684817]

[3] Jacobs SA, Roobrouck VD, Verfaillie CM, Van Gool SW. Immunological characteristics of human mesenchymal stem cells and multipotent adult progenitor cells. Immunol Cell Biol 2013; 91(1): 32-9.
[http://dx.doi.org/10.1038/icb.2012.64] [PMID: 23295415]

[4] Via AG, Frizziero A, Oliva F. Biological properties of mesenchymal stem cells from different sources. Muscles Ligaments Tendons J 2012; 2(3): 154-62.
[PMID: 23738292]

[5] Zhang J, Huang X, Wang H, *et al.* The challenges and promises of allogeneic mesenchymal stem cells for use as a cell-based therapy. Stem Cell Res Ther 2015; 6: 234.
[http://dx.doi.org/10.1186/s13287-015-0240-9] [PMID: 26620426]

[6] Parolini O, Alviano F, Bagnara GP, *et al.* Concise review: isolation and characterization of cells from human term placenta: outcome of the first international Workshop on Placenta Derived Stem Cells. Stem Cells 2008; 26(2): 300-11.
[http://dx.doi.org/10.1634/stemcells.2007-0594] [PMID: 17975221]

[7] Kim D-W, Staples M, Shinozuka K, Pantcheva P, Kang SD, Borlongan CV. Wharton's jelly-derived mesenchymal stem cells: phenotypic characterization and optimizing their therapeutic potential for clinical applications. Int J Mol Sci 2013; 14(6): 11692-712.
[http://dx.doi.org/10.3390/ijms140611692] [PMID: 23727936]

[8] Kim N, Cho S-G. Clinical applications of mesenchymal stem cells. Korean J Intern Med 2013; 28(4): 387-402.
[http://dx.doi.org/10.3904/kjim.2013.28.4.387] [PMID: 23864795]

[9] Wang L, Ott L, Seshareddy K, Weiss ML, Detamore MS. Musculoskeletal tissue engineering with human umbilical cord mesenchymal stromal cells. Regen Med 2011; 6(1): 95-109.
[http://dx.doi.org/10.2217/rme.10.98] [PMID: 21175290]

[10] Zhu SF, Zhong ZN, Fu XF, *et al.* Comparison of cell proliferation, apoptosis, cellular morphology and ultrastructure between human umbilical cord and placenta-derived mesenchymal stem cells. Neurosci Lett 2013; 541: 77-82.
[http://dx.doi.org/10.1016/j.neulet.2013.03.018] [PMID: 23523648]

[11] Guillot PV, De Bari C, Dell'Accio F, Kurata H, Polak J, Fisk NM. Comparative osteogenic

transcription profiling of various fetal and adult mesenchymal stem cell sources. Differentiation 2008; 76(9): 946-57.
[http://dx.doi.org/10.1111/j.1432-0436.2008.00279.x] [PMID: 18557767]

[12] Balasubramanian S, Thej C, Venugopal P, *et al.* Higher propensity of Wharton's jelly derived mesenchymal stromal cells towards neuronal lineage in comparison to those derived from adipose and bone marrow. Cell Biol Int 2013; 37(5): 507-15.
[http://dx.doi.org/10.1002/cbin.10056] [PMID: 23418097]

[13] Lee HJ, Jung J, Cho KJ, Lee CK, Hwang SG, Kim GJ. Comparison of *in vitro* hepatogenic differentiation potential between various placenta-derived stem cells and other adult stem cells as an alternative source of functional hepatocytes. Differentiation 2012; 84(3): 223-31.
[http://dx.doi.org/10.1016/j.diff.2012.05.007] [PMID: 22885322]

[14] Wagner W, Ho AD. Mesenchymal stem cell preparations-comparing apples and oranges. Stem Cell Rev 2007; 3(4): 239-48.
[http://dx.doi.org/10.1007/s12015-007-9001-1] [PMID: 18074246]

[15] Wang Youwei, Han Zhi-bo, Song Yong-ping, Zhong Chao Han. Safety of mesenchymal stem cells for clinical application. Stem Cells Int 2012; 2012: 652034.

[16] Tarte K, Gaillard J, Lataillade JJ, *et al.* Clinical-grade production of human mesenchymal stromal cells: occurrence of aneuploidy without transformation. Blood 2010; 115(8): 1549-53.
[http://dx.doi.org/10.1182/blood-2009-05-219907] [PMID: 20032501]

[17] Division on Earth and Life Studies; Institute of Medicine; National Academy of Sciences. Stem Cell Therapies: Opportunities for Ensuring the Quality and Safety of Clinical Offerings: Summary of a Joint Workshop. Washington (DC): National Academies Press (US); Comparative Regulatory and Legal Frameworks. Available from: http://www.ncbi.nlm.nih.gov/books/NBK223198/ 2014.

[18] Lazarus HM, Haynesworth SE, Gerson SL, Rosenthal NS, Caplan AI. *ex vivo* expansion and subsequent infusion of human bone marrow-derived stromal progenitor cells (mesenchymal progenitor cells): implications for therapeutic use. Bone Marrow Transplant 1995; 16(4): 557-64.
[PMID: 8528172]

[19] Horwitz EM, Prockop DJ, Fitzpatrick LA, *et al.* Transplantability and therapeutic effects of bone marrow-derived mesenchymal cells in children with osteogenesis imperfecta. Nat Med 1999; 5(3): 309-13.
[http://dx.doi.org/10.1038/6529] [PMID: 10086387]

[20] Horwitz EM, Gordon PL, Koo WK, *et al.* Isolated allogeneic bone marrow-derived mesenchymal cells engraft and stimulate growth in children with osteogenesis imperfecta: Implications for cell therapy of bone. Proc Natl Acad Sci USA 2002; 99(13): 8932-7.
[http://dx.doi.org/10.1073/pnas.132252399] [PMID: 12084934]

[21] Le Blanc K, Götherström C, Ringdén O, *et al.* Fetal mesenchymal stem-cell engraftment in bone after in utero transplantation in a patient with severe osteogenesis imperfecta. Transplantation 2005; 79(11): 1607-14.
[http://dx.doi.org/10.1097/01.TP.0000159029.48678.93] [PMID: 15940052]

[22] Gnecchi M, Zhang Z, Ni A, Dzau VJ. Paracrine mechanisms in adult stem cell signaling and therapy. Circ Res 2008; 103(11): 1204-19.

[http://dx.doi.org/10.1161/CIRCRESAHA.108.176826] [PMID: 19028920]

[23] Chen L, Tredget EE, Wu PY, *et al.* Paracrine factors of mesenchymal stem cells recruit macrophages and endothelial lineage cells and enhance wound healing. PLoS One 2008; 3(4): 1886.

[24] Badiavas EV, Falanga V. Treatment of chronic wounds with bone marrow-derived cells. Arch Dermatol 2003; 139(4): 510-6.
[http://dx.doi.org/10.1001/archderm.139.4.510] [PMID: 12707099]

[25] Chen JS, Wong VW, Gurtner GC. Therapeutic potential of bone marrow-derived mesenchymal stem cells for cutaneous wound healing. Front Immunol 2012; 3: 192.
[http://dx.doi.org/10.3389/fimmu.2012.00192] [PMID: 22787462]

[26] Dash NR, Dash SN, Routray P, Mohapatra S, Mohapatra PC. Targeting nonhealing ulcers of lower extremity in human through autologous bone marrow-derived mesenchymal stem cells. Rejuvenation Res 2009; 12(5): 359-66.
[http://dx.doi.org/10.1089/rej.2009.0872] [PMID: 19929258]

[27] Yoshikawa T, Mitsuno H, Nonaka I, *et al.* Wound therapy by marrow mesenchymal cell transplantation. Plast Reconstr Surg 2008; 121(3): 860-77. [PubMed: 18317135].
[http://dx.doi.org/10.1097/01.prs.0000299922.96006.24] [PMID: 18317135]

[28] Liew A, O'Brien T. Therapeutic potential for mesenchymal stem cell transplantation in critical limb ischemia. Stem Cell Res Ther 2012; 3(4): 28.
[http://dx.doi.org/10.1186/scrt119] [PMID: 22846185]

[29] Gupta PK, Chullikana A, Parakh R, *et al.* A double blind randomized placebo controlled phase I/II study assessing the safety and efficacy of allogeneic bone marrow derived mesenchymal stem cell in critical limb ischemia. J Transl Med 2013; 11: 143.
[http://dx.doi.org/10.1186/1479-5876-11-143] [PMID: 23758736]

[30] Altaner C, Altanerova V, Cihova M, *et al.* Characterization of mesenchymal stem cells of "no-options" patients with critical limb ischemia treated by autologous bone marrow mononuclear cells. PLoS One 2013; 8(9): e73722.
[http://dx.doi.org/10.1371/journal.pone.0073722] [PMID: 24069226]

[31] Zoja C, Garcia PB, Rota C, *et al.* Mesenchymal stem cell therapy promotes renal repair by limiting glomerular podocyte and progenitor cell dysfunction in adriamycin-induced nephropathy. Am J Physiol Renal Physiol 2012; 303(9): F1370-81.
[http://dx.doi.org/10.1152/ajprenal.00057.2012] [PMID: 22952284]

[32] El-Ansary M, Saadi G, Abd El-Hamid SM. Mesenchymal stem cells are a rescue approach for recovery of deteriorating kidney function. Nephrology (Carlton) 2012; 17(7): 650-7.
[http://dx.doi.org/10.1111/j.1440-1797.2012.01622.x] [PMID: 22640266]

[33] Perico N, Casiraghi F, Gotti E, *et al.* Mesenchymal stromal cells and kidney transplantation: pretransplant infusion protects from graft dysfunction while fostering immunoregulation. Transpl Int 2013; 26(9): 867-78.
[http://dx.doi.org/10.1111/tri.12132] [PMID: 23738760]

[34] Tan J, Wu W, Xu X, *et al.* Induction therapy with autologous mesenchymal stem cells in living-related kidney transplants: a randomized controlled trial. JAMA 2012; 307(11): 1169-77.

[http://dx.doi.org/10.1001/jama.2012.316] [PMID: 22436957]

[35] Zhu X-Y, Lerman A, Lerman LO. Concise review: mesenchymal stem cell treatment for ischemic kidney disease. Stem Cells 2013; 31(9): 1731-6.
[http://dx.doi.org/10.1002/stem.1449] [PMID: 23766020]

[36] Stephen C. Hypoxia and Inflammatory Injury in Human Renovascular Hypertension : Phase 1 Trial of Mesenchymal Stem Cell Therapy. Available from: https://clinicaltrials.gov/ct2/show/study/NCT0226-6394

[37] Amin MA, Sabry D, Rashed LA, *et al.* Short-term evaluation of autologous transplantation of bone marrow-derived mesenchymal stem cells in patients with cirrhosis: Egyptian study. Clin Transplant 2013; 27(4): 607-12.
[http://dx.doi.org/10.1111/ctr.12179] [PMID: 23923970]

[38] Jang YO, Kim YJ, Baik SK, *et al.* Histological improvement following administration of autologous bone marrow-derived mesenchymal stem cells for alcoholic cirrhosis: a pilot study. Liver Int 2014; 34(1): 33-41.
[http://dx.doi.org/10.1111/liv.12218] [PMID: 23782511]

[39] Wang L, Li J, Liu H, *et al.* Pilot study of umbilical cord-derived mesenchymal stem cell transfusion in patients with primary biliary cirrhosis. J Gastroenterol Hepatol 2013; 28 (Suppl. 1): 85-92.
[http://dx.doi.org/10.1111/jgh.12029] [PMID: 23855301]

[40] Zhang Z, Lin H, Shi M, *et al.* Human umbilical cord mesenchymal stem cells improve liver function and ascites in decompensated liver cirrhosis patients. J Gastroenterol Hepatol 2012; 27 (Suppl. 2): 112-20.
[http://dx.doi.org/10.1111/j.1440-1746.2011.07024.x] [PMID: 22320928]

[41] Wang L, Han Q, Chen H, *et al.* Allogeneic bone marrow mesenchymal stem cell transplantation in patients with UDCA-resistant primary biliary cirrhosis. Stem Cells Dev 2014; 23(20): 2482-9.
[http://dx.doi.org/10.1089/scd.2013.0500] [PMID: 24835895]

[42] Shi M, Zhang Z, Xu R, *et al.* Human mesenchymal stem cell transfusion is safe and improves liver function in acute-on-chronic liver failure patients. Stem Cells Transl Med 2012; 1(10): 725-31.
[http://dx.doi.org/10.5966/sctm.2012-0034] [PMID: 23197664]

[43] Kharaziha P, Hellström PM, Noorinayer B, *et al.* Improvement of liver function in liver cirrhosis patients after autologous mesenchymal stem cell injection: a phase I-II clinical trial. Eur J Gastroenterol Hepatol 2009; 21(10): 1199-205.
[http://dx.doi.org/10.1097/MEG.0b013e32832a1f6c] [PMID: 19455046]

[44] El-Ansary M, Abdel-Aziz I, Mogawer S, *et al.* Phase II trial: undifferentiated *versus* differentiated autologous mesenchymal stem cells transplantation in Egyptian patients with HCV induced liver cirrhosis. Stem Cell Rev 2012; 8(3): 972-81.
[http://dx.doi.org/10.1007/s12015-011-9322-y] [PMID: 21989829]

[45] Chen SL, Fang WW, Ye F, *et al.* Effect on left ventricular function of intracoronary transplantation of autologous bone marrow mesenchymal stem cell in patients with acute myocardial infarction. Am J Cardiol 2004; 94(1): 92-5.
[http://dx.doi.org/10.1016/j.amjcard.2004.03.034] [PMID: 15219514]

[46] Oswald J, Boxberger S, Jørgensen B, *et al.* Mesenchymal stem cells can be differentiated into endothelial cells *in vitro*. Stem Cells 2004; 22(3): 377-84.
[http://dx.doi.org/10.1634/stemcells.22-3-377] [PMID: 15153614]

[47] Planat-Benard V, Silvestre JS, Cousin B, *et al.* Plasticity of human adipose lineage cells toward endothelial cells: physiological and therapeutic perspectives. Circulation 2004; 109(5): 656-63.
[http://dx.doi.org/10.1161/01.CIR.0000114522.38265.61] [PMID: 14734516]

[48] Chen SL, Fang WW, Qian J, *et al.* Improvement of cardiac function after transplantation of autologous bone marrow mesenchymal stem cells in patients with acute myocardial infarction. Chin Med J (Engl) 2004; 117(10): 1443-8.
[PMID: 15498362]

[49] Williams AR, Trachtenberg B, Velazquez DL, *et al.* Intramyocardial stem cell injection in patients with ischemic cardiomyopathy: functional recovery and reverse remodeling. Circ Res 2011; 108(7): 792-6.
[http://dx.doi.org/10.1161/CIRCRESAHA.111.242610] [PMID: 21415390]

[50] Shabbir A, Zisa D, Lin H, *et al.* Activation of host tissue trophic factors through JAK-STAT3 signaling: a mechanism of mesenchymal stem cell-mediated cardiac repair. Am J Physiol Heart Circ Physiol 2010; 299(5): H1428-38.
[http://dx.doi.org/10.1152/ajpheart.00488.2010] [PMID: 20852053]

[51] Connolly JF, Guse R, Tiedeman J, Dehne R. Autologous marrow injection as a substitute for operative grafting of tibial nonunions. Clin Orthop Relat Res 1991; (266): 259-70.
[PMID: 2019059]

[52] Hendrich C, Franz E, Waertel G, Krebs R, Jäger M. Safety of autologous bone marrow aspiration concentrate transplantation: initial experiences in 101 patients. Orthop Rev (Pavia) 2009; 1(2): e32.
[http://dx.doi.org/10.4081/or.2009.e32] [PMID: 21808691]

[53] Shafiee A, Soleimani M, Chamheidari GA, *et al.* Electrospun nanofiber-based regeneration of cartilage enhanced by mesenchymal stem cells. J Biomed Mater Res A 2011; 99(3): 467-78.
[http://dx.doi.org/10.1002/jbm.a.33206] [PMID: 21887742]

[54] Tay LX, Ahmad RE, Dashtdar H, *et al.* Treatment outcomes of alginate-embedded allogenic mesenchymal stem cells *versus* autologous chondrocytes for the repair of focal articular cartilage defects in a rabbit model. Am J Sports Med 2012; 40(1): 83-90.
[http://dx.doi.org/10.1177/0363546511420819] [PMID: 21917609]

[55] Buda R, Vannini F, Cavallo M, Grigolo B, Cenacchi A, Giannini S. Osteochondral lesions of the knee: a new one-step repair technique with bone-marrow-derived cells. J Bone Joint Surg Am 2010; 92(92) (Suppl. 2): 2-11.
[http://dx.doi.org/10.2106/JBJS.J.00813] [PMID: 21123588]

[56] Centeno CJ, Busse D, Kisiday J, Keohan C, Freeman M, Karli D. Increased knee cartilage volume in degenerative joint disease using percutaneously implanted, autologous mesenchymal stem cells. Pain Physician 2008; 11(3): 343-53.
[PMID: 18523506]

[57] Jones E, McGonagle D. Human bone marrow mesenchymal stem cells *in vivo*. Rheumatology

(Oxford) 2008; 47(2): 126-31.
[http://dx.doi.org/10.1093/rheumatology/kem206] [PMID: 17986482]

[58] Jäger M, Jelinek EM, Wess KM, *et al.* Bone marrow concentrate: a novel strategy for bone defect treatment. Curr Stem Cell Res Ther 2009; 4(1): 34-43.
[http://dx.doi.org/10.2174/157488809787169039] [PMID: 19149628]

[59] Longo UG, Lamberti A, Maffulli N, Denaro V. Tissue engineered biological augmentation for tendon healing: a systematic review. Br Med Bull 2011; 98: 31-59.
[http://dx.doi.org/10.1093/bmb/ldq030] [PMID: 20851817]

[60] Steinert AF, Rackwitz L, Gilbert F, Nöth U, Tuan RS. Concise review: the clinical application of mesenchymal stem cells for musculoskeletal regeneration: current status and perspectives. Stem Cells Transl Med 2012; 1(3): 237-47.
[http://dx.doi.org/10.5966/sctm.2011-0036] [PMID: 23197783]

[61] Centeno CJ, Schultz JR, Cheever M, Robinson B, Freeman M, Marasco W. Safety and complications reporting on the re-implantation of culture-expanded mesenchymal stem cells using autologous platelet lysate technique. Curr Stem Cell Res Ther 2010; 5(1): 81-93.
[http://dx.doi.org/10.2174/157488810790442796] [PMID: 19951252]

[62] Haller MJ, Viener HL, Wasserfall C, Brusko T, Atkinson MA, Schatz DA. Autologous umbilical cord blood infusion for type 1 diabetes. Exp Hematol 2008; 36(6): 710-5.
[http://dx.doi.org/10.1016/j.exphem.2008.01.009] [PMID: 18358588]

[63] Dave SD, Vanikar AV, Trivedi HL, *et al.* Autologous adipose tissue derived insulin-secreting mesenchymal stem cell transplantation in late onset of autoimmune mediated diabetes mellitus (LADA). IJBR 2013; 4(3)

[64] Dave SD, Vanikar AV, Trivedi HL. *Ex vivo* generation of glucose sensitive insulin secreting mesenchymal stem cells derived from human adipose tissue. Indian J Endocrinol Metab 2012; 16(16) (Suppl. 1): S65-9.
[http://dx.doi.org/10.4103/2230-8210.94264] [PMID: 22701849]

[65] Hu J, Yu X, Wang Z, *et al.* Long term effects of the implantation of Wharton's jelly-derived mesenchymal stem cells from the umbilical cord for newly-onset type 1 diabetes mellitus. Endocr J 2013; 60(3): 347-57.
[http://dx.doi.org/10.1507/endocrj.EJ12-0343] [PMID: 23154532]

[66] Liu X, Zheng P, Wang X, *et al.* A preliminary evaluation of efficacy and safety of Wharton's jelly mesenchymal stem cell transplantation in patients with type 2 diabetes mellitus. Stem Cell Res Ther 2014; 5(2): 57.
[http://dx.doi.org/10.1186/scrt446] [PMID: 24759263]

[67] Siegel G, Kluba T, Hermanutz-Klein U, Bieback K, Northoff H, Schäfer R. Phenotype, donor age and gender affect function of human bone marrow-derived mesenchymal stromal cells. BMC Med 2013; 11(1): 146.
[http://dx.doi.org/10.1186/1741-7015-11-146] [PMID: 23758701]

[68] Sun L, Wang D, Liang J, *et al.* Umbilical cord mesenchymal stem cell transplantation in severe and refractory systemic lupus erythematosus. Arthritis Rheum 2010; 62(8): 2467-75.
[http://dx.doi.org/10.1002/art.27548] [PMID: 20506343]

[69] Wang D, Li J, Zhang Y, *et al.* Umbilical cord mesenchymal stem cell transplantation in active and refractory systemic lupus erythematosus: a multicenter clinical study. Arthritis Res Ther 2014; 16(2): R79.
[http://dx.doi.org/10.1186/ar4520] [PMID: 24661633]

[70] Onken JE, *et al.* Successful outpatient treatment of refractory Crohn's disease using adult mesenchymal stem cells. American College of Gastroenterology Conference, Las Vegas, NV, Abstract 121.

[71] García-Olmo D, Herreros D, De-La-Quintana P, *et al.* Adipose-derived stem cells in Crohn's-rectovaginal fistula. Case Rep Med 2010.

[72] Ciccocioppo R, Bernardo ME, Sgarella A, *et al.* Autologous bone marrow-derived mesenchymal stromal cells in the treatment of fistulising Crohn's disease. Gut 2011; 60(6): 788-98.
[http://dx.doi.org/10.1136/gut.2010.214841] [PMID: 21257987]

[73] Chamberlain G, Fox J, Ashton B, Middleton J. Concise review: mesenchymal stem cells: their phenotype, differentiation capacity, immunological features, and potential for homing. Stem Cells 2007; 25(11): 2739-49.
[http://dx.doi.org/10.1634/stemcells.2007-0197] [PMID: 17656645]

[74] Duijvestein M, Vos AC, Roelofs H, *et al.* Autologous bone marrow-derived mesenchymal stromal cell treatment for refractory luminal Crohn's disease: results of a phase I study. Gut 2010; 59(12): 1662-9.
[http://dx.doi.org/10.1136/gut.2010.215152] [PMID: 20921206]

[75] Yoo J, Kim HS, Hwang DY. Stem cells as promising therapeutic options for neurological disorders. J Cell Biochem 2013; 114(4): 743-53.
[http://dx.doi.org/10.1002/jcb.24427] [PMID: 23097262]

[76] Anghileri E, Marconi S, Pignatelli A, *et al.* Neuronal differentiation potential of human adipose-derived mesenchymal stem cells. Stem Cells Dev 2008; 17(5): 909-16.
[http://dx.doi.org/10.1089/scd.2007.0197] [PMID: 18564036]

[77] Yamout B, Hourani R, Salti H, *et al.* Bone marrow mesenchymal stem cell transplantation in patients with multiple sclerosis: a pilot study. J Neuroimmunol 2010; 227(1-2): 185-9.
[http://dx.doi.org/10.1016/j.jneuroim.2010.07.013] [PMID: 20728948]

[78] Karussis D, Kassis I, Kurkalli BG, Slavin S. Immunomodulation and neuroprotection with mesenchymal bone marrow stem cells (MSCs): a proposed treatment for multiple sclerosis and other neuroimmunological/neurodegenerative diseases. J Neurol Sci 2008; 265(1-2): 131-5.
[http://dx.doi.org/10.1016/j.jns.2007.05.005] [PMID: 17610906]

[79] Venkataramana NK, Kumar SK, Balaraju S, *et al.* Open-labeled study of unilateral autologous bone-marrow-derived mesenchymal stem cell transplantation in Parkinson's disease. Transl Res 2010; 155(2): 62-70.
[http://dx.doi.org/10.1016/j.trsl.2009.07.006] [PMID: 20129486]

[80] Hayashi T, Wakao S, Kitada M, *et al.* Autologous mesenchymal stem cell-derived dopaminergic neurons function in parkinsonian macaques. J Clin Invest 2013; 123(1): 272-84.
[http://dx.doi.org/10.1172/JCI62516] [PMID: 23202734]

[81] Le Blanc K, Rasmusson I, Sundberg B, *et al.* Treatment of severe acute graft-*versus*-host disease with

third party haploidentical mesenchymal stem cells. Lancet 2004; 363(9419): 1439-41. [http://dx.doi.org/10.1016/S0140-6736(04)16104-7] [PMID: 15121408]

[82] Ringdén O, Uzunel M, Rasmusson I, *et al.* Mesenchymal stem cells for treatment of therapy-resistant graft-*versus*-host disease. Transplantation 2006; 81(10): 1390-7. [http://dx.doi.org/10.1097/01.tp.0000214462.63943.14] [PMID: 16732175]

[83] Kebriaei P, Isola L, Bahceci E, *et al.* Adult human mesenchymal stem cells added to corticosteroid therapy for the treatment of acute graft-*versus*-host disease. Biol Blood Marrow Transplant 2009; 15(7): 804-11. [http://dx.doi.org/10.1016/j.bbmt.2008.03.012] [PMID: 19539211]

[84] Kurtzberg J, Prockop S, Teira P, *et al.* Allogeneic human mesenchymal stem cell therapy (remestemcel-L, Prochymal) as a rescue agent for severe refractory acute graft-*versus*-host disease in pediatric patients. Biol Blood Marrow Transplant 2014; 20(2): 229-35. [http://dx.doi.org/10.1016/j.bbmt.2013.11.001] [PMID: 24216185]

[85] Behre G, Theurich S, Weber T, Christopeit M. Reply to 'The correlation between cotransplantation of mesenchymal stem cells and higher recurrence rates in hematologic malignancy patients: outcome of a pilot clinical study' by Ning *et al.* Leukemia 2009; 23(1): 178. [http://dx.doi.org/10.1038/leu.2008.150] [PMID: 18548101]

[86] Kim DW, Chung YJ, Kim TG, Kim YL, Oh IH. Cotransplantation of third-party mesenchymal stromal cells can alleviate single-donor predominance and increase engraftment from double cord transplantation. Blood 2004; 103(5): 1941-8. [http://dx.doi.org/10.1182/blood-2003-05-1601] [PMID: 14592834]

[87] Fouillard L, Bensidhoum M, Bories D, *et al.* Engraftment of allogeneic mesenchymal stem cells in the bone marrow of a patient with severe idiopathic aplastic anemia improves stroma. Leukemia 2003; 17(2): 474-6. [http://dx.doi.org/10.1038/sj.leu.2402786] [PMID: 12592355]

[88] Cortinovis M, Casiraghi F, Remuzzi G, Perico N. Mesenchymal stromal cells to control donor-specific memory T cells in solid organ transplantation. Curr Opin Organ Transplant 2015; 20(1): 79-85. [http://dx.doi.org/10.1097/MOT.0000000000000145] [PMID: 25563995]

[89] Zuk PA, Zhu M, Mizuno H, *et al.* Multilineage cells from human adipose tissue: implications for cell-based therapies. Tissue Eng 2001; 7(2): 211-28. [http://dx.doi.org/10.1089/107632701300062859] [PMID: 11304456]

[90] Kassis I, Zangi L, Rivkin R, *et al.* Isolation of mesenchymal stem cells from G-CSF-mobilized human peripheral blood using fibrin microbeads. Bone Marrow Transplant 2006; 37(10): 967-76. [http://dx.doi.org/10.1038/sj.bmt.1705358] [PMID: 16670702]

[91] Gronthos S, Mankani M, Brahim J, Robey PG, Shi S. Postnatal human dental pulp stem cells (DPSCs) *in vitro* and *in vivo*. Proc Natl Acad Sci USA 2000; 97(25): 13625-30. [http://dx.doi.org/10.1073/pnas.240309797] [PMID: 11087820]

[92] Hennrick KT, Keeton AG, Nanua S, *et al.* Lung cells from neonates show a mesenchymal stem cell phenotype. Am J Respir Crit Care Med 2007; 175(11): 1158-64. [http://dx.doi.org/10.1164/rccm.200607-941OC] [PMID: 17332484]

[93] Cho NH, Park YK, Kim YT, Yang H, Kim SK. Lifetime expression of stem cell markers in the uterine endometrium. Fertil Steril 2004; 81(2): 403-7.
[http://dx.doi.org/10.1016/j.fertnstert.2003.07.015] [PMID: 14967381]

[94] Williams JT, Southerland SS, Souza J, Calcutt AF, Cartledge RG. Cells isolated from adult human skeletal muscle capable of differentiating into multiple mesodermal phenotypes. Am Surg 1999; 65(1): 22-6.
[PMID: 9915526]

[95] Fukuchi Y, Nakajima H, Sugiyama D, Hirose I, Kitamura T, Tsuji K. Human placenta-derived cells have mesenchymal stem/progenitor cell potential. Stem Cells 2004; 22(5): 649-58.
[http://dx.doi.org/10.1634/stemcells.22-5-649] [PMID: 15342929]

[96] Lindenmair A, Hatlapatka T, Kollwig G, *et al.* Mesenchymal stem or stromal cells from amnion and umbilical cord tissue and their potential for clinical applications. Cells 2012; 1(4): 1061-88.
[http://dx.doi.org/10.3390/cells1041061] [PMID: 24710543]

[97] Al-Nbaheen M, Vishnubalaji R, Ali D, *et al.* Human stromal (mesenchymal) stem cells from bone marrow, adipose tissue and skin exhibit differences in molecular phenotype and differentiation potential. Stem Cell Rev 2013; 9(1): 32-43.
[http://dx.doi.org/10.1007/s12015-012-9365-8] [PMID: 22529014]

[98] Im GI, Shin YW, Lee KB. Do adipose tissue-derived mesenchymal stem cells have the same osteogenic and chondrogenic potential as bone marrow-derived cells? Osteoarthritis Cartilage 2005; 13(10): 845-53.
[http://dx.doi.org/10.1016/j.joca.2005.05.005] [PMID: 16129630]

[99] Zanini C, Bruno S, Mandili G, *et al.* Differentiation of mesenchymal stem cells derived from pancreatic islets and bone marrow into islet-like cell phenotype. PLoS One 2011; 6(12): e28175.
[http://dx.doi.org/10.1371/journal.pone.0028175] [PMID: 22194812]

[100] de Girolamo L, Lucarelli E, Alessandri G, *et al.* Mesenchymal stem/stromal cells: a new "cells as drugs" paradigm. Efficacy and critical aspects in cell therapy. Curr Pharm Des 2013; 19(13): 2459-73.
[http://dx.doi.org/10.2174/1381612811319130015] [PMID: 23278600]

[101] Zheng YB, Gao ZL, Xie C, *et al.* Characterization and hepatogenic differentiation of mesenchymal stem cells from human amniotic fluid and human bone marrow: a comparative study. Cell Biol Int 2008; 32(11): 1439-48.
[http://dx.doi.org/10.1016/j.cellbi.2008.08.015] [PMID: 18782626]

[102] Pittenger MF, Mackay AM, Beck SC, *et al.* Multilineage potential of adult human mesenchymal stem cells. Science 1999; 284(5411): 143-7.
[http://dx.doi.org/10.1126/science.284.5411.143] [PMID: 10102814]

[103] Kitagawa YK, Kobori M, Toriyama K, Kamei Y, Torii S. History of discovery of human adipose-derived stem cells and their clinical applications. Jpn J Plast Reconstr Surg 2006; 49: 1097-104.

[104] Roubelakis MG, Pappa KI, Bitsika V, *et al.* Molecular and proteomic characterization of human mesenchymal stem cells derived from amniotic fluid: comparison to bone marrow mesenchymal stem cells. Stem Cells Dev 2007; 16(6): 931-52.
[http://dx.doi.org/10.1089/scd.2007.0036] [PMID: 18047393]

[105] Vishnubalaji R, Al-Nbaheen M, Kadalmani B, Aldahmash A, Ramesh T. Comparative investigation of the differentiation capability of bone-marrow- and adipose-derived mesenchymal stem cells by qualitative and quantitative analysis. Cell Tissue Res 2012; 347(2): 419-27. [http://dx.doi.org/10.1007/s00441-011-1306-3] [PMID: 22287041]

[106] Rebelatto CK, Aguiar AM, Moretão MP, *et al.* Dissimilar differentiation of mesenchymal stem cells from bone marrow, umbilical cord blood, and adipose tissue. Exp Biol Med (Maywood) 2008; 233(7): 901-13. [http://dx.doi.org/10.3181/0712-RM-356] [PMID: 18445775]

[107] Ramkisoensing AA, Pijnappels DA, Askar SF, *et al.* Human embryonic and fetal mesenchymal stem cells differentiate toward three different cardiac lineages in contrast to their adult counterparts. PLoS One 2011; 6(9): e24164. [http://dx.doi.org/10.1371/journal.pone.0024164] [PMID: 21931658]

[108] Baksh D, Yao R, Tuan RS. Comparison of proliferative and multilineage differentiation potential of human mesenchymal stem cells derived from umbilical cord and bone marrow. Stem Cells 2007; 25(6): 1384-92. [http://dx.doi.org/10.1634/stemcells.2006-0709] [PMID: 17332507]

[109] Yan ZJ, Hu YQ, Zhang HT, *et al.* Comparison of the neural differentiation potential of human mesenchymal stem cells from amniotic fluid and adult bone marrow. Cell Mol Neurobiol 2013; 33(4): 465-75. [http://dx.doi.org/10.1007/s10571-013-9922-y] [PMID: 23478940]

[110] Stern-Straeter J, Bonaterra GA, Juritz S, *et al.* Evaluation of the effects of different culture media on the myogenic differentiation potential of adipose tissue- or bone marrow-derived human mesenchymal stem cells. Int J Mol Med 2014; 33(1): 160-70. [http://dx.doi.org/10.3892/ijmm.2013.1555] [PMID: 24220225]

[111] Melief SM, Geutskens SB, Fibbe WE, Roelofs H. Multipotent stromal cells skew monocytes towards an anti-inflammatory interleukin-10-producing phenotype by production of interleukin-6. Haematologica 2013; 98(6): 888-95. [http://dx.doi.org/10.3324/haematol.2012.078055] [PMID: 23349310]

[112] Jin HJ, Bae YK, Kim M, *et al.* Comparative analysis of human mesenchymal stem cells from bone marrow, adipose tissue, and umbilical cord blood as sources of cell therapy. Int J Mol Sci 2013; 14(9): 17986-8001. [http://dx.doi.org/10.3390/ijms140917986] [PMID: 24005862]

[113] Di Trapani M, Bassi G, Ricciardi M, *et al.* Comparative study of immune regulatory properties of stem cells derived from different tissues. Stem Cells Dev 2013; 22(22): 2990-3002. [http://dx.doi.org/10.1089/scd.2013.0204] [PMID: 23819720]

[114] Balasubramanian S, Venugopal P, Sundarraj S, Zakaria Z, Majumdar AS, Ta M. Comparison of chemokine and receptor gene expression between Wharton's jelly and bone marrow-derived mesenchymal stromal cells. Cytotherapy 2012; 14(1): 26-33. [http://dx.doi.org/10.3109/14653249.2011.605119] [PMID: 22091833]

[115] Schu S, Nosov M, O'Flynn L, *et al.* Immunogenicity of allogeneic mesenchymal stem cells. J Cell Mol Med 2012; 16(9): 2094-103.

[http://dx.doi.org/10.1111/j.1582-4934.2011.01509.x] [PMID: 22151542]

[116] Tögel F, Cohen A, Zhang P, Yang Y, Hu Z, Westenfelder C. Autologous and allogeneic marrow stromal cells are safe and effective for the treatment of acute kidney injury. Stem Cells Dev 2009; 18(3): 475-85.
[http://dx.doi.org/10.1089/scd.2008.0092] [PMID: 18564903]

[117] Auletta JJ, Bartholomew AM, Maziarz RT, *et al.* The potential of mesenchymal stromal cells as a novel cellular therapy for multiple sclerosis. Immunotherapy 2012; 4(5): 529-47.
[http://dx.doi.org/10.2217/imt.12.41] [PMID: 22642335]

[118] Cooper K, Viswanathan C. Establishment of a Mesenchymal Stem Cell Bank. Stem Cells Int 2011.
[http://dx.doi.org/10.4061/2011/905621]

[119] Madonna R, Renna FV, Cellini C, *et al.* Age-dependent impairment of number and angiogenic potential of adipose tissue-derived progenitor cells. Eur J Clin Invest 2011; 41(2): 126-33.
[http://dx.doi.org/10.1111/j.1365-2362.2010.02384.x] [PMID: 20874854]

[120] Kalaszczynska I, Ferdyn K. Wharton's jelly derived mesenchymal stem cells: future of regenerative medicine? Recent findings and clinical significance. BioMed res int 2015.

[121] Hermann A, List C, Habisch HJ, *et al.* Age-dependent neuroectodermal differentiation capacity of human mesenchymal stromal cells: limitations for autologous cell replacement strategies. Cytotherapy 2010; 12(1): 17-30.
[http://dx.doi.org/10.3109/14653240903313941] [PMID: 19878082]

[122] Chen H, Liu X, Zhu W, *et al.* SIRT1 ameliorates age-related senescence of mesenchymal stem cells *via* modulating telomere shelterin. Front Aging Neurosci 2014; 6: 103.
[http://dx.doi.org/10.3389/fnagi.2014.00103] [PMID: 24917814]

[123] Yuan H-F, Zhai C, Yan X-L, *et al.* SIRT1 is required for long-term growth of human mesenchymal stem cells. J Mol Med 2012; 90(4): 389-400.
[http://dx.doi.org/10.1007/s00109-011-0825-4] [PMID: 22038097]

[124] Phadnis SM, Ghaskadbi SM, Hardikar AA, Bhonde RR. Mesenchymal stem cells derived from bone marrow of diabetic patients portrait unique markers influenced by the diabetic microenvironment. Rev Diabet Stud 2009; 6(4): 260-70.
[http://dx.doi.org/10.1900/RDS.2009.6.260] [PMID: 20043038]

[125] Stanko P, Kaiserova K, Altanerova V, Altaner C. Comparison of human mesenchymal stem cells derived from dental pulp, bone marrow, adipose tissue, and umbilical cord tissue by gene expression. Biomed Pap Med Fac Univ Palacky Olomouc Czech Repub 2014; 158(3): 373-7.
[http://dx.doi.org/10.5507/bp.2013.078] [PMID: 24145770]

[126] El-Badri N, Ghoneim MA. Mesenchymal stem cell therapy in diabetes mellitus: progress and challenges. J Nucleic Acids 2013; (6): 194858.

[127] Lee HJ, Wu J, Chung J, Wrathall JR. SOX2 expression is upregulated in adult spinal cord after contusion injury in both oligodendrocyte lineage and ependymal cells. J Neurosci Res 2013; 91(2): 196-210.
[http://dx.doi.org/10.1002/jnr.23151] [PMID: 23169458]

[128] Sart S, Ma T, Li Y. Preconditioning stem cells for *in vivo* delivery. Biores Open Access 2014; 3(4):

137-49.
[http://dx.doi.org/10.1089/biores.2014.0012] [PMID: 25126478]

[129] Hsiao ST, Dilley RJ, Dusting GJ, Lim SY. Ischemic preconditioning for cell-based therapy and tissue engineering. Pharmacol Ther 2014; 142(2): 141-53.
[http://dx.doi.org/10.1016/j.pharmthera.2013.12.002] [PMID: 24321597]

[130] Pendergrass KD, Boopathy AV, Seshadri G, *et al.* Acute preconditioning of cardiac progenitor cells with hydrogen peroxide enhances angiogenic pathways following ischemia-reperfusion injury. Stem Cells Dev 2013; 22(17): 2414-24.
[http://dx.doi.org/10.1089/scd.2012.0673] [PMID: 23544670]

[131] Sart S, Tsai AC, Li Y, Ma T. Three-dimensional aggregates of mesenchymal stem cells: cellular mechanisms, biological properties, and applications. Tissue Eng Part B Rev 2014; 20(5): 365-80.
[http://dx.doi.org/10.1089/ten.teb.2013.0537] [PMID: 24168395]

[132] Ye Z, Zhou Y, Cai H, Tan W. Myocardial regeneration: Roles of stem cells and hydrogels. Adv Drug Deliv Rev 2011; 63(8): 688-97.
[http://dx.doi.org/10.1016/j.addr.2011.02.007] [PMID: 21371512]

[133] Bieback K, Kinzebach S, Karagianni M. Translating research into clinical scale manufacturing of mesenchymal stromal cells. Stem Cells Int 2010.
[http://dx.doi.org/10.4061/2010/193519]

[134] Sensebé L, Bourin P. Producing MSC according GMP: process and controls. Biomed Mater Eng 2008; 18(4-5): 173-7.
[PMID: 19065017]

[135] Chen B, Hu J, Liao L, *et al.* Flk-1+ mesenchymal stem cells aggravate collagen-induced arthritis by up-regulating interleukin-6. Clin Exp Immunol 2010; 159(3): 292-302.
[http://dx.doi.org/10.1111/j.1365-2249.2009.04069.x] [PMID: 20002448]

[136] Zhao Q, Ren H, Han Z. Mesenchymal stem cells: Immunomodulatory capability and clinical potential in immune diseases. J Cell Immunother 2015.

SUBJECT INDEX

A

B

Atta-ur-Rahman & Shazia Anjum (Eds.)

C

F

G

H

I

N

O

P

Q

R

S

T

U

V

W

Y

www.ingramcontent.com/pod-product-compliance
Lightning Source LLC
LaVergne TN
LVHW070948210926
848482LV00002B/138